ELECTRICAL INTERACTIONS IN MOLECULAR BIOPHYSICS

ELECTRICAL INTERACTIONS IN MOLECULAR BIOPHYSICS

An Introduction

RAYMOND GABLER

Millipore Corporation
Bedford, Massachusetts

1978

ACADEMIC PRESS   New York   San Francisco   London
A Subsidiary of Harcourt Brace Jovanovich, Publishers

ACADEMIC PRESS, INC.
111 Fifth Avenue, New York, New York 10003

United Kingdom Edition published by
ACADEMIC PRESS, INC. (LONDON) LTD.
24/28 Oval Road, London NW1 7DX

Library of Congress Cataloging in Publication Data

Gabler, Raymond.
    Electrical interactions in molecular biophysics.

    Includes bibliographies.
    1. Biological chemistry. 2. Biological physics.
3. Molecular biology. 4. Electrostatics.
I. Title. [DNLM: 1. Molecular biology. 2. Bio-
physics. QT34 G115m]
QP514.2.G3   574.1'92   77-6595
ISBN 0-12-271350-8

PRINTED IN THE UNITED STATES OF AMERICA

CONTENTS

4   Dipole Moments of Biological Macromolecules

5   Types of Molecular Interactions

PREFACE

This book is designed to act as an introduction to the subject of electrical interactions between biomolecules. As such, it encompasses two subjects, molecular biology and physics. The text is written to give the reader an appreciation of the fact that at the fundamental level, all biochemical or molecular biological reactions are electrical in nature, and as such must conform with the laws of physics and electrostatics. In this same respect, no new laws of physics have to be postulated to explain the interactions of biomolecules. The text also illustrates the wide variety of ways in which biomolecules can interact with one another, and how use is made of the electrical properties of biomolecules in order to obtain information about their structure and function. The principles of electrostatics are used to explain some of the basic units of structure on a molecular level.

Although both biochemistry and physics are covered, the reader need not have prior experience in either. The introductory chapters on both subjects are written to give a simple and basic understanding of the background material necessary to understand the rest of the text. Familiarity with the elements of differential and integral calculus is, however, assumed. It is possible to take a concurrent course in calculus, but this is thought to be more difficult. In this respect then, this text could be used by any person in the fields of biochemistry, molecular biology, physics, or microbiology.

The book is organized in such a fashion that the chapters generally build on the materials in the previous chapters. Mathematics is kept to a minimum, although Chapters 4 and 7 will require special attention. Chapter 1 presents the concepts and structures of biochemistry needed to understand the rest of the text. Emphasis is placed on knowing how to recognize different structural biochemical groups and how they are used as building blocks in forming molecules. Very few chemical reactions are considered, so the reader need not memorize a great deal. Chapter 2 covers the basics of elementary electrostatics. This chapter lays the foundation for the formulation of the physical theories presented in the other chapters. Both Chapters 1 and 2 assume no prior experience in these subjects. Chapter 3 builds on Chapters

1 and 2 and considers dielectrics and dipole moments. This is the first chapter in which biochemistry and physics are mixed. Chapter 4 deals with the dipole moments of biomolecules, how the dipole moment is measured, and what it means in terms of the physical structure of a molecule. In Chapter 5, equations are derived to describe strength and other properties for a wide variety of different electrical interactions. Biological examples are given to illustrate the importance and occurrence of each. Chapter 6 is devoted entirely to van der Waals forces, which are another form of electrical interaction. These unique and generally unappreciated forces are described in general terms and several examples are given to illustrate their importance. Chapter 7 is also devoted to one subject, Debye–Hückel theory. This theory seeks to describe the much more realistic situation of electrical interactions in a solution containing counterions. Chapter 8 treats water and water structure from a physical standpoint and indicates water's role in the overall scheme of molecular biology. Finally, Chapter 9 describes experimental techniques that rely upon the electrical properties of biomolecules. This chapter also discusses what types of information can be obtained from each experimental form and emphasizes that the electrical nature of biomolecules can be used to yield information of importance to research workers and also to physicians in dealing with disease.

The electrical nature of biomolecules and their interactions can be an exceedingly complex subject, and this book is written to give the reader a fundamental overview of the basics, and perhaps to spark an interest in further study of the physical reasons for biomolecules behaving as they do.

ACKNOWLEDGMENTS

I want to thank Dr. Norman Ford for reading the manuscript and supplying me with many useful comments and criticisms. I would also like to thank my wife Rose Mary for proofreading the manuscript, suggesting changes, and for enduring my many nights of writing and rewriting.

Thanks are also extended to Drs. Edward Westhead and Kenneth Langley for numerous helpful conversations, and to Dr. Irwin Bendet for initially introducing me to biophysics.

1

BIOCHEMISTRY

INTRODUCTION

The science of biophysics is concerned with the use of physical principles and techniques to solve biological problems. Biophysics is an interdisciplinary science that requires a knowledge of physics, biology, and chemistry. It is the type of discipline where various fields of knowledge overlap, and only by considering them all can a complete and organized overview be obtained. Molecular biophysics is that branch dealing with the structure and function of the basic biomolecules that are responsible for the different forms of life systems. Even though biology is usually thought of as a discipline completely independent of physics, when one examines biological reactions, on either a macroscopic or submicroscopic level, it is known that biomolecules, cells, etc. must obey the fundamental physical laws. Just because an entity is involved with a life system does not exempt it from the rules of physical behavior exhibited by the rest of the universe. In considering the specific case of molecular behavior it should be emphasized that at the most basic and fundamental level all interactions are electrical in nature, and that structural and functional aspects of biological macromolecules are determined by electrical laws. The fact that the basic electrical laws cannot be applied with certainty to all cases of interest is an indication of the complexity of biomolecular systems or of the inadequacy of our understanding of the physical universe.

This textbook is an attempt to make the reader more aware of the importance of electrical interactions, and in particular to emphasize their application to molecular biology. It is an effort to illustrate how basic biological processes can be governed, influenced, and controlled by physical forces.

This chapter will cover the basic biochemical information the reader should be familiar with before proceeding to other chapters. To those who are familiar with biochemistry, this will be a review. To those who have never had exposure to biochemistry, this chapter represents the minimum knowledge that should be mastered. The emphasis here will be on the ability to recognize different types of chemical compounds that are important in biophysics and biochemistry. Very few actual chemical reactions will be discussed in detail since this knowledge is not going to be needed to any large extent. The types of electrical interactions a biomolecule can enter into and the strength of these interactions are ultimately determined by the chemical makeup of the molecule itself. It is therefore important that the reader be familiar with the various parts of a biomolecule from a chemical standpoint. The material in this chapter is of a general nature, and any amplifications of details that are necessary will be made throughout the text as they are needed. For those who are interested in learning more or who would like to see another presentation, several organic chemistry and biochemistry textbooks are listed at the end of the chapter.

BIOELEMENTS

Before getting into the biochemical aspects directly it is instructive to take a look at the basic building blocks of all biological molecules, namely, the bioelements. Any molecule, whether large or small, organic or inorganic, etc. will have physical and chemical properties that are determined by the individual atoms composing it. Biological molecules tend to be thought of as being rather special because they are involved in a process called life, but they too are constructed from a collection of atoms. In examining a wide variety of these molecules it has been noticed that they are built from a preponderence of a certain few atoms, rather than from a wide distribution of all the elements seen in the periodic chart. That this is true indicates that this small group of elements apparently possesses certain properties that make them especially useful to the construction of biological molecules. These bioelements, then, deserve a special look to see just what properties they have that makes them so special.

One might naïvely think that the percentages of elements found in bio-molecules would parallel the percentages of elements found in the earth's crust. This seems natural since the more abundant an element, the more likely it will be incorporated into a growing system of molecules. However, this would be true only if all the elements possessed the same properties, and none had electrical configurations that would favor one element over another. The percentages for some of the more common elements found in the earth's crust are shown in Table 1-1. In terms of biological molecules the four most commonly found elements are hydrogen (H), carbon (C), oxygen (O), and nitrogen (N). In all of nature these four elements make up well over 90% of all living matter. Two other elements that are extremely important are phosphorus (P) and sulfur (S). Ions that are important in biological processes are sodium (Na^+), potassium (K^+), magnesium (Mg^{+2}),

TABLE 1-1

Common Elements in the Earth's Crust and Their Relative Abundance[a]

Element	Percent	Element	Percent
Oxygen	47	Carbon	0.2
Silicon	28	Potassium	2.5
Aluminum	8	Titanium	0.4
Sodium	2.5	Hydrogen	0.2
Calcium	3.5	Nickel	0.2
Iron	4.5	Copper	0.002
Magnesium	2.5	Zinc	0.001
Phosphorus	0.1		

[a] Data from "Van Nostrand's Scientific Encyclopedia," fifth ed. Van Nostrand, Princeton, New Jersey, 1976.

calcium (Ca^{+2}), and chloride (Cl^-). Also encountered, but usually in very small or trace amounts, are manganese (Mn), aluminum (Al), vanadium (V), iron (Fe), cobalt (Co), copper (Cu), molybdenum (Mo), iodine (I), silicon (Si), zinc (Zn), and boron (B). So, these 22 elements just listed are the ones that are most commonly found in biological systems. Considering that there are well over one hundred elements in the periodic table, it is easily seen that the incorporation of atoms into biomolecules has been restricted to a fairly small group.

If we now take a look at the four most common elements and their placement in the periodic table (Fig. 1-1), several things are noticed. Being the simplest of all elements, hydrogen is in the first period, while C, N, and O are in the second period. This means that the most common bioelements are among the lightest. The electronic structures of these atoms are shown schematically in Fig. 1-2 where each dot represents an electron in the outer shell. More will be said in Chapter 8 concerning the geometric arrangement of the orbiting electrons. For our purposes, it is enough to note that the number of electrons in the outer orbits has a large role in determining an element's ability to form chemical bonds with other elements. The inner electrons are more protected and hence cannot react as well with other atoms. The number of electrons an atom has surrounding it is also important from the standpoint of stability. It is well known that atoms with certain numbers of electrons tend to possess greater stability than those that do not have these special configurations. For the lighter elements two or eight electrons surrounding a nucleus in the outer shell is an advantage in terms of stability. In this sense we see that H requires one additional electron to attain a stable configuration, while O, N, and C require two, three, and four electrons, respectively. So, these elements represent the four lightest

Period	Group I	II						Transition elements						III	IV	V	VI	VII	VIII
1	1 H 1.00797																		2 He 4.0026
2	3 Li 6.939	4 Be 9.0122												5 B 10.811	6 C 12.01115	7 N 14.0067	8 O 15.9994	9 F 18.9984	10 Ne 20.183
3	11 Na 22.9898	12 Mg 24.312												13 Al 26.9815	14 Si 28.086	15 P 30.9738	16 S 32.064	17 Cl 35.453	18 Ar 39.948
4	19 K 39.102	20 Ca 40.08	21 Sc 44.956	22 Ti 47.90	23 V 50.942	24 Cr 51.996	25 Mn 54.9380	26 Fe 55.847	27 Co 58.9332	28 Ni 58.71	29 Cu 63.54	30 Zn 65.37		31 Ga 69.72	32 Ge 72.59	33 As 74.9216	34 Se 78.96	35 Br 79.909	36 Kr 83.80
5	37 Rb 85.47	38 Sr 87.62	39 Y 88.905	40 Zr 91.22	41 Nb 92.906	42 Mo 95.94	43 Tc (99)	44 Ru 101.07	45 Rh 102.905	46 Pd 106.4	47 Ag 107.870	48 Cd 112.40		49 In 114.82	50 Sn 118.69	51 Sb 121.75	52 Te 127.60	53 I 126.9044	54 Xe 131.30
6	55 Cs 132.905	56 Ba 137.34	57-71 *	72 Hf 178.49	73 Ta 180.948	74 W 183.85	75 Re 186.2	76 Os 190.2	77 Ir 192.2	78 Pt 195.09	79 Au 196.967	80 Hg 200.59		81 Tl 204.37	82 Pb 207.19	83 Bi 208.980	84 Po (210)	85 At (210)	86 Rn (222)
7	87 Fr (223)	88 Ra (227)	(89-103) †	(104)	(105)														

*Lanthanide elements	57 La 138.91	58 Ce 140.12	59 Pr 140.907	60 Nd 144.24	61 Pm (145)	62 Sm 150.35	63 Eu 151.96	64 Gd 157.25	65 Tb 158.924	66 Dy 162.50	67 Ho 164.930	68 Er 167.26	69 Tm 168.934	70 Yb 173.04	71 Lu 174.97
†Actinide elements	89 Ac (227)	90 Th 232.038	91 Pa (231)	92 U 238.03	93 Np (237)	94 Pu (242)	95 Am (243)	96 Cm (245)	97 Bk (249)	98 Cf (249)	99 Es (254)	100 Fm (252)	101 Md (256)	102 No (254)	103 Lw (257)

Key:

26 ———— Atomic number (Z)

Fe ———— Element symbol

55.847 ———— Atomic mass of the naturally occurring isotopic mixture; for the elements that are naturally radioactive, the numbers in parentheses are mass numbers of the most stable isotopes of these elements.

Fig. 1-1   Periodic table of the elements.

$$\text{H} \bullet \qquad \text{C} \, \vdots \qquad \bullet \, \text{N} \, \vdots \qquad \vdots \, \text{O} \, \vdots$$

Fig. 1-2 Schematic electronic arrangement of hydrogen, carbon, nitrogen, and oxygen atom. Each dot represents an electron in the outer shell.

atoms that require the fewest number of additional electrons to form stable electronic configurations.

At this point the reader may well wonder why a stable configuration of electrons is important, how these additional electrons are acquired, and what this all has to do with molecular biology. The stability question is rather easy to understand, in that biological systems would not be of much lasting value if they were not stable, but could literally fall apart at any time. This implies that the atoms composing the systems must be stable. Biological systems do not last forever; they are constantly undergoing an evolutionary process both on the macroscopic and molecular level. But they do possess stability of structure for significant periods of time. In general the additional electrons needed for stability are gained by having two atoms come together to form a bond in such a manner that both atoms share mutual electrons so both atoms feel they have the full complement required to form a stable configuration. This particular type of association is called a covalent bond and is the most frequently found bond in biological systems. Naturally, the number of other atoms one atom may combine with depends on how many additional electrons are needed to complete a stable set. For instance, H needs one electron to complete its stable configuration; so two hydrogen atoms may form a covalent bond with each other where each atom shares its own single electron plus its neighbor's. Both hydrogens then think they have a total of two electrons. In this instance the covalent bond is represented as H:H or H—H where the dash represents two shared electrons. In this configuration both H atoms are happy, and we have formed molecular hydrogen H_2. Usually, but not always, an atom will be able to donate one electron to be shared in a covalent bond for every atom with which it is interacting. In this case the covalent bond is known as a single bond.

Another type of bond that is also encountered is the ionic bond. In this type of union one atom actually donates an electron to its partner. The first atom then acquires a positive charge, and the receiving atom becomes negatively charged. The two atoms are then held together by electrostatic forces. An example of this is seen in the union between Na and Cl, shown below in equation form:

$$\bullet \text{Na} \longrightarrow \text{Na}^+ + \bullet$$

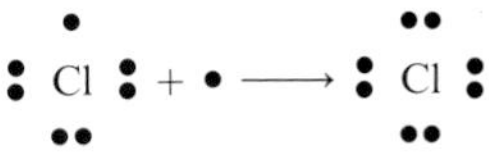

$$\vdots \, \text{Cl} \, \vdots + \bullet \longrightarrow \vdots \, \text{Cl} \, \vdots$$

$$\text{Cl}^- + \text{Na}^+ \longrightarrow \text{NaCl}$$

Again, the ultimate result is to form a stable configuration of electrons around both atoms. In this case the chloride atom ends up with eight electrons in the outer orbit, as does the sodium atom.

Of the two types the covalent bond is by far the more important in biological systems since it is responsible for the majority of bonds needed to form biomolecules. One advantage it has in this respect is that it has less susceptibility to being broken by the presence of water, whereas NaCl, for example, will easily dissociate in water. The exact way in which a covalent bond is formed between two atoms depends on the geometric shape of the electron orbits involved, where this information ultimately comes from an analysis using quantum mechanics. This analysis can best be described by stating that quantum mechanics can predict the probability of an electron being found in a particular location. By plotting the positions of high probability a geometric shape will evolve that describes the overall shape of a particular electron's orbit. In this light, a covalent bond is formed when an electron orbit from one atom overlaps an electron orbit from another atom. The strength of the bond depends on the amount of overlap existing between the two orbits; the more overlap the stronger the bond.

The apparent reason for nature's selection of the smaller elements in forming biological systems has to do with two facts. The first is that the smaller elements usually form the tightest, most stable bonds. The second is that only the lighter atoms tend to be able to form multiple bonds. A multiple bond is one in which a participating atom can donate more than one electron to be shared in conjunction with a neighboring atom. Multiple bonds are usually represented by two or three lines between atomic symbols, depending on whether a double or triple bond is being represented, e.g., $C{=}C$, $C{\equiv}C$. The mechanics of multiple bonds are more complicated than single bonds, but that is of little interest to us here. The advantages of stability have already been discussed, but what are the advantages of multiple bonds? One characteristic of a multiple bond is that it is stronger than a single bond, but the main property that it gives to a potential biological system is variety. By being able to choose not only from a variety of different elements but also from a variety of potential bonding schemes, nature is able to construct a myriad of structures with each having different physical and chemical properties. This is very important from an evolutionary standpoint in that nature can then pick and choose from a wider reservoir in order to find the most advantageous structure for a particular function. Without variety the system suffers. An analogy can be made with a coach's problem of fielding the best ball team possible. If the coach has a large group of candidates from which to choose, then the probability of ending up with a good team is better than if the group is small. It was realized in the early

1920s that of all the elements the ability to form multiple bonds was almost entirely confined to C, N, and O.

When one studies organic chemistry or biochemistry, it is frequently stated that these subjects are concerned with the chemical compounds of carbon. This is not strictly accurate, although the importance of carbon is seen almost immediately in these subjects. Of the elements so far mentioned, it is not entirely untrue to state that carbon has properties that are particularly important to life systems. It has the ability to form single, double, and triple bonds with itself and also with O and N. It can combine with H, N, O, and S; and a carbon atom can associate via covalent bonds with up to four other atoms. These properties thus confer upon carbon compounds the potential of forming the backbone or foundation of an immense variety of chemical structures (see Fig. 1-3). When one considers the total number of biomolecules found in all living matter, the numbers become astronomical; it is only through the many possible arrangements between carbon and other atoms that this variety is possible. So, in dealing with the biochemical aspects of this textbook, carbon will play a major role; and it should be emphasized again that the importance of carbon and the other major bioelements is due to the electronic structure of their atoms. This is a physical, and not a biological, characteristic.

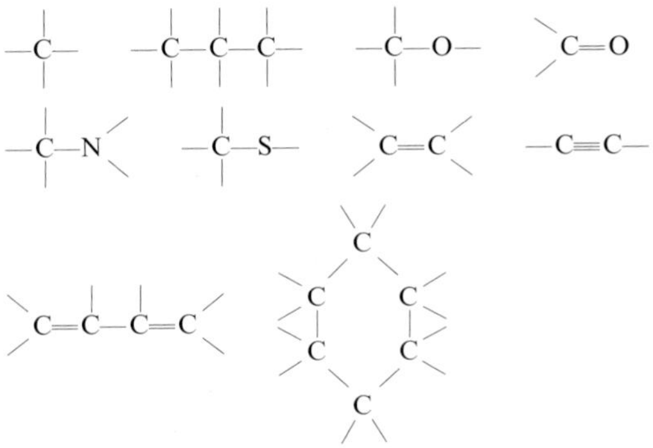

Fig. 1-3 Some of the different possible chemical combinations of carbon. Each line represents a possible or actual covalent bond.

GENERAL BIOCHEMISTRY

Now that the basic reasons for the preponderance of certain elements in biosystems have been discussed, we can proceed to consider some of the biochemistry that will be useful for this book. As one might expect, the biochemical compounds or groups that will be discussed are going to be

constructed primarily from H, C, N, and O. Because of the number of electrons in its outer orbit, carbon can form four single covalent bonds with
other atoms, nitrogen can form three, oxygen can form two, and hydrogen
one. The simplest combination of carbon and hydrogen is methane CH_4
where four covalent bonds satisfy both the carbon and the four hydrogens.
The number of bonds a certain atom can form is called its valency; and when
all those bonds are engaged, valency is said to be satisfied. In stable compounds the participating atoms almost always have their valencies satisfied.
If the valency of a particular atom is not fulfilled, then it is usually a very
reactive species and will readily combine with another atom in order to
satisfy its valency requirement. Combinations of H with N, O, and S are
called ammonia, water, and hydrogen sulfide respectively. These are shown
in Fig. 1-4.

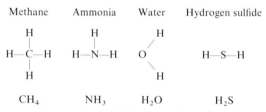

Fig. 1-4   Some simple molecules in which hydrogen is combined with C, N, O, or S.

Hydrocarbons constitute a large group of chemical compounds composed
only of C and H. If only single covalent bonds are present, the hydrocarbon
is termed saturated; if multiple bonds are present, the hydrocarbon is
termed unsaturated. Some typical hydrocarbons are shown in Table 1-2.
It should be noticed that hydrocarbons can be either straight chained or
branched, which leads to almost unlimited configurational possibilities.
When a hydrocarbon (or any other compound) has alternating single and
double bonds, it is called conjugated; this type of compound has chemical
and physical properties that are slightly different from an identical compound without conjugation. Saturated hydrocarbons usually have a name
that ends in *ane*, while those with one double bond end in *ene*. The first
part of the hydrocarbon's name usually, but not always, indicates how many
carbon atoms are present. The n in front of the saturated hydrocarbon's
name stands for normal and indicates that the hydrocarbon is unbranched.
In naming the unsaturated hydrocarbons shown in Table 1-2 the numbers
preceding the names refer to the position of the double bond(s) with respect
to one end of the chain. The system is such that the lowest numbers possible
are used. For example, 1-butene could just as well be called 3-butene; but

the convention is to use the lowest number, so 1-butene is usually used. Whenever a hydrocarbon is substituted or has one of its hydrogens replaced by another element, the number of the carbon atom(s) bearing the new element(s) precedes the name of the compound to identify where the non-hydrogen atom is located. For instance, 1, 2-dichlorethane is drawn as

$$
\begin{array}{ccc}
& Cl & Cl \\
& | & | \\
H\text{---} & C\text{---}C & \text{---}H \\
& | & | \\
& H & H
\end{array}
$$

For the simple examples shown in Table 1-2, the naming or nomenclature system is rather straightforward; however, this is not always the case. The reader should notice that two hydrocarbons can have identical empirical formulas yet have quite distinct structures. It should also be remembered that the compounds shown in Table 1-2 are two-dimensional representations of three-dimensional structures.

For just about every type or group of organic compounds, there is a systematic method of naming the compound. Specific rules of nomenclature have been agreed upon by international committees to ensure uniformity, and some of these rules have been previously described for naming hydrocarbons. More detail on these nomenclature rules will not be presented in this textbook; the interested reader is referred to a book on organic chemistry or biochemistry. In this book most chemical compounds that are discussed in the text will be illustrated in a figure and named to avoid confusion. However, the reader should be able to recognize different chemical groups as belonging to one classification or another.

As a chemical group, hydrocarbons tend to have certain common physical properties allowing them to interact with themselves and other molecules in a similar fashion. The main hydrocarbon property that is of interest to us is their nonpolar nature (having little or no permanent dipole moment); this property dictates the types of electrical interactions in which they can participate. It will be seen in later chapters that the major electrical interaction involving hydrocarbons is known as the van der Waals force. We shall also see that the nonpolar nature of hydrocarbons will explain such chemical properties of hydrocarbons as their inability to be dissolved in aqueous solvents and their preference for solvents such as benzene and chloroform. In future chapters our interest will not be necessarily with hydrocarbons per se, but rather in the hydrocarbon portions of more complex molecules and in the hydrocarbonlike properties of those molecules. It turns out that the physical properties of a molecule can be understood by considering the separate distinct chemical groups present in the molecule. The method of considering the properties of a large molecule in terms of its

TABLE 1-2

Typical Bonding Arrangements Seen in Saturated and Unsaturated Hydrocarbons

Saturated hydrocarbons			Unsaturated hydrocarbons		
Empirical formula	Structural formula	Name	Empirical formula	Structural formula	Name
CH_4	H–C–H (with H above and H below)	Methane	C_2H_4	H₂C=CH₂	Ethylene
C_2H_6	H–C–C–H (with H's)	Ethane	C_3H_6	H₂C=CH–CH₃	Propylene
C_3H_8	H–C–C–C–H (with H's)	Propane	C_4H_8	H₂C=CH–CH₂–CH₃	1-Butene

2-Butene

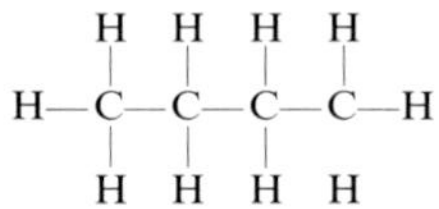

C_4H_8

1-Pentene

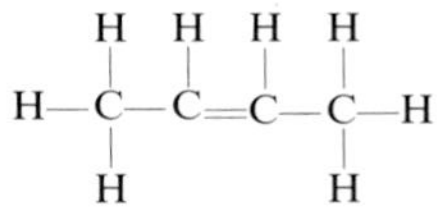

C_5H_{10}

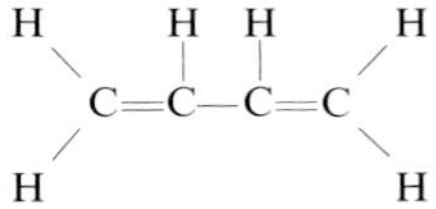

1,3-Butadiene (conjugated)

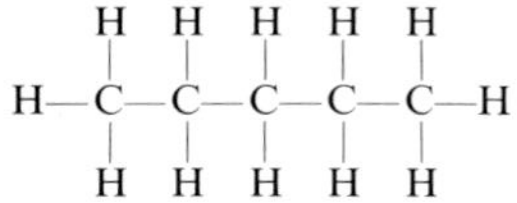

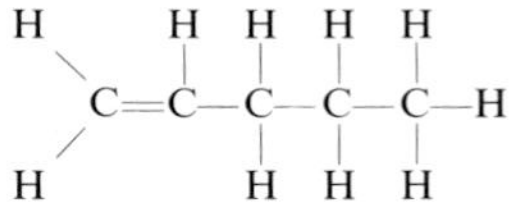

C_4H_6

n-Butane

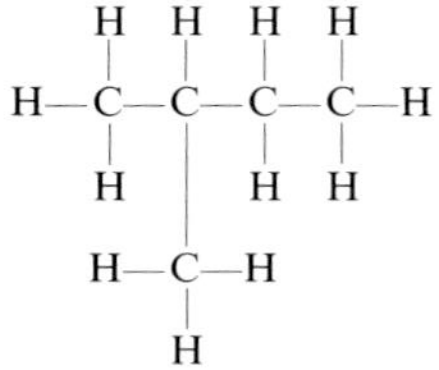

C_4H_{10}

n-Pentane

C_5H_{12}

Isopentane

C_5H_{12}

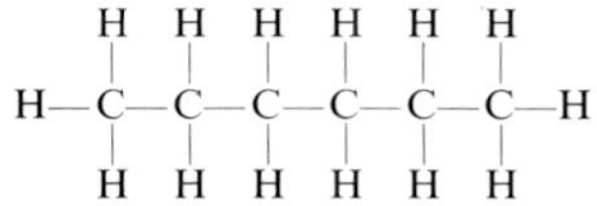

n-Hexane

C_6H_{14}

constituent chemical groups is a general one, and this approach will be taken quite often.

When considering large chemical structures, it becomes practical to single out specific portions or groups within the whole structure and consider the naming of the whole structure, and also its characteristics, in terms of these smaller groups. Some of these specific groups that are commonly encountered are shown in Table 1-3. By combining one or several of these groups with a carbon in a hydrocarbon, new classes of chemical compounds are formed with physical and chemical characteristics that are quite different from the parent hydrocarbon. For instance, when a methyl group in butane is replaced by a carboxyl group, the compound butyric acid is formed, which is more soluble in water than butane. From its name, butyric acid must have acidic properties; and it must be the carboxyl group that is acidic because butane itself is not. (More will be discussed about organic

TABLE 1-3

Common Biochemical Function Groupings

Empirical formula	Structural formula	Name	Empirical formula	Structural formula	Name
$-CH_3$	H | $-C-H$ | H	Methyl	$-NH_2$	H / $-N$ \ H	Amine
$-OH$	$-O-H$	Hydroxyl	$-SH$	$-S-H$	Sulfhydryl or mercaptan
$-COOH$	O // $-C$ \ OH	Carboxyl	$-CO$	$-C=O$	Carbonyl
$-CONH_2$	O // $-C$ \ NH_2	Amide	$-SS-$	$-S-S-$	Disulfide
$-C_6H_5$	H H | | C$-$C // \\\\ $-C$ C$-$H \ C$=$C | | H H	Benzyl			

acids in a short while.) The chemistry of each group shown in Table 1-3 is a chapter in itself; the interested reader is again referred to one of the text-books on organic chemistry or biochemistry listed at the chapter's end. For now, it is satisfactory that the reader be aware of the various groups and is able to recognize them.

The hydrocarbons shown in Table 1-3 are all seen to have a definite beginning and a definite end, or several ends depending on the degree of branching. Chemical compounds can also form ring structures in which there is no beginning or end; it turns out that ring structures play an ex-tremely important role in biophysical chemistry. Some common ring struc-tures are shown in Fig. 1-5. The reader should notice that many of the structures shown in this figure have conjugated or partially conjugated configurations, and that some are pure hydrocarbons while others are not.

This book will be concerned primarily with only two types of biological macromolecules, proteins and nucleic acids, with the emphasis being on proteins. It should not be assumed however that other types of biomolecules do not use electrical forces in their interactions. This restriction merely reflects the facts that proteins and nucleic acids have been the most widely studied and that today the largest reservoir of knowledge surrounds them. Also, they are among the most abundant of biomolecules and their impor-tance is unquestioned. The next sections will consider the detailed chemical composition of these molecules, and it will be seen that it is here that the very nature of their electrical forces originate. Throughout the text brief mention will be made of other types of biostructures, and at these times any pertinent chemical information will be given.

PROTEINS

Proteins are a group of biomolecules that not only play a wide variety of functional roles, but are virtually found everywhere in living systems. Viable cells are known to have up to 50% of their dry weight constituted by proteins of one sort or another. Cells are constantly synthesizing new protein molecules to replace those that have been "used up"; hence it is necessary for organisms to be continually digesting the needed raw materials to be used for this purpose. Pure proteins contain carbon, hydrogen, oxygen, nitrogen, and sometimes sulfur. Proteins frequently contain other atoms than the five just mentioned, although these are usually present in very small quantities. Before going into the detailed chemical makeup of proteins, it will be profitable to first briefly describe their role in nature. In this respect the diversity of functions that proteins can perform is best illustrated by some examples.

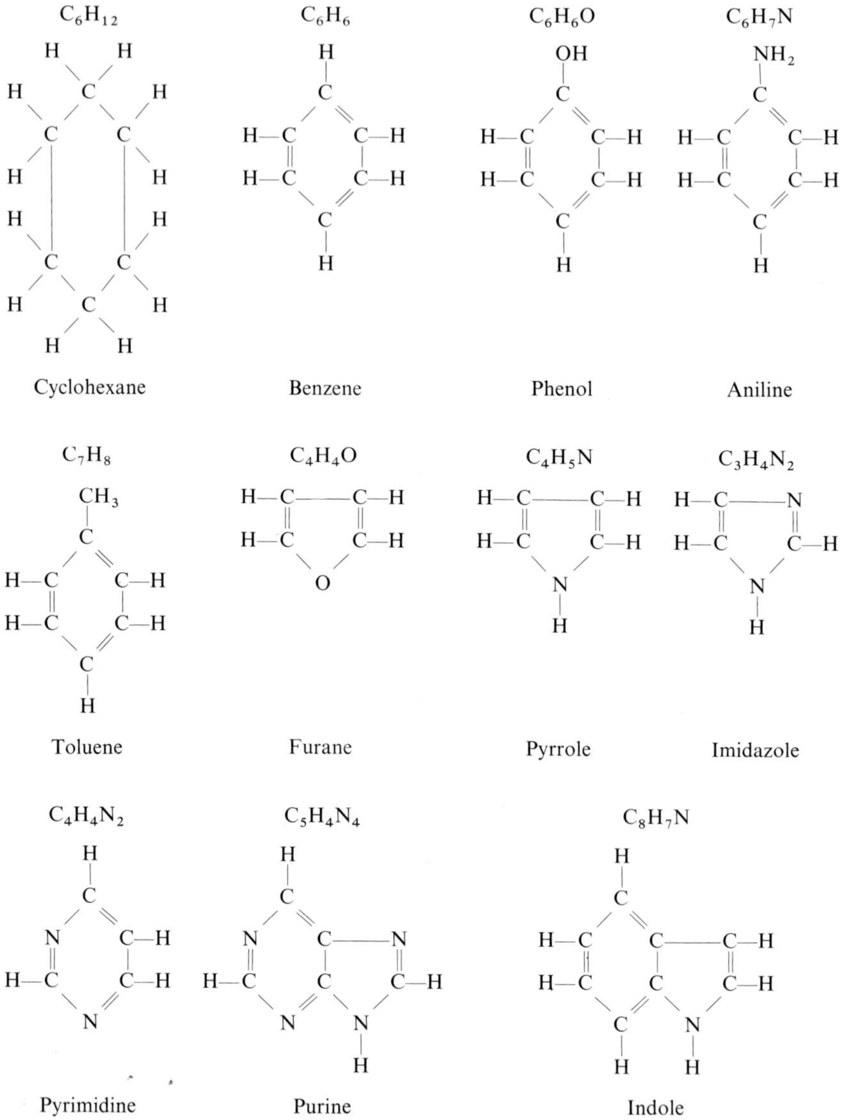

Fig. 1-5 Commonly encountered biochemical ring structures.

By far the largest subgroup of proteins is made up of the enzymes. Enzymes are molecules that help a particular chemical reaction to completion without becoming permanently involved with any of the reactants; i.e., they are essentially catalysts. Most biochemical reactions do not occur spontaneously (or at least at a rapid enough rate to be useful) at temperatures encountered in life systems; hence enzymes are necessary to speed up chemical reactions in an organism's metabolism. Without enzymes, life systems would be hopelessly lost. The type of reactions that enzymes catalyze vary from such seemingly simple tasks as transferring an electron from one molecule to another, to completely destroying another protein or a nucleic acid. It should also be mentioned that enzymes do not catalyze thermodynamically unfavorable reactions; i.e., they do not help a reaction to completion that would not normally be completed on its own, but at a much slower rate. Enzymes do not alter the change in energy of the system between the reactants and the products. They just make the reaction go faster than it would normally if left on its own. Enzymes work at concentrations that are very small compared to the reactants; and enzymes show an extreme degree of selectivity with, in general, each individual biochemical reaction requiring a specific enzyme. In size, enzymes can range from small to very large. The molecular weights of enzymes can range from 10,000 to 20,000 to over 500,000 daltons.*

Another large class of proteins are the structural proteins. These proteins have the function of helping to form the structural framework around which an organism is built. Examples here are collagen which is found in tendons and cartilage, elastin which is found in elastic connective tissue, and keratin which is found in skin, hair, and feathers. Other structural proteins are found in the membranes that surround many cells or subdivide cells into compartments.

Some hormones are proteins, or mostly protein, in nature. The best known example is insulin which regulates the metabolism of glucose. People suffering from diabetes have a deficiency of insulin and must have it regularly supplied to them.

Protective proteins, such as antibodies, guard the host organism against invading foreign molecules or cells. Antibodies and the immune system help keep the organism healthy from a perpetual onslaught of diseases and metabolic disorders. Fibrinogen and thrombin are protective proteins that are involved with blood clotting in vertebrates and help prevent excessive bleeding.

Proteins are also used as carriers. For example, hemoglobin transports oxygen from the lungs to individual cells via the bloodstream. Serum albumin is another blood protein thought to transport many different kinds of materials. Myoglobin also carries oxygen, but does so primarily in muscle cells. Whales and other air-breathing animals that swim submerged for long periods usually have large stores of myoglobin.

* A dalton is a unit of molecular weight and it is equal to the mass of one hydrogen atom.

Proteins such as myosin and actin, when organized properly, make up muscle fibers and are involved in muscular contraction. Ovalbumin in egg whites and casein and β-lactoglobulin in milk are used as nourishment for growing organisms. Some extremely poisonous substances are proteins, as is the toxin that is secreted from the bacterium *Clostridium botulinum*, and can cause a fatal case of food poisoning. Snake venoms are frequently enzymes that destroy critical molecules involved in a victim's metabolism.

So you can see that the distribution and function of proteins in nature is quite widespread and that their importance to living systems is critical. With such a diversity in function, one would expect that the chemical make-up of proteins should be capable of achieving a wide variety of structures. Now let us look at the chemical composition of these ubiquitous molecules to see just how they are built.

Proteins, as all large biological molecules, are composed of an assemblage of smaller parts joined together in a well-organized way. In the case of proteins these smaller parts are called amino acids. These amino acids are covalently bonded together to form long polymer chains, and the resulting structure is termed a protein. An amino acid has the following general structural formula:

$$
\begin{array}{ccc}
H & H & O \\
\diagdown & | & \diagup\!\diagup \\
& N\!-\!C\!-\!C & \\
\diagup & | & \diagdown \\
H & R & OH
\end{array}
$$

where R represents a side chain, which is a chemical grouping that distinguishes one amino acid from another. In all there are about 20 different amino acids (Rs) commonly found in biological situations with others occurring less frequently. The amino acids can be subclassified according to the electrical properties, or polarity, of their side chains, although this should not necessarily be obvious to the reader at present. There are three groups in this method of classification: hydrophobic or nonpolar, polar but uncharged at neutral pH, and charged at neutral pH. Figure 1-6 shows the 20 commonly occurring amino acids along with their letter abbreviations according to these three groups.

After looking at the structures in Fig. 1-6 one thing is quite apparent. Except for proline, an amine group and a carboxyl group are common to all amino acids, and since the COOH group is the characteristic feature of an organic acid, the name amino acid is used. The reader should be reminded that the structures in Fig. 1-6 are only two-dimensional representations of the amino acids. In reality it turns out that the three-dimensional structure of amino acids is extremely important. For every amino acid structure except glycine, there exists the possibility of another structure, identical except that it is a mirror image of the first. Mirror image amino acids do

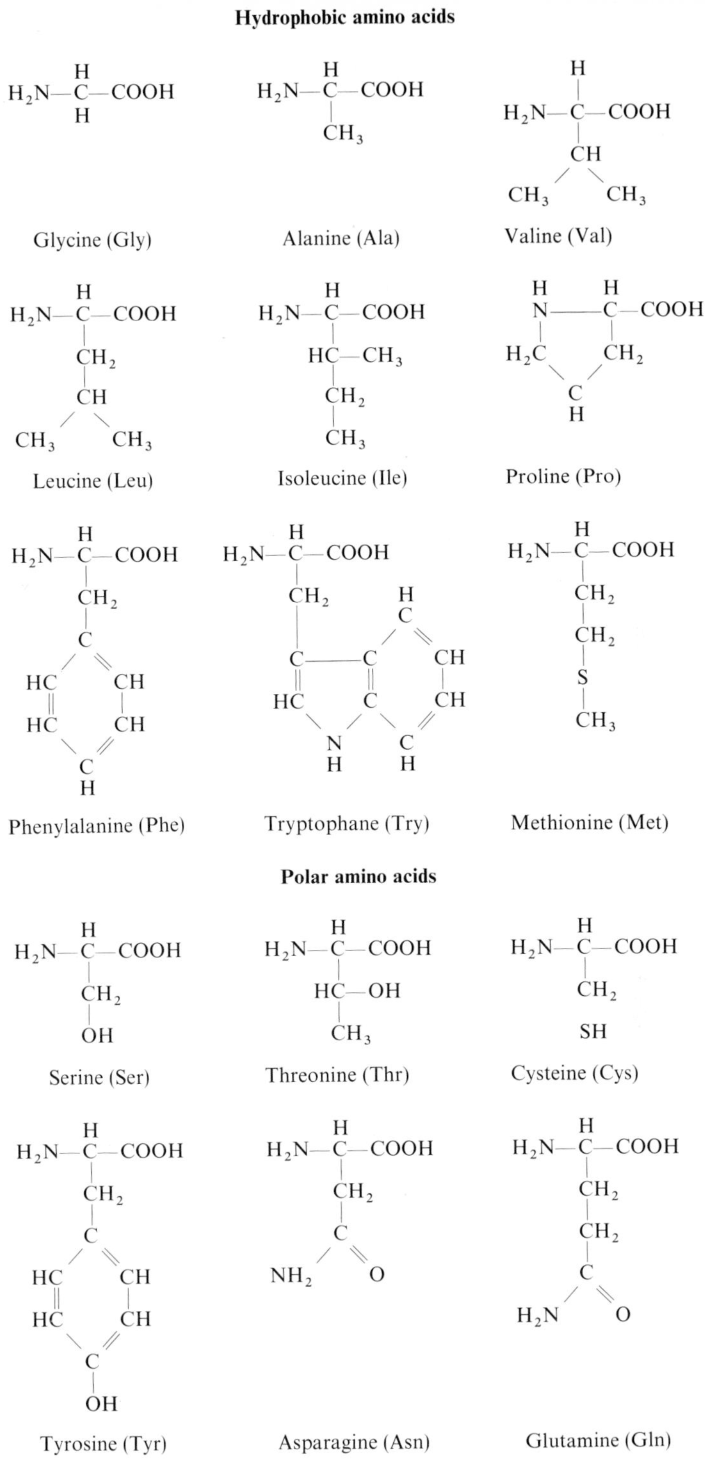

Hydrophobic amino acids
Glycine (Gly)
Alanine (Ala)
Valine (Val)
Leucine (Leu)
Isoleucine (Ile)
Proline (Pro)
Phenylalanine (Phe)
Tryptophane (Try)
Methionine (Met)
Polar amino acids
Serine (Ser)
Threonine (Thr)
Cysteine (Cys)
Tyrosine (Tyr)
Asparagine (Asn)
Glutamine (Gln)

Fig. 1-6 The common amino acids are subdivided into groups based on their electrical properties.

Charged amino acids

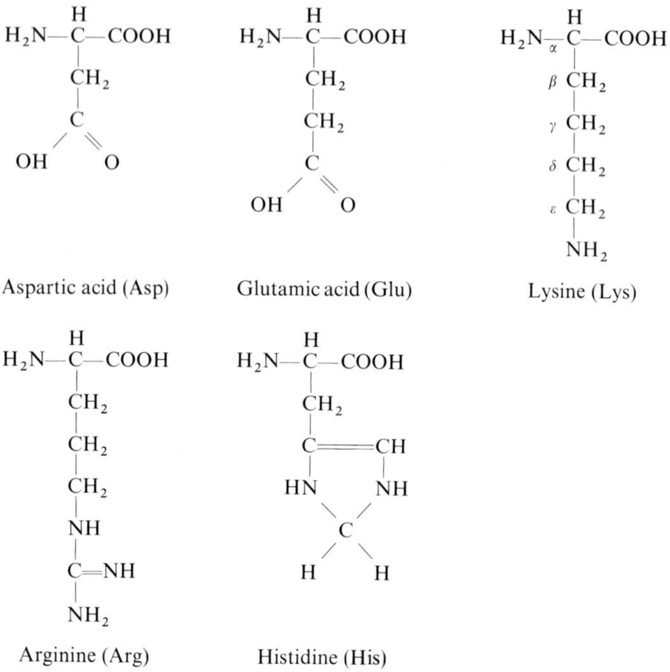

$$H_2N-\overset{\overset{\displaystyle H}{|}}{C}-COOH$$

$$\begin{array}{c} CH_2 \\ | \\ C \\ \diagup \diagdown \\ OH \quad\ \ O \end{array}$$

$$H_2N-\overset{\overset{\displaystyle H}{|}}{C}-COOH$$

$$\begin{array}{c} CH_2 \\ | \\ CH_2 \\ | \\ C \\ \diagup \diagdown \\ OH \quad\ \ O \end{array}$$

$$H_2N-\underset{\alpha}{\overset{\overset{\displaystyle H}{|}}{C}}-COOH$$

$$\begin{array}{c} \beta\ CH_2 \\ | \\ \gamma\ CH_2 \\ | \\ \delta\ CH_2 \\ | \\ \varepsilon\ CH_2 \\ | \\ NH_2 \end{array}$$

Aspartic acid (Asp)   Glutamic acid (Glu)   Lysine (Lys)

$$H_2N-\overset{\overset{\displaystyle H}{|}}{C}-COOH$$

$$\begin{array}{c} CH_2 \\ | \\ CH_2 \\ | \\ CH_2 \\ | \\ NH \\ | \\ C=NH \\ | \\ NH_2 \end{array}$$

$$H_2N-\overset{\overset{\displaystyle H}{|}}{C}-COOH$$

$$\begin{array}{c} CH_2 \\ | \\ C=\!=\!=\!CH \\ | \qquad\ | \\ HN \qquad NH \\ \diagdown\ \diagup \\ C \\ \diagup\ \diagdown \\ H \qquad H \end{array}$$

Arginine (Arg)   Histidine (His)

Fig. 1-6   (continued)

exist in nature and are called stereoisomers. The metabolic processes in living systems can usually distinguish between mirror image amino acid stereoisomers; in fact, except in rare situations, only one type of mirror image amino acid is used in all biochemical reactions.*

One nomenclature system for naming the various atoms of the side chains is illustrated in the case of lysine in Fig. 1-6. The carbon next to the carboxyl group is called the α carbon, the first atom in the side chain is the β atom, the next is the γ atom, and so forth. Only the main atoms are classified in this way. The hydrogens are ignored. Because of this system, the amino acids in life systems are also called α amino acids because it is the α carbon that is attached to the amine group. Another system calls the carbon atom of the carboxyl group number 1, the α carbon number 2, the β atom number 3, etc. These systems of nomenclature make it easier to identify a particular atom in the amino acid and also to name a substituted variety

---

* Amino acids are designated L or D depending on whether they rotate a beam of polarized light to the left or right, respectively. The L and D amino acids are essentially mirror images of one another. Except in rare situations, only L amino acids are used in biochemical reactions.

The main chemical reaction of amino acids that is of interest to us is peptide bond formation. When amino acids combine to form proteins, an amine group and a carboxyl group form a covalent bond between them with the subsequent elimination of a molecule of water. This is shown in Fig. 1-7. In this way amino acids can be linked to form polymer chains, or polypeptides, of almost any length. Given the fact that there are 20 amino acids to work with, an unlimited number of chains can be imagined in which either the sequence or number of amino acids is different; hence we can explain the chemical basis for the wide diversity of proteins. By having different chemical compositions or sequences, proteins can be formed to have totally different structures and functions. Just to give a small example of the diversity of possible sequences, let us consider the following. Suppose we were to make up a protein containing only five amino acid residues. Let us assume that each of the 20 amino acids can be used only once. How many possible pentapeptides are there? Permutation theory tells us the answer is $20 \times 19 \times 18 \times 17 \times 16$ or 1.8×10^6. Now let us suppose that we relax our original restriction that an amino acid can be used only once. Now, how many possibilities are there? The answer is 20^5 or 3.2×10^6. By relaxing our original restriction the possible number of different pentapeptides has only doubled, not increased by an order of magnitude. This latter situation is much closer to the way nature operates. In either case the possible forms our pentapeptide can take are enormous. This also means that there will be an extremely wide diversity of possible physical and chemical properties generated by the various combinations of amino acids. Even though the situation just described is impressive in terms of the possibilities, it is unrealistic because it is seldom that a protein will have as few as five amino acids. Proteins typically have 50, 75, 100, or more amino acids. Just for comparison, let us calculate the number of possible sequences for a protein containing 75 amino acids. Assuming each residue can be used any number of times, it turns out that there are over 3×10^{97} possibilities. So, by increasing the number of residues from 5 to 75 the number

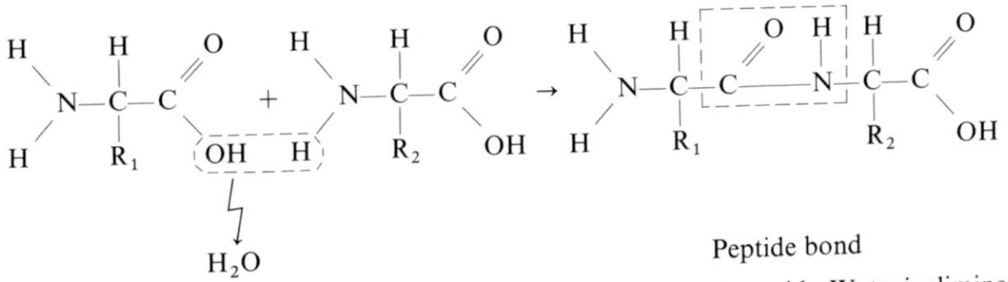

Fig. 1-7 Formation of a peptide covalent bond between two amino acids. Water is eliminated in this process.

of possible sequences has increased by 10^{91}. It is very difficult to comprehend the magnitude of these numbers; however, the possible diversity of proteins is awesome. Just as the existence of multiple bonds increases the possible variations in simple chemical structures, the sequence of amino acids accounts for the wide diversity of structure and function in proteins. Every different sequence of amino acids is going to have a unique set of physical and chemical properties. Not all possible sequences are going to be found useful, but at least the potential is there.

The physical characteristics of amino acids that are of most interest to us in studying electrical interactions are their ionization properties. The fact that free amino acids can form strong crystals and have large dielectric constants and dipole moments (Chapter 3) has led people to the conclusion that in neutral aqueous solutions amino acids are charged and are not really structured as shown in Fig. 1-6. In aqueous solvents amino acids are capable of several reactions that involve a change in charge: The NH_2 group can pick up a hydrogen ion (H^+) from an aqueous solvent, and the COOH group can discharge a hydrogen ion to the solvent. These reactions and the resulting amino acid structures are shown in Fig. 1-8. The double arrows in this figure are used to indicate that the reactions shown are reversible, that they can go both ways. When reactions (a) or (b) are in equilibrium, they are in a dynamic equilibrium, meaning that the rate of the forward reaction is equal to the rate of the reverse. In dynamic equilibrium all macroscopic quantities are stable with time; but on a submicroscopic level there is a considerable amount of fluctuation about an average value. If it were possible to follow the path of an individual H^+ ion, it would be seen that in reaction (a), for example, it is constantly combining with and being released from the amine group, as are many other H^+ ions. The net macroscopic concentration of NH_3^+ at equilibrium however is constant. This concept of dynamic equilibrium is an important one and should be thoroughly understood by the reader.

The fact that the amine and carboxyl groups of amino acids can acquire an electrical charge might then explain the presence of charges on proteins and the fact that proteins undergo electrical interactions. The flaw in this

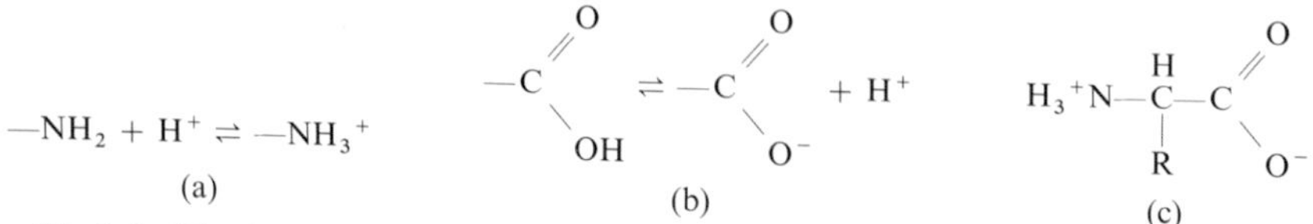

Fig. 1-8 The ionization properties of the NH_2 and COOH groups of an amino acid. (a) Amino group picking up a hydrogen ion. (b) Acid group losing a hydrogen ion. (c) Structure of an amino acid with both groups charged.

argument however is that in a protein both the NH_2 and COOH group are involved in peptide binding and hence are not free to ionize in the manner shown above. The terminal NH_2 on one end of the protein and the COOH on the other end may ionize in this fashion, but these two potential charges certainly cannot explain the fact that some proteins have net charges of over 40 electron charges. The polypeptide backbone, then, is not the major source of charge on proteins. The only part of the amino acid that is left to generate a charge is the side chain. Looking at Fig. 1-6, there are five amino acids listed in the group, charged at neutral pH; it is these amino acid side chains that are responsible for the explicit charges that show up on proteins. The COOH and NH_2 groups in the side chains are able to undergo a reaction similar to those shown in Fig. 1-8, thus giving the individual side chain a positive or negative charge. Under certain conditions of pH the net charge on a protein can be positive, negative, or zero, depending on the number and kinds of amino acids in the protein.

pH AND IONIZATION OF AMINO ACIDS

Using a somewhat nonrigorous definition, the pH of a solution is defined as

$$pH = -\log[H^+] \qquad (1-1)$$

where $[H^+]$ represents the free hydrogen ion concentration in solution in moles per liter. Pure water at 25°C has a neutral pH of 7.0; acidic solutions have lower pH, and alkaline or basic solutions have higher pH. Acidic solutions then tend to have high concentrations of H^+. Because a logarithm is used in the definition, a change in one pH unit means that the $[H^+]$ has changed by a factor of 10; so when the pH is reduced from 4 to 3, $[H^+]$ changes from 10^{-4} to 10^{-3} moles/liter.

For water, the H^+ concentration can be calculated using the law of mass action. This principle, which was first stated in the 1860s, says that if you have reactants A and B being changed to products C and D,

$$A + B \rightarrow C + D \qquad (1-2)$$

then at equilibrium the ratio of concentrations

$$K_{eq} = \frac{[C][D]}{[A][A]} \qquad (1-3)$$

is equal to a constant K_{eq}. The value of K_{eq} will vary for different temperatures. Water dissociates very slightly according to

$$H_2O \rightarrow H^+ + OH^- \qquad (1-4)$$

and the value of K_{eq} has been measured to be 1.8×10^{-16} at 25°C. We then have

$$K_{eq} = \frac{[H^+][OH^-]}{[H_2O]} = 1.8 \times 10^{-16} \qquad (1\text{-}5)$$

Since water has a molar concentration of $[H_2O] = 55.5\ M$ (1000/18), it can be found that the product $[H^+][OH^-] = 10^{-14}$; and since $[H^+] = [OH^-]$, we have $[H^+] = 10^{-7}\ M$. Using the definition, the pH of water is calculated to be 7.0.

The carboxyl group COOH is generally considered acidic because under certain conditions it can dissociate a H^+ ion to the solution as shown in Fig. 1-8b. The NH_2 group is considered basic because it can pick up a H^+ ion from solution. These descriptions follow what is known as the Bronsted definition of acids and bases where an acid is any substance that can donate a proton (hydrogen ion), and a base is a substance that can accept a proton. The easier a substance can donate a proton, the more acidic it will be.

At low pHs, where there are a lot of H^+ ions in solution, the amine group in the side chains of lysine, argine, asparagine, and glutamine form the NH_3^+ ion. The N involved in the double bond in the side chain of histidine can also pick up a H^+ and become charged. Therefore, because of the high concentration of hydrogen ions in solution at acidic pH, reaction (a) in Fig. 1-8 is pushed to the right. This means that some hydrogen ions find themselves spending more time associated with an NH_3^+ group than free in solution. As the concentration of H^+ ions in solution is reduced by raising the pH, fewer NH_2 groups will be protonated. Under the same conditions of low pH, reaction (b) is pushed to the left. At low pH values, the COOH group remains uncharged. At high pH, where the concentration of hydrogen ions in solution is low, just the opposite happens; the carboxyl group tends to be negatively charged, whereas the NH_2 group is neutral. It should be remarked that even at low pH, where NH_3^+ dominates, because of the dynamic equilibrium aspects of the reaction, there is still the statistical possibility for a COO^- group to be present. This probability is lower, the lower the pH. The concentration of COO^- ions present at a particular pH can be calculated using the law of mass action if a value of K_{eq} for the reaction is known. By adjusting the pH of a solution then, it is possible to have a dominant charge group present. Table 1-4 shows the dominant species present at low and high pH. As the transition from low to high pH takes place, or vice versa, the ratio of charged groups changes. This is usually a reversible phenomenon; e.g., NH_2 may be protonated and deprotonated many times as the pH changes.

In order to facilitate talking about the electrical properties of biomolecules several definitions will now be presented. The pK of an ionizing group is

TABLE 1-4

Forms Taken by the Amino and Carboxyl Groups at Low and High pH Values[a]

Low pH	High pH
$-NH_3{}^+$	$-NH_2$

[a] At low values of the pH the amino group is charged, whereas the carboxyl group is charged at high pH values.

that pH at which equal molar concentrations of both forms of the group exist in solution. For instance, the pK of the COOH group of the free amino acid glycine is 2.34. This means that at a pH of 2.34, there are equal concentrations of

The pK of the $NH_3{}^+$ group is 9.6, so at this pH there are equal concentrations of

The isoionic point* is that pH at which there is no net charge on the molecule. For free glycine, this corresponds to the pH of around 6.0 where we have the species

The NH_2 and COOH groups in the side chains of the amino acids also have characteristic pK values; some of these values are given in Table 1-5. When

* Sometimes called the isoelectric point.

TABLE 1-5

**Typical pK Values of Side Chain Groups
for Different Amino Acid Side Chains**[a]

Amino acid	pK of side chain
Aspartic acid	3.65
Glutamic acid	4.25
Histidine	6.0
Lysine	10.58
Arginine	12.48

[a] Data from "Handbook of Biochemistry," 2nd ed. (H. A. Saber, ed.). Chemical
Rubber Publ. Co., Cleveland, Ohio, 1970.

these amino acid side chains are incorporated into proteins, pK values can be altered due to the influence of neighboring side chains. In principle it is possible to calculate the charge on a protein if the pK values of all its ionizable amino acids are known.

It is seen that proteins can acquire an electrical charge, that this charge depends on the amino acid composition, and that the charge can be varied depending on the pH of the solution. It should be briefly remarked that other factors such as temperature and protein shape can also influence the charge on a protein, although the pH is the dominant factor.

PROTEIN STRUCTURE

While we are still discussing proteins, it is convenient to define several terms that are related to protein structure and to which we shall refer many times in subsequent chapters. It has been seen that proteins are constructed of an assemblage of amino acids, but the question now is, What shape or conformation does the string of amino acids assume? Is it straight and very rigid like an uncooked piece of spaghetti, is it very compact and tangled like cooked spaghetti, or just what does it look like? The answer to this question is different for every protein, and it is one of the basic questions in which biophysicists and biochemists are interested. In general it is extremely difficult to determine a protein's detailed shape; to date only a few proteins have been completely mapped. In order to be able to discuss the concept of structure, scientists have devised a system whereby they describe different "levels" of structure in an effort to fully understand the three-dimensional size and shape. Definitions and illustrations of these various "levels" of structure are given below.

Primary structure refers to the kinds of amino acids that comprise the protein, the number of each, and the sequence from one end of the polypeptide chain to another in which they occur. It also refers to the number of polypeptide chains that make up the protein. The primary structure imposes a fundamental limit on all other levels of structure, much as the number and kinds of structural components can determine limitations in the size and shape of a building. Figure 1-9 shows the primary structure for the protein insulin. This sequence of amino acids was deciphered by the British biochemist Fredrick Sanger in the period 1944–1954; and for his work, Sanger received the Nobel prize. Insulin was the first protein to have its amino acid sequence established. The reader should note that insulin has two polypeptide chains which are held together by S—S bonds.

Secondary structure refers to the local conformation of the polypeptide backbone. It is concerned with whether or not successive amino acids have a regular or periodic three-dimensional relationship to one another. One frequently encountered example of secondary structure in proteins is the α-helix in which the polypeptide backbone follows the shape of a helix. Figures 5-21a and 5-21b show a picture of the α-helix that was first proposed by Linus Pauling and R. Corey in the early 1950s. There are also other forms of secondary structure that are commonly found in nature. The entire length of a polypeptide chain does not necessarily have to have the same secondary structure, and in fact most proteins have several types of secondary structure depending on where one looks along the chain. There are also situations in which a polypeptide chain has no regular secondary structure. Table 1-6 shows the relative percentage of α-helix for several different proteins.

Tertiary structure refers to the overall way in which the polypeptide chain folds onto itself. Even if a polypeptide chain has no regular secondary structure, it is still possible for the chain to fold or bend back on itself to form a compact structure. Consider that the individual strands in a piece of rope have a helical shape and that it is quite usual to see the rope pass back on itself when tied in a knot. The exact three-dimensional way in which one section of the peptide chain is related to another is described in its tertiary structure. Figure 1-10 shows a representation of the tertiary structure of the protein myoglobin. The sausagelike strands are the polypeptide chain, and individual short sections have α-helical secondary structure. Proteins that have a very compact or "knotlike" tertiary structure are called globular. Proteins that have no rigid secondary or tertiary structure, but occupy all stereochemically accessible conformations on a statistical basis, are said to be in a random coil conformation.

Quaternary structure refers to the spatial relationship of the separate peptide chains if two or more separate chains are part of the protein. Many enzymes and large bio-molecules are made up of subunits, and only when

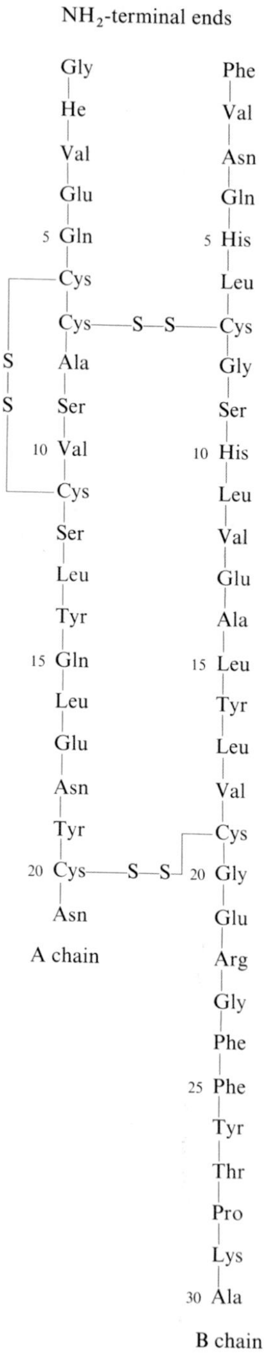

NH$_2$-terminal ends

```
        Gly                    Phe
         |                      |
         He                    Val
         |                      |
        Val                    Asn
         |                      |
        Glu                    Gln
         |                      |
    5  Gln               5   His
         |                      |
  ┌─── Cys                   Leu
  │      |                      |
  │    Cys──S─S──Cys
  │      |                      |
  S    Ala                    Gly
  |      |                      |
  S    Ser                    Ser
  │      |                      |
  │ 10 Val            10  His
  │      |                      |
  └─── Cys                   Leu
         |                      |
        Ser                    Val
         |                      |
        Leu                    Glu
         |                      |
        Tyr                    Ala
         |                      |
   15 Gln             15  Leu
         |                      |
        Leu                    Tyr
         |                      |
        Glu                    Leu
         |                      |
        Asn                    Val
         |                   ┌── Cys
        Tyr                  │    |
   20 Cys──S─S─┘ 20  Gly
         |                      |
        Asn                    Glu
                                |
     A chain                   Arg
                                |
                               Gly
                                |
                               Phe
                                |
                          25  Phe
                                |
                               Tyr
                                |
                               Thr
                                |
                               Pro
                                |
                               Lys
                                |
                          30  Ala
```

B chain

Fig. 1-9 The primary structure for the protein insulin. There are two polypeptide chains in the protein that are held together by two disulfide bridges.

26

Percentage Helix of Some Different Proteins[a]

Protein	Percent helix	Molecular weight
Myoglobin	77	16,900
Lysozyme	29	14,400
RNase	18	13,600
Insulin	31	5700
Hemoglobin	72	64,500
Papain	21	20,700

[a] Data reprinted with permission from G. Fasman, *Biochemistry* **13**, 222 (1974). Copyright by the American Chemical Society.

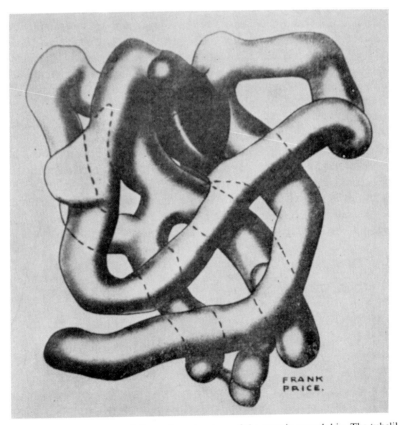

Fig. 1-10 A representation of the tertiary structure of the protein myoglobin. The tubelike portion of the molecule represents the polypeptide backbone. The flat structure in the right center is a porphorin group that actually combines with the oxygen. The tertiary structure is concerned with the manner in which the polypeptide backbone folds on itself (Kendrew *et al.*, 1958).

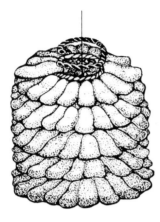

Fig. 1-11   The quaternary structure of tobacco mosaic virus shows how the protein subunits are related to one another. The individual subunits are pure protein, and the helically wound strand is the RNA of the virus (Stanford, 1975).

all subunits are present in the correct order is there any biological activity. Figure 1-11 is a model of tobacco mosaic virus which has an outer protein coat containing well over 2000 individual protein subunits. These protein subunits all fit together in a tight array to form the outer protective coating of the virus. The physical and chemical relationship of one subunit to another is described by the quaternary structure. Viruses are not the only biomolecules to have subunits. Many enzymes typically have on the order of two to four subunits and are inactive without all subunits in the proper place. Quaternary structure is concerned with the exact manner in which subunits interact with one another.

Starting with the quaternary structure which describes the gross or overall view, one successively looks at more and more details of the protein as the progression to primary structure is made. It is one of the prime tenets of molecular biology that biological function depends heavily on molecular structure; very often even a minute change in structure, at any level, will completely deactivate a protein's biological role. In the chapters to follow we shall look at how electrical interactions help determine conformation.

Another parameter used to describe macromolecules is the molecular weight, which is the sum of the atomic weights in grams of all the atoms in the structure. Table 1-7 gives some typical molecular weights of several macromolecules. To get a feel for the size of biological molecules, molecular weights of over one million are generally considered very large, whereas, molecules with molecular weights below 20,000 are considered small.

NUCLEIC ACIDS

Nucleic acids, like proteins, are widely distributed throughout nature. By far, the most commonly known nucleic acid is deoxyribonucleic acid

TABLE 1-7

Several Examples of Proteins or Protein–Nucleic Acid Structures and Their Respective Molecular Weights

Macromolecule	Molecular weight (daltons)
Tobacco mosaic virus	40,000,000
Ribosome	2,500,000
Alcohol dehydrogenase	151,000
Serum albumin	68,500
Hemoglobin	65,000
β-Lactoglobin	37,100
Myoglobin	16,700
Insulin	5733

(DNA) which forms the genetic blueprint that is passed from generation to generation in organisms. However, this is not the only role nucleic acids play. Nucleic acids can also serve as coenzymes or cofactors in enzymatic or metabolic reactions. These cofactors are compounds that are not permanently associated with an enzyme, but are necessary for the enzyme to carry out its biological function. Other nucleic acids, messenger ribonucleic acids (RNAs), are used to transfer specific parts of the genetic information in the DNA to the site of protein synthesis; they act as a template from which proteins are constructed. Still another type of ribonucleic acid, transfer ribonucleic acid, acts to capture free amino acids in the cytoplasm of the cell so that these amino acids can be incorporated into proteins. Nucleic acids are also incorporated into the structure of various biological macromolecules like ribosomes. Ribosomes are the site where nascent proteins are actually made in the cell, and nucleic acids play a significant role in their structure. So, as with proteins, nucleic acids occupy an important role in molecular biology; and they are found in a wide variety of places.

Nucleic acids are constructed of many repeating building blocks. These repeat units themselves are called nucleotides and are composed of three parts: the base, the sugar, and phosphoric acid.

Considering both ribonucleic acids (RNA) and deoxyribonucleic acids (DNA), there are a total of five commonly encountered bases. These bases are shown in Fig. 1-12. It will be noticed that each base is classified as either a purine or a pyrimidine structure (see Fig. 1-5). To be strictly correct, it should be mentioned that there are other structures these bases may take and the fact that a particular compound is drawn in a specific way on paper does not mean it exists that way, and only that way, in solution. Drawn as in Fig. 1-12, the bases clearly show their conjugated or partially conjugated structures. The five bases shown also tend to be planar, meaning the ring structures are fairly flat. The bases are the variable portion of the DNA

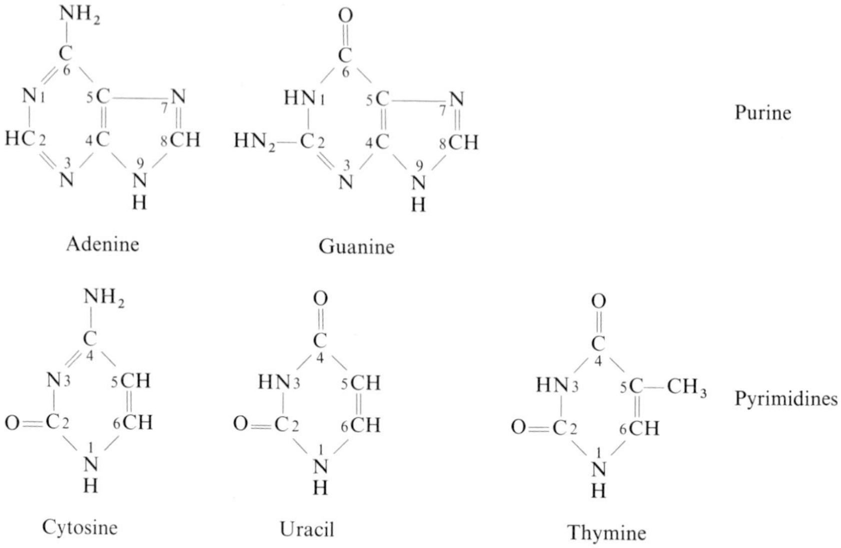

Fig. 1-12 The commonly found nucleic acid bases. Thymine is found only in DNA, while uracil is found only in RNA. Adenine and guanine are known as purines; and cytosine, uracil, and thymine are known as pyrimidines. The numbers that appear in the bases are for identifying specific atoms and helping to name the compounds.

structure, and as such are analogous to the amino acid side chains of proteins. Compared to the number of amino acids, it is seen that the nucleotide bases are rather limited in scope. The numbers that appear in the bases in Fig. 1-12 are for the purpose of identifying the various atoms of the ring when naming substituted derivatives.

The second component of a nucleotide is the sugar moiety. The two types of sugars commonly found in nucleic acids are shown in Fig. 1-13.

This type of pictorial representation for the sugars is named after the English chemist W. H. Haworth. The four-carbon ring is situated in a plane perpendicular to the page with the thick line between carbons 2' and 3' being closest to the observer. The groups that are attached to the ring are situated above and below that plane. The numbers are used to identify the various carbons in the ring. They are primed to distinguish them from equivalent numbers of the nucleotide bases. Both sugars shown are pentoses, which indicates that they have a total of five carbons. Deoxyribose, as the name implies, differs from ribose by the fact that it lacks an oxygen atom at the 2' carbon. Again, the two sugars can be drawn in different structural arrangements, but the ones shown here are the most commonly encountered.

The last member of the nucleotide complex, phosphoric acid, is shown in Fig. 1-14. Under certain conditions of pH, phosphoric acid can dissociate

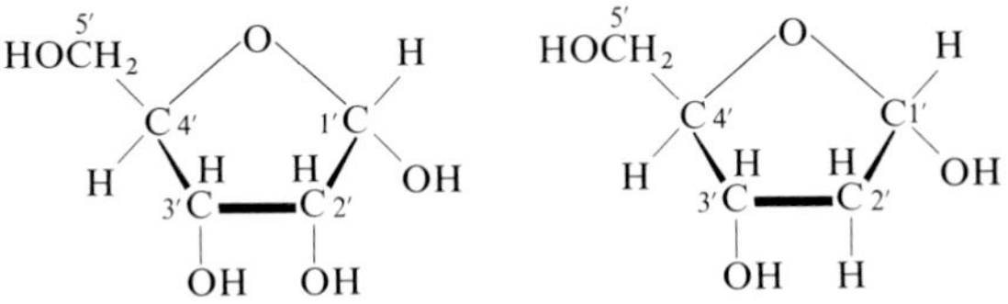

Fig. 1-13   The two types of sugars commonly found in nucleic acids. This way of drawing the sugars was devised by the English chemist W. H. Haworth. The four-carbon ring is situated in a plane perpendicular to the plane of the paper with the 2′ and 3′ carbons closest to the reader. The groups attached to the carbons in the ring are either above or below that plane. The numbers are used to identify specific atoms. The primes are used to differentiate the numbers from those for the nucleotide bases.

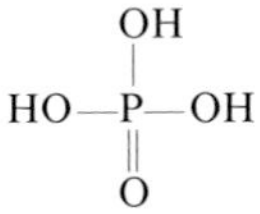

Fig. 1-14   A diagram of phosphoric acid. Notice that it has a potential of three dissociatable hydrogens. Phosphoric acid gives DNA and RNA their acidic properties.

to the solution one, two or three hydrogen ions, and it is for this reason that it is called an acid. The phosphoric acid component of DNA and RNA gives these structures their acidic properties. It should be realized that when an H^+ is dissociated, phosphoric acid acquires a negative charge. It can therefore have a maximum charge of -3. In nature, phosphoric acid is usually found with only one or two hydrogens dissociated. This is due to the fact that phosphoric acid is usually covalently bonded in a nucleic acid structure, and all of its ionizable hydrogens are not free to dissociate. This is made clearer in Figs. 1-15 and 1-16. Sometimes the symbol Ⓟ is used to represent phosphoric acid.

A nucleotide is formed by the attachment of phosphoric acid via an ester linkage to the oxygen on the 5′ carbon of the sugar, and by a glycosyl linkage of the 1′ carbon of the sugar to either the N-1 atom of a pyrimidine or the N-9 atom of a purine. This is shown in Fig. 1-15. An ester bond, or linkage, involves phosphorus or carbon being doubly bonded to one oxygen and singly bonded to another oxygen which in turn is bonded to another group. If phosphoric acid is absent from the structures shown in Fig. 1-15, the resulting compound is called a nucleoside, and more specifically adenosine and deoxyuridine, respectively. In a nucleotide phosphoric acid can dissociate a maximum of two hydrogen ions, to obtain a net of two electronic charges.

Nucleic acids are formed from nucleotides by having the phosphate group on the 5′ carbon of the sugar form another ester bond with the OH group

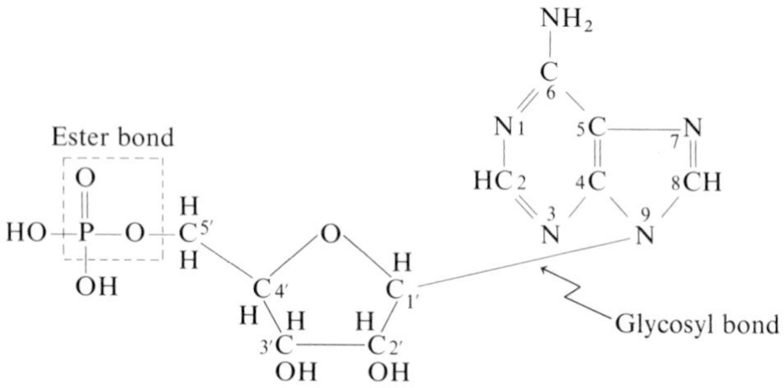

NH$_2$

Ester bond

O

HO—P—O—C$_{5'}$

OH

H

H

O

H

C$_{4'}$

C$_{1'}$

H H   H

$_{3'}$C———C$_{2'}$

OH   OH

N$_1$ $_5$C———N$_7$

HC$_2$ $_4$C $_8$CH

$_3$N   N$_9$

Glycosyl bond

Adenylic acid
(a ribose nucleotide)

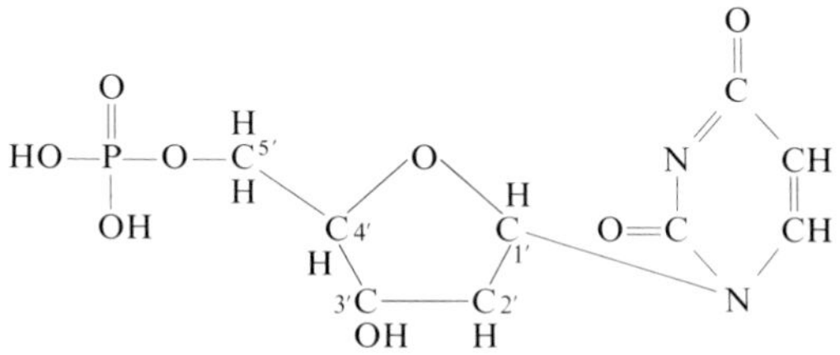

O

HO—P—O—C$_{5'}$

OH

H

H

O

H

C$_{4'}$

C$_{1'}$

H

$_{3'}$C———C$_{2'}$

OH   H

O

C

N   CH

O=C   CH

N

Deoxyuridylic acid
(a deoxyribose nucleotide)

Fig. 1-15   The manner in which a base, a sugar, and phosphoric acid join together to form a nucleotide. The sugar joins with phosphoric acid via an ester bond and with the base via a glycosyl bond.

on the 3' carbon of a sugar on another nucleotide. This is shown in Fig. 1-16. The sugar–phosphate part of the nucleic acid is called the backbone, and the order of bases is called the sequence. The nucleic acid chain has a polarity, or a specific direction, which is designated by the location of the terminal phosphates. In Fig. 1-16, the top of the figure is the 5' end, while the bottom is the 3' end. One spatial configuration that nucleic acids may assume (double helix) will be briefly discussed in another section. Since the number of nucleotide bases is considerably less than the 20 amino acids, it is necessary for the chromosomes of organisms to have very long base sequences in order to contain the needed information. To give an example of size, the molecular weight of the chromosome in the bacterium *E. coli* is on the order of 10^9 daltons. This is much larger than the molecular weight of a common protein.

The ionization properties of the nucleotides follow the same general principles as the amino acids. The phosphoric group is strongly acidic and can dissociate two H$^+$ ions if it is complexed only with a free nucleotide, or one H$^+$ ion if it is involved in a diester bond. Therefore, at neutral pHs,

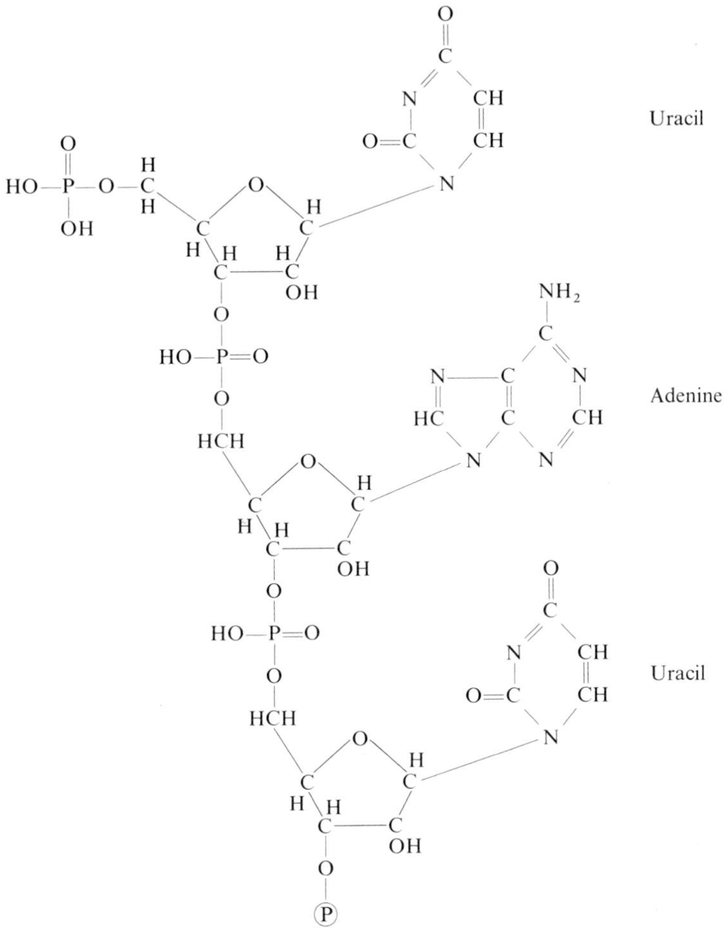

Fig. 1-16 The manner in which several nucleotides join together to form a nucleic acid chain or polymer. The separate nucleotides are joined together via phosphodiester bonds joining the 3′ carbon of one sugar with the 5′ carbon of the next. In this fashion very long chains can be constructed. The sequence of this chain is UAU.

the phosphoric acid group contributes one negative charge per base to DNA and RNA. The bases themselves all have one ionizable site, except guanine which has two; however, the details of this will not concern us here.

SUMMARY

Biological molecules are mainly composed of a rather small group of atomic elements. These bioelements are so used because of their electronic configurations and their ability to form not only different types of bonds but also strong bonds. Certain frequently encountered groups of elements are given common names, and large chemical structures are named in terms of these fundamental groups. One chemical group, the hydrocarbons, are composed of only carbon and hydrogen. This group can be straight-chained, branched, or form ring structures. They can also have multiple bonds. The hydrocarbons can form the backbone for a wide variety of other types of compounds. Proteins and nucleic acids are two very common types of biological molecules and are found in many places and have a variety of functions. They are made up of basic building blocks called amino acids or nucleotides, respectively; these chemical building blocks can acquire charges depending on the pH, thus giving the overall structure a net electrical charge.

GENERAL REFERENCES

Asimov, I. (1962). "The Genetic Code." New American Library, New York.
Conn, E. E., and Stumpf, P. K. (1967). "Outlines of Biochemistry." Wiley, New York.
Dickerson, R. E., and Geis, I. (1973). "The Structure and Action of Proteins." W. A. Benjamin, Menlo Park, California.
Freifelder, David (1975). "Physical Biochemistry." Freeman, San Francisco, California.
Lehninger, A. L. (1970). "Biochemistry." Worth, New York.
Morrison, R. T., and Boyd, R. N. (1966). "Organic Chemistry," 2nd ed. Allyn and Bacon, Boston.
Steiner, R. F. (1965). "The Chemical Foundations of Molecular Biology." Van Nostrand, Princeton, New Jersey.

REFERENCES

Crick, F. H. C. (1954). The structure of hereditary material, *Scientific American* October, p. 54.
Doty, Paul (1957). Proteins, *Scientific American* September, p. 173.
Fasman, G. (1974). *Biochemistry* **13**, 222.
Kendrew, J. C., Bodo, G., Dintzis, H. M., Parrish, R. G., Wyckoff, H, and Phillips, D. C. (1958). Three-dimensional model of the myoglobin molecule obtained by x-ray analysis, *Nature* **181**, 662.
Philips, D. C. (1966). The three dimensional structure of an enzyme molecule, *Scientific American* November, p. 78.
Stanford, A. L. (1975). "Foundations of Biophysics." Academic Press, New York.

2

ELECTROSTATICS

INTRODUCTION

In this chapter a basic outline of electrostatics will be presented. To those who have already taken a course in physics, parts of this material should be a review of topics already seen. To those who are unfamiliar with electrostatics, this chapter represents the minimum that should be known before proceeding to other chapters. The topics presented here form a basis upon which future treatments will be based. Just as it is necessary to understand the basic biochemistry of macromolecules, so too is it necessary to have a command over the basic experimental and theoretical concepts of electrostatics before electrical properties and interactions of macromolecules can be discussed. For those who would like additional readings in electrostatics, several references are listed at the end of the chapter.

It was known as far back as 600 B.C. that various pieces of matter could attract or repel one another by rubbing the first object and then holding it close to, but not touching, the second object. The generalized concept that matter could interact through some type of force other than gravitional is thus a fairly old one; however, it remained until the latter part of the nineteenth century and the beginning of the twentieth century for the formalism to describe electrical interactions to be developed. It was at this time that the basic model of the atom was proposed and its various components (electron, proton, neutron) were actually detected and characterized. In the days of Ben Franklin, electricity was thought to have been a continuous quantity, but the atomic models of J. J. Thompson and Niels Bohr proposed that the units of the atom and electrical charge were actually discrete entities. It was also in the latter part of the nineteenth century that J. C. Maxwell developed his famous set of four equations that gave a unifying viewpoint to electrical and magnetic phenomena. From a historical standpoint this era was a very exciting time since it was then that the basic units of electrical charge were finally verified by experiment and then ceased to be merely a figment in physicists' imaginations. It was with this general background that the scientific attack on the solution properties of biomolecules began in the 1920s and 1930s. In this endeavor many of the basic theoretical interpretations were based upon the fundamental principles of electrostatic theory, and it is these principles that will be presented here.

COULOMB'S LAW

Even though the electron and proton were not characterized until the twentieth century, in 1785 a French scientist named Charles Augustin de Coulomb performed a series of experiments in which he investigated the basic nature of electrical forces between charged particles. Using a device known as a torsional balance, which is shown schematically in Fig. 2-1, Coulomb placed various amounts of electric charge on balls a and b thus causing them to attract or repel one another depending on the charge. Since ball a was free to move, and in the process twist the fine wire supporting the cross arm, Coulomb could have measured the angle θ through which a given charge caused the wire to twist. An equivalent procedure, and the one actually used by Coulomb, was to keep the two balls a fixed distance apart by manually twisting the supporting wire himself as the charges caused repulsion or attraction of the two balls. From these experiments Coulomb deduced that the fundamental electrical force acting between the balls was directly proportional to the amount of charge on both balls and inversely

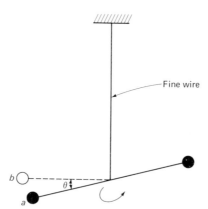

Fig. 2-1   A representation of the torsional balance used by Coulomb. A small rod supporting two balls is suspended by a thin wire so that the rod may rotate. One of the balls a may be charged. Another ball b may also be charged and brought near ball a. The electrical forces will then cause the balls to separate or come closer together. The strength of a particular electrical force could be determined by how much the suspension wire twisted or conversely by how much force had to be applied to twist the wire to keep the two balls a fixed distance apart.

proportional to the square of the distance separating them. He also concluded that like charges repelled one another, whereas unlike charges attracted each other. These observations were to become some of the most basic principles in electrostatics. From the mechanical laws previously derived by Newton it was also necessary that the electrical force act along a line between the two balls. Coulomb's observations can be expressed mathematically as

$$F \propto q_1 q_2 / r^2 \qquad (2\text{-}1)$$

where q_1 and q_2 represent the magnitude of the two charges, and r is the distance of separation. The symbol $\propto$ means proportional. This relationship can be transformed into an equality by introducing a constant of proportionality k:

$$F = k q_1 q_2 / r^2 \qquad (2\text{-}2)$$

Equation (2-2) is then a mathematical statement of Coulomb's law which holds only for the special case where charges q_1 and q_2 are in a vacuum and can be represented by points in space. The reason for this latter restriction is due to certain mathematical contradictions that arise otherwise. The form Coulomb's law takes when the charges are immersed in a medium will be discussed in the next chapter. The reader should realize that Coulomb's law is one of the most fundamental and important in electrostatics and that it is based upon experimental fact.

UNITS

In order to obtain useful quantitative information from Coulomb's law and other expressions to follow, it is necessary to adopt a system of units in which charge, force, and distance are measured. There are presently two systems of units currently in use to describe electrical interactions: electrostatic (or CGS) and MKS. In the electrostatic system the units of mass, length, and time are the gram, centimeter, and second, respectively. The unit of force is the dyne (abbreviated, dyn) which has units of gram centimeters per second squared (g cm/sec^2).* The unit of charge is called the statcoulomb or electrostatic unit (esu), and it is defined such that the constant of proportionality (k) in Coulomb's law is equal to 1. A statcoulomb,

---

* The unit of force may be derived by considering Newton's second law of motion. Essentially it says that the net force on an object is equal to the product of its mass times the acceleration, or $F = ma$. If the unit of mass is the gram and the units of acceleration are centimeters per second squared, then the units of force must be gram centimeters per second squared, one of which is defined to be 1 dyn.

then, is that charge which repels an equal charge of the same sign with a force of 1 dyn when the two charges are separated by a distance of 1 cm. For this system of units, Coulomb's law takes the form

$$F = q_1q_2/r^2 \qquad \text{(Coulomb's law using electrostatic units)} \qquad (2\text{-}3)$$

This system of units is older than the MKS system and is frequently used by people in the biophysical sciences, although others sometimes prefer the MKS system of units. Electrostatic units will be used almost exclusively in this textbook.

One reason to prefer using the MKS system of units is that these units are directly related to commonly encountered electrical terms, such as ampere (A) and volt (V). In the MKS system the units of length, mass, and time are the meter, kilogram, and second, respectively. The unit of force is the newton (N), which has units of kilogram meters per second squared (kg m/sec^2). The unit of charge is the coulomb (C), which is defined as the amount of charge that in one second passes through the cross section of a conductor that is carrying a current of one ampere. Since the MKS units are defined in this way, the constant of proportionality k in Coulomb's law is not equal to unity, but rather takes on a value that must be determined experimentally. This value of k has been determined, and it is found that

$$k = 8.987 \times 10^9 \quad \text{N m}^2/\text{C}^2 \qquad (2\text{-}4)$$

Table 2-1 summarizes the units of each system.

Because of its convenience for the development of other equations, physicists frequently write Coulomb's constant k as

$$k = 1/4\pi\varepsilon_\circ \qquad (2\text{-}5)$$

where $\varepsilon_\circ$ is called the permittivity of empty space and is equal to

$$\varepsilon_\circ = 8.85 \times 10^{-12} \quad \text{C}^2/\text{N m}^2 \qquad (2\text{-}6)$$

TABLE 2-1

Summary of Units for the CGS and MKS Systems

	Electrostatic	MKS
Mass	gram (g)	kilogram (kg)
Length	centimeter (cm)	meter (m)
Time	second (sec)	second (sec)
Force	dyne (dyn)	newton (N)
Energy	erg or kilocalorie (kcal)	joule (J)
Charge	statcoulomb (statC) or esu	coulomb (C)
k	1	8.987×10^9

Hence Coulomb's equation in the MKS system becomes

$$F = \frac{1}{4\pi\varepsilon_\circ} \frac{q_1 q_2}{r^2} \qquad (2\text{-}7)$$

The units of charge may be changed from one system of units to another by the conversion factor

$$1 \quad C = 3 \times 10^9 \quad \text{statC} \qquad (2\text{-}8)$$

The electric charge on an electron or proton is given as

$$e^- = 1.6 \times 10^{-19} \text{ C (MKS)} \quad \text{or} \quad e^- = 4.8 \times 10^{-10} \text{ statC (esu)} \quad (2\text{-}9)$$

Although the reader may not realize it, one coulomb of charge is an extraordinary amount of charge. One may gain a feeling for its size by realizing that a typical lightning bolt transfers on the order of only 20–30 C of charge to the earth. It should also be emphasized that Coulomb's law is a statement of experimental fact and that it is really independent of which system of units are used. The two systems of units were developed purely as a matter of convenience, and they in no way alter the basic physics of the situation.

Coulomb's initial measurements were rather crude compared to those possible with modern technology, but his basic observations were quite correct. In fact, it has been verified that the exponent of the distance dependence of electrical force is equal to 2.0 ± 0.000000002. It is interesting to realize that Newton's law of gravity also has an inverse squared distance dependence and that Coulomb's law bears a striking resemblance to it:

$$\begin{array}{ll} \text{Newton's law of gravity} & F = Gm_1 m_2/r^2 \\ \text{Coulomb's electrical law} & F = kq_1 q_2/r^2 \end{array} \qquad (2\text{-}10)$$

In Newton's law of gravity G is a constant of proportionality and m_1 and m_2 are the masses of the interacting bodies. Written as in Eqs. (2-10), it is seen that mass and charge occupy the same position of importance in the two types of interactions.

A relatively simple application of Coulomb's law to a biochemical phenomenon can be seen by considering the different pK values of aspartic and glutamic acid. Table 1-5 shows that the side chain carboxyl of aspartic acid has a pK that is almost $\frac{1}{2}$ pH unit lower than that of glutamic acid, yet they both have a very similar structure. Why is this so? The answer has to do with the physical length of the respective side chains. Aspartic acid's side chain is shorter, hence its side chain carboxyl group is physically closer to the NH_3^+ attached to the α carbon. The positive charge of the NH_3^+ group therefore tends to repel the H^+ ion of the carboxyl more on the aspartic acid side chain than on the glutamic acid side chain. In other words, the positive charge of the NH_3^+ encourages the H^+ to dissociate. From Coulomb's law, the closer the two charges are to begin with, the stronger is the repulsive force between the two ions; hence a lower pH is needed for the H^+ to dissociate for the amino acid with the shorter side chain.

VECTORS AND SCALARS

Before continuing directly with electrostatics, it is first necessary to consider a few mathematical points.

The generalized concept of force is one that has several subtle facets. The quantity called force, and electrical force is a special case, is generally classified from a purely mathematical point of view as a vector. A vector is defined as a quantity that has a magnitude and also a direction; it takes a knowledge of both for a completely adequate description. It is not only necessary to know the strength of a force, but also its direction of action. The symbol → over a letter or symbol is used to designate a vector quantity in handwritten equations; boldface symbols are used in this text. A scalar quantity, on the other hand, has a magnitude with no specific direction; i.e., a number is a scalar, as are mass, time, and temperature. The difference between vector and scalar quantities is an important concept in electrostatics, and the manipulation of both types of quantities should be thoroughly understood. For those who are unfamiliar with vectors, Appendix A is provided to cover some of their most useful properties. The formulas given so far for Coulomb's law have indicated only the strength of the force, although it was stated that the direction of a repulsive or attractive force between two particles lies on the line connecting the two charges. This then defines the direction for the force of interaction between only two electrical bodies. The obvious question now is, How does one find the resultant force when a number of charges are involved? This is best illustrated by an example.

Consider the situation depicted in Fig. 2-2, where it is required to find the resultant force on charge q_D. By resultant force is meant the strength of this force and also its direction. In Fig. 2-2 we have four charges, and it is desired to find the resultant force acting on charge q_D due to the other three charges. The magnitude of each charge is in esu units as shown in the diagram. To find the resultant force acting on charge q_D, it is necessary to consider all the individual forces acting on q_D. These are also shown in the diagram. The directions of these forces were partially determined by the fact that like charges repel and unlike charges attract. The individual forces are also along the straight line joining the two charges under consideration. To find the resultant force, it is convenient to construct a rectangular coordinate

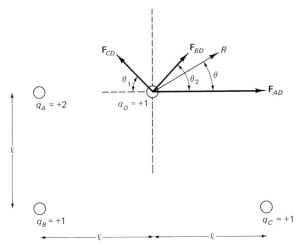

Fig. 2-2 Four charges located in a plane. It is desired to find the resultant force of three charges on the fourth. Each pair of charges will generate a force between them, and the resultant effect is found by vectorially summing all of the individual forces. The individual forces between any two charges is indicated by an arrow and a subscripted **F**. The angle θ indicates the angle with respect to the horizontal axis of the resultant force (**R**). A rectangular coordinate system is shown as dotted lines.

system with its origin at q_D. The problem then reduces to finding the resultant of three vectors (see Appendix A). Because of symmetry, we have $\theta_1 = \theta_2 = 45°$. Table 2-2 summarizes the components of the forces. The algebraic sums of the x components and of the y components are equal to the x and y components of the resultant force, respectively, and are given by

$$R_x = \sum F_x = (2)l^{-2}, \qquad R_y = \sum F_y = (\sin 45°)l^{-2} \qquad (2\text{-}11)$$

The magnitude of the resultant force is

$$R = (R_x^2 + R_y^2)^{1/2} = \left(\frac{4}{l^4} + \frac{\sin^2 45°}{l^4}\right)^{1/2} = \frac{(4.5)^{1/2}}{l^2} \qquad (2\text{-}12)$$

TABLE 2-2

Summary of Force Components for the Situation in Fig. 2-2

Force	x component	y component
F_{CD}	$-\frac{1}{2}l^{-2}\cos 45°$	$\frac{1}{2}l^{-2}\sin 45°$
F_{BD}	$\frac{1}{2}l^{-2}\cos 45°$	$\frac{1}{2}l^{-2}\sin 45°$
F_{AD}	$(2)l^{-2}$	0
	sum $= 2l^{-2}$	sum $= (\sin 45°)l^{-2}$

and θ is given by

$$\theta = \tan^{-1}\left(\frac{R_y}{R_x}\right) = \tan^{-1} 0.353 \simeq 19° \qquad (2\text{-}13)$$

Therefore, the resultant force is at an angle of 19° with respect to the horizontal. Since q was given in esu units, and if l is measured in centimeters, then the units of F will be dynes. This example illustrates the general procedure for deducing the net result when a number of electrical forces are influencing the body of interest. The reader should thoroughly understand the principles involved since they will be used again and again.

ELECTRIC FIELD

In developing the formalism to describe how an electric charge influences its surroundings, physicists have introduced the concept of the electric field. The essence of this idea is that a charged body perturbs all space around it by the very fact that it has a charge, and that this perturbation of space is described by an electric field. It is this electric field that interacts with other charges and actually exerts a force on them. Charged particles act on one another through an electric field. It was once thought that two charged particles could interact with each other directly and instantaneously, but this view is no longer accepted. In field theory the field acts as an intermediary, and changes of the field are propagated at the speed of light, not instantaneously. This important fact has been verified experimentally. The electric field is operationally defined at a specific point in space by placing a positive test charge q at the point in question and measuring the force exerted on this charge. The electric field is then defined as

$$\mathbf{E} = \mathbf{F}/q \qquad (2\text{-}14)$$

Since force $\mathbf{F}$ is a vector quantity, the electric field $\mathbf{E}$ must also be a vector. A very rigorous definition of $\mathbf{E}$ would be similar to Eq. (2-14) except in the limit where q approaches zero magnitude. The units of $\mathbf{E}$ are newtons per coulombs in the MKS system, and dynes per esu in the esu system. The direction $\mathbf{E}$ takes is the same as the direction of the force experienced by the test charge q. Using Eqs. (2-3) and (2-14), the magnitude of the field of a single point of charge q is given using the esu system of units as

$$E = q/r^2 \qquad (2\text{-}15)$$

In this case $\mathbf{E}$ points radially outward if q is positive or radially inward if q is negative. This is shown in Fig. 2-3.

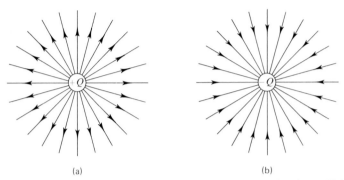

(a)   (b)

Fig. 2-3   The direction of the electric field for a positive and a negative charge. If the charge of interest is positive, the field lines radiate outward; if negative, they are pointed radially inward. The direction of the electric field at a point is the same as the direction of the force acting on a small positively charged test charge that is occupying that point.

Now that we have developed the mathematical formalism to calculate electric fields and forces, it is appropriate to ask whether or not electrical forces are really powerful enough to warrant spending so much time on them. Alternatively, the question is whether or not electrical forces between biological macromolecules are significant. The fact that this book has thickness indicates that they are, but it is useful to make some sample calculations. To get more of a feeling for the strength of electrical fields, let us calculate the strength of the field generated by an electron at a distance of 10 Å (angstroms) or 10×10^{-8} cm. We have

$$E = -(4.8 \times 10^{-10} \quad esu)/(10^{-7} \quad cm)^2 = -4.8 \times 10^{+4} \quad dyn/esu$$

This is quite a large field, and in fact it is enormous. This point will again be amplified later in this chapter when the unit of potential difference (volt) is introduced. If it is now desired to calculate the force exerted on another electron placed 10 Å from the first one, Eq. (2-14) is to be used. This force is calculated to be

$$F = -(4.8 \times 10^4 \quad dyn/esu)(4.8 \times 10^{-10} \quad esu) = 2.3 \times 10^{-5} \quad dyn$$

This is to be compared to a force of 5.5×10^{-48} dyn for the force of gravity acting between two electrons separated by 10 Å. Similarly, the weight of the electron itself can be calculated to be equal to $mg = 8.9 \times 10^{-25}$ dyn where g is the acceleration due to gravity and m is the electron mass. As another comparison, let us calculate both the electrical and gravitational forces acting between two typical biomolecules. Human serum albumin has a molecular weight of about 69,000 daltons and is capable of becoming highly charged. One of its roles is the transportation of metabolites through the blood stream. Let us assume it has 10 electronic charges while the carried metabolite has 1 electronic charge and a molecular weight of 500 daltons. From these data it can be calculated that the electrical force of attraction is 9.2×10^{-4} dyn, while the gravitational force is equal to 2.6×10^{-33} dyn, for a distance of separation equal to 5 Å. From these examples we see that electrical forces can be very large compared to gravitational forces, so one would expect that electrical forces between biomolecules will be significant.

To find the resultant electric field at a specific point due to a collection of charges, it is necessary to vectorially add the contributions from each individual source in a fashion similar to the example in the previous section. Mathematically, this is written as

$$\mathbf{E}_{tot} = \sum_i \mathbf{E}_i \qquad (2\text{-}16)$$

where the $\mathbf{E}_i$ are the individual fields. If there is a continuum of charges instead of discrete charges, then the summation of Eq. (2-16) is replaced by an integration. An example best illustrates this concept.

Consider that we have an infinite rod whose length has a continuous charge density γ (charge per unit length); it is required to find the value of the electric field a distance x away from the rod (Fig. 2-4). The approach

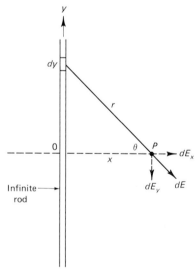

Fig. 2-4   An infinite rod whose length is uniformly covered with charge. The charge per unit length is denoted as γ, and x is the distance perpendicular to the rod. It is desired to find the value of $\mathbf{E}$ at a point P that is x units away from the rod. The variable r denotes the distance between an arbitrary infinitesimal length of the rod dy and the point of interest. The dotted line represents the x axis of a cartesian coordinate system, whereas the rod itself is the y axis.

taken will be to find the field due to the infinitesimal charge in length dy of the rod, and to then add all such contributions vectorially. The infinitesimal charge in the length of rod dy is given by $dq = \gamma\, dy$, so that the magnitude of the infinitesimal field due to dq is

$$dE = \frac{dq}{r^2} = \frac{\gamma\, dy}{r^2} \qquad (2\text{-}17)$$

The components of dE along the x and y directions are given by

$$dE_x = dE \cos\theta = \gamma \cos\theta\, dy/r^2, \qquad dE_y = dE \sin\theta = \gamma \sin\theta\, dy/r^2 \quad (2\text{-}18)$$

Since our rod is assumed to be of infinite length, the component of the field parallel to the rod will be zero, i.e., $E_y = 0$. This is due to the fact that corresponding elements of charge are symmetrically located on either side of the origin, with the y components of each element's field canceling one another out. Knowing this, we now have to worry only about the x component. The x component is given by

$$E_x = \int_{-\infty}^{\infty} \frac{\gamma \cos\theta}{r^2}\, dy \qquad (2\text{-}19)$$

but by simple trigonometry we know

$$\cos\theta = \frac{y}{(x_2 + y^2)^{1/2}}, \qquad r^2 = x^2 + y^2 \qquad (2\text{-}20)$$

so

$$E_x = 2\gamma \int_0^{\infty} \frac{y\, dy}{(x^2 + y^2)^{3/2}} \qquad (2\text{-}21)$$

where instead of integrating y from $-\infty$ to ∞, the limits are taken as 0 to ∞, and the whole integral is multiplied by 2. The integral in Eq. (2-21) can be looked up in a table of integrals, which shows

$$\int_0^{\infty} \frac{y\, dy}{(x^2 + y^2)^{3/2}} = -\frac{1}{(x^2 + y^2)^{1/2}} \bigg|_0^{\infty} = \frac{1}{x} \qquad (2\text{-}22)$$

The quantity x here is treated like a constant, y is the only variable. The final answer is

$$E = E_x = 2\gamma/x \qquad (2\text{-}23)$$

The answer then states that the electric field depends only on the quantities x and γ, not on y at all. Any point P that is x units away from the rod has the same electric field regardless of where you are along its length. This is true only because the rod is assumed to have infinite length. One may also wonder what the value of this calculation is since infinite rods are seldom encountered. One case where it is useful comes about when considering the electric field (or force) on a particle that is located very close to a wire. In this case, for all practical purposes, the particle is "seeing" a wire of infinite length. The farther out the particle moves, the less precise is this approximation.

In the situation where a detailed charge distribution is unknown, and only the electric field at a specific point is known, the force acting on a charge q can be found from the relation $\mathbf{F} = q\mathbf{E}$.

GAUSS'S LAW

Calculating the direction and magnitude of an electric field can be a somewhat cumbersome and tedious procedure, as was just demonstrated in the previous example. For certain situations, however, an appropriate application of Gauss's law can dramatically shorten this procedure. Gauss's law is useful not only for this purpose, but in the chapter on the Debye–Huckel theory (Chapter 7) it will be used to derive an important equation of classical electrodynamics. Before considering Gauss's law directly, the idea of lines of force must be introduced.

The concept of lines of force originated with the English experimenter Michael Faraday as a convenient way of visualizing the shape and strength of an electric field. These lines of force are imaginary and are drawn in a manner such that the number of lines passing through a unit cross-sectional area represents the magnitude of the field, while the tangent to a line of force at a point represents the direction of the field at that point. Figure 2-5a shows the appropriate lines of force for two equal, but oppositely charged particles. The arrows indicate the direction a small positive test charge would be pushed if placed on any line. The high density of lines near each charge, and in the region between the charges, indicates regions of high electric field. Far from the charges, the lines are relatively far apart because the field is weaker. The lines in Fig. 2-5a also give a feeling for the inhomogeneity of the field in this particular case.

The flux of an electric field is defined as the number of lines of force that pass at right angles through an arbitrary surface. Referring to Fig. 2-5b, the reader can see that the lines of force passing through a surface S surrounding a positive charge are pointing outward, and this direction is defined as positive. Qualitatively, one can visualize the situation where the flux crossing a surface enclosing a charge will depend on the amount of charge within the surface. This is illustrated in Fig. 2-5b, where the dotted lines represent surfaces that have equal radii, and hence equal area in all three

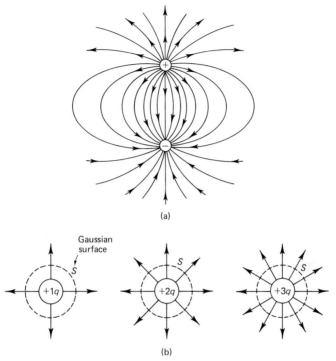

(a)

Gaussian
surface

(b)

Fig. 2-5 The lines of force for some simple situations. (a) The lines of force for two equal but oppositely charged particles. The lines of force represent the strength and direction of the electric field at any point. The larger the number of lines that pass through a cross-sectional area, the larger is the strength of the field. This is illustrated in (b) where three different particles have charges of different magnitudes. The dotted lines represent gaussian surfaces S that are all of the same radius. The number of force lines that cross the surfaces S are proportional to the charges contained within. The arrows on the force lines indicate the tangential direction a small test charge would be pushed if placed at that point.

cases. The flux passing through the surface surrounding the charge of $3q$ is obviously three times greater than the corresponding flux for the charge of $1q$. It seems likely, then, that a quantitative expression could be derived relating flux through a closed surface to the net electrical charge contained within that surface. This can be done, and the result is known as Gauss's law. Before writing this law it is first necessary to develop a mathematical description of flux.

Consider the situation depicted in Fig. 2-6 where a surface is drawn in the path of the lines of force. Intuitively, one would expect that a mathematical expression for flux could be written as a product of the electric field strength E and the area of the surface S. However, this is not quite adequate since the lines of force cross the arbitrary surface S at many different

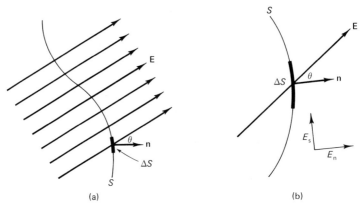

Fig. 2-6   The line of force passing through an arbitrary surface S. The area ΔS is an infinitesimally small subsurface, and **n** is a vector of unit length pointing at a right angle to ΔS. (b) An enlarged view of the situation in (a). The two components of the electric field **E** are designated as E_s and E_n, being respectively parallel to and perpendicular to the surface ΔS. The angle θ defines the direction of **E** relative to **n**. $E_s = E \sin \theta$ and $E_n = E \cos \theta$.

angles. To solve this problem, let the surface be subdivided into many very small subsurfaces ΔS, and let the orientation of each subsurface be defined by a vector **n** whose magnitude is unity and whose direction is always pointing perpendicularly outward from the surface. Figure 2-6b shows an enlargement of Fig. 2-6a. The electric field vector passing through ΔS can be decomposed into mutually perpendicular components that are parallel to the vector **n** (E_n) or parallel to the subsurface (E_s). The amount of **E** that actually passes through the subsurface is E_n, or $E \cos \theta$. The component E_s actually passes parallel to the surface and does not pass through it; hence, this component cannot be considered in the definition of flux. The flux through ΔS is then defined as

$$\text{flux through } \Delta S = E \cos \theta \, \Delta S \qquad (2\text{-}24)$$

which can be rewritten using the vector dot product notation as

$$\text{flux} = \mathbf{E} \cdot \Delta S \, \mathbf{n} = \mathbf{E} \cdot \Delta \mathbf{S} \qquad (2\text{-}25)$$

where $\Delta \mathbf{S}$ is the vector formed by multiplying the scalar ΔS by the vector **n**. The flux passing through the entire surface is found by adding all the contributions from the individual areas ΔS, or

$$\text{total flux} \equiv \phi = \sum_i \mathbf{E}_i \cdot \Delta \mathbf{S}_i \qquad (2\text{-}26)$$

If it is assumed that in the limit as the subsurface area shrinks to zero and that the limit of the right-hand side of Eq. (2-26) exists, then the summation in Eq. (2-26) can be replaced by an integral. We then have

$$\phi = \lim_{\Delta S \to 0} \sum_i \mathbf{E}_i \cdot \Delta \mathbf{S}_i = \oint_S \mathbf{E} \cdot d\mathbf{S} \qquad (2\text{-}27)$$

where the $\oint_S$ sign indicates that the integral is to be evaluated over the entire surface area S. Equation (2-27) is then the mathematical definition of flux. If the angle θ between $\mathbf{E}$ and $\mathbf{n}$ is less than $90°$, the flux is positive; if θ is greater than $90°$, the flux is negative; and if $\theta = 90°$, there is no flux.

Gauss's law can now be written mathematically as

$$\oint_S \mathbf{E} \cdot d\mathbf{S} = 4\pi q_s \qquad (2\text{-}28)$$

This equation states that the net flux through any closed surface S is equal to 4π times the net charge q_s contained within the closed surface. If the MKS system of units is to be used, the 4π factor on the right-hand side of Eq. (2-28) does not appear. The total flux through the surface is independent of the spatial arrangement of the charges; and if there are no charges, then the net flux is zero.

The advantages of using Gauss's law in calculating $\mathbf{E}$ can be seen in situations where the integral in Eq. (2-28) can be easily evaluated. These situations occur where the charges are arranged with a high degree of symmetry and a convenient surface (gaussian surface) can be constructed. The enclosing gaussian surface can be of any shape or size, but usually a clever geometric choice makes the integral in Eq. (2-28) particularly simple. Two examples will be given to illustrate the application of Gauss's law.

Consider the single positive charge shown in Fig. 2-7 where the lines of force are directed radially outward; it is required to determine the electric field at a distance r from the charge. A clever geometry for a gaussian surface is a spherical surface of radius r. By choosing this geometry, the lines of force always cross the surface at right angles so that the product $\mathbf{E} \cdot d\mathbf{S}$ reduces to $E dS \cos 0° = E dS$. Since the spherical surface is everywhere equidistant from the charge q, E is constant over the surface. Gauss's law now states that

$$\oint_S \mathbf{E} \cdot d\mathbf{S} = E \oint_S dS = 4\pi q \qquad (2\text{-}29)$$

E is brought out in front of the integral because it is a constant. The integral itself is just the summation of all infinitesimal surface areas dS on the spherical surface; this is just equal to the total area of the spherical surface:

$$\oint_S dS = 4\pi r^2 \qquad (2\text{-}30)$$

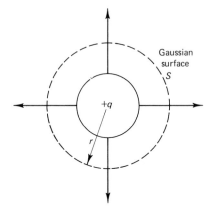

Gaussian
surface
S

Fig. 2-7 A gaussian surface shown surrounding a charge q. The surface is everywhere perpendicular to the lines of force, and is also equidistant from the charge.

So Gauss's law finally gives us

$$E \oint_S dS = 4\pi r^2 = 4\pi q \qquad \text{or} \qquad E = q/r^2 \qquad (2\text{-}31)$$

which we know to be true from previous considerations. The key to the application of Gauss's law was in finding a geometrical shape where the lines of force were always perpendicular to it so the dot product reduced to a simple scalar product.

A somewhat more complex situation is shown in Fig. 2-8 where there is an infinitely long rod having a linearly uniform positive charge density γ. It is required to find the electric field at a distance r from the rod. This problem has already been solved by direct integration using the definition of the electric field; however, using that method it was necessary to solve a non-trivial integral. A simpler solution can be achieved via Gauss's law by constructing a cylindrical gaussian surface around the rod as shown in Fig. 2-8. By symmetry arguments it can again be argued that the electric field is everywhere perpendicular to the rod and also to the cylindrical surface; hence the lines of force are perpendicular to the cylindrical surface also. There is no flux through either the top or bottom surface because the lines of force are parallel to these parts of the cylinder. Gauss's law then states

$$\oint_S \mathbf{E} \cdot d\mathbf{S} = E \oint_S dS = E2\pi r l \qquad (2\text{-}32)$$

$$E2\pi r l = 4\pi q = 4\pi l \gamma \qquad (2\text{-}33)$$

so that

$$E = 2\gamma/r \qquad (2\text{-}34)$$

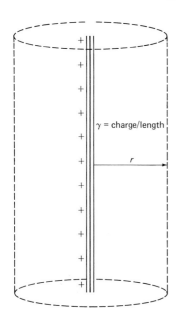

γ = charge/length

r

Fig. 2-8 A gaussian surface surrounding a rod that has a uniform charge per unit length of γ. The gaussian surface, except for its ends, is every-where equidistant from the rod.

which is identical to the previously derived expression. The reader should be aware that Eq. (2-33) would have resulted regardless of the shape of the gaussian surface chosen; it just happens that in the example shown the choice of a cylinder makes for a particularly speedy solution. The results of this example will be used in Chapter 5.

ELECTRIC POTENTIAL

In describing the electric field so far we have used the electric vector **E** which has all the properties of a vector. Alternatively, it is equally possible to describe an electric field in terms of a scalar quantity called the electric potential V. The decision of which one to choose generally depends on the particular situation, and it can be shown that knowledge of either the potential or the field implies a knowledge of the other. The reason for introducing the concept of potential lies in the advantage of simplifying the mathematics of some situations. The scalar nature of the potential makes it unnecessary to worry about components and vector addition.

In describing the electric potential it is usual to speak only of a potential difference rather than to refer to an absolute potential. The definition of the potential difference between two points A and B is defined as

$$V_B - V_A = W_{AB}/q \qquad (2-35)$$

where W_{AB} is the amount of work required to move a test charge q from point A to point B. W_{AB} can be positive, negative, or zero. If point B is at a higher potential than point A, W_{AB} is positive. If point B is at a lower potential, then W_{AB} is negative. If A and B have the same potential, $W_{AB} = 0$. A reference point at infinity is arbitrarily defined to have zero potential; hence Eq. (2-35) becomes

$$V_B - V_\infty = W_B/q \qquad (2\text{-}36)$$

where now W_B is the work involved in moving q in from infinity to point B. If work must be performed against an electric force to bring q in from infinity, then W is positive. If the electrical forces are such that the test charge is pulled in from infinity, then W is negative. This convention is arbitrary, and arises because infinity has been defined to have zero potential.

Since work can be defined as the scalar product of force times distance

$$W = \mathbf{F} \cdot \mathbf{r} \qquad (2\text{-}37)$$

the units of electric potential can be derived from its definition as newton meters per coulomb = joules per coulomb (N M/C = J/C) in the MKS system and dyme centimeters per esu = ergs per esu (dyn cm/esu = ergs/esu) in the esu system. One joule per coulomb is also defined as a volt (V). Since work is a scalar quantity, the value of W_B in Eq. (2-36) is independent of the path; i.e., the same value of W_B will be obtained regardless of how the charge q is brought in from infinity. If this were not true, then the point B would not have a single unique potential value, which the concept of potential requires. This is illustrated in Fig. 2-9. The collection of points all having the same potential (with respect to infinity implied) is called an equipotential surface. For the simple case of a single point charge, these surfaces are a set of concentric spheres with the charge being their center. Since all points on the same surface have the same potential, no work is required to move a

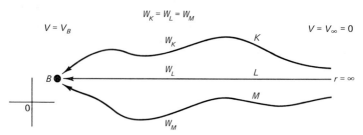

Fig. 2-9   The work W necessary to bring a charge in from infinity to a point B is independent of the path taken. If this were not so, then the potential at point B would not have a unique value, which is a requirement of the concept of potential. The origin of an arbitrary coordinate system is shown on the left. The potential at infinity is defined as equal to zero. K, L, and M are different paths.

test charge from one point to another on that surface, even if the path deviates from the surface itself (Fig. 2-10). Equipotential surfaces are always at right angles to the electric field vector $\mathbf{E}$; for if they were not, then there would be a component of $\mathbf{E}$ lying parallel to the surface and hence work would be required to move a test charge along that equipotential surface. This would be a contradiction to the properties of an equipotential surface.

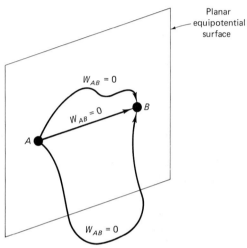

Fig. 2-10   The work needed to move a charge around on an equipotential surface is zero. Regardless of the path, it takes no net work to move a charge from point A to point B, even if the path goes off the surface. If the path leaves the surface and returns, half the path will be going against the electric field and the other half will be going with the electric field, making the net work equal to zero.

RELATION BETWEEN ELECTRIC POTENTIAL AND FIELD

The statement was made earlier in this chapter that the electric potential and electric field were related to one another. In this section we shall elucidate exactly what this relationship is. The approach taken will be to calculate the difference in potential between two points as a function of the electric field that is present. This will be done first for a simple case, and then for a more general case.

For the first case, consider the situation depicted in Fig. 2-11a, where we have a simple uniform electric field as shown. Suppose that we wish to find the potential difference $V_B - V_A$ between points A and B. To move a positive test charge from point A to point B requires work because the charge is going against the direction of the electric field. The work involved is given

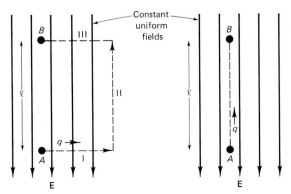

Fig. 2-11 The work required to move a test charge q from point A to B is independent of path. In part (a) charge q is moved directly from A to B and the total work involved is $-Eql$. In part (b) charge q is moved using a three-step path. When the charge is moved along steps I and III, no work is required because these paths are along equipotential surfaces. The only path that requires work is along step II. The amount of work required here is again equal to $-Eql$.

by the product of force times distance

$$W_{AB} = Fl \tag{2-38}$$

and the force F can be equated to

$$F = -Eq \tag{2-39}$$

where the minus sign indicates $\mathbf{F}$ and $\mathbf{E}$ are in opposite directions. We then have

$$W_{AB} = -Eql \tag{2-40}$$

Utilizing the definition of potential difference, V_{AB} is given as

$$V_B - V_A = W_{AB}/q = -El \tag{2-41}$$

So, the potential difference is equal to the simple product of the strength of the electric field times the distance the charge q travels.

To consider a more general case where $\mathbf{E}$ is not uniform, and the path from A to B is not a straight line, it is necessary to use the more rigorous definition of W given by Eq. (2-37).

Work is given as the dot product between the force and the distance through which the force acts. The vector $\mathbf{r}$ is called the displacement vector (Fig. 2-12) and it describes the direction and distance from point A to B. Equation (2-37) then states that the amount of work a general force $\mathbf{F}$ does is proportional to that component of $\mathbf{F}$ that is parallel to $\mathbf{r}$ ($F \cos \theta$), so that if $\mathbf{F}$ is perpendicular to $\mathbf{r}$, it can do no work in the direction of $\mathbf{r}$. A general

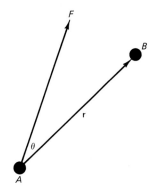

Fig. 2-12   The displacement vector **r** describes the direction and distance from point A to point B. The component of **F** that is parallel to **r** is given by $F \cos \theta$. If $\theta = 90°$, the force **F** can do no work in the direction of **r**.

relationship between **F** and V may now be given with the aid of Fig. 2-13, where we have a nonuniform field; it is required to find the potential difference between points A and B. The work required to move the test charge from A to B may be found by calculating the infinitesimal work needed to move the charge q over the infinitesimal distance $d\mathbf{r}$, and then to sum all infinitesimal work elements. We have

$$dW = \mathbf{F} \cdot d\mathbf{r} \tag{2-42}$$

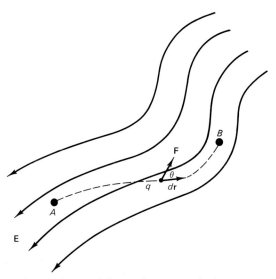

Fig. 2-13   A test charge q is moved from points A to B in the presence of a nonuniform electric field. At any one instant, the force **F** will make an angle θ with respect to the direction of the path. The total work involved can be calculated by summing the individual increments of work dW along the path.

where $\mathbf{F}$ is a general external force needed to move the test charge q in a direction against the field $\mathbf{E}$. The total work can now be given as

$$W_{AB} = \int_A^B \mathbf{F} \cdot d\mathbf{r} = -q \int_A^B \mathbf{E} \cdot d\mathbf{r} \tag{2-43}$$

So the potential difference is given as

$$V_B - V_A = \frac{W_{AB}}{q} = -\int_A^B \mathbf{E} \cdot d\mathbf{r} \tag{2-44}$$

Equation (2-44) then lets us calculate the potential difference between any two points A and B as long as the electric field $\mathbf{E}$ is known. The potential difference between points A and B is independent of the path taken to go between the two points, so in evaluating the integral in Eq. (2-44) one usually chooses a path to make the accompanying mathematics simple. Again, an example best illustrates this.

To illustrate the use of Eq. (2-44) and to show that the potential difference is independent of path, let us once more look at Fig. 2-11. In part (a) the path between points A and B is taken as a straight line, and application of Eq. (2-44) gives

$$V_B - V_A = -\int_A^B \mathbf{E} \cdot d\mathbf{r} = -E \int_A^B dr = -El \tag{2-45}$$

which is exactly as was previously derived for the potential difference. The integral $\int_A^B dr$ is just the summation of all the infinitesimal line segments on the straight line between points A and B; hence, it is just the length of the line. Suppose instead we had desired to take q along the dotted line path shown in part (b). What then is $V_B - V_A$? This can be calculated in three separate parts, and the total found by summing the parts:

$$\text{part I} \qquad \Delta V = -\int \mathbf{E} \cdot d\mathbf{r} = 0 \qquad \text{because } E \text{ is } \perp \text{ to } d\mathbf{r}$$

$$\text{part II} \qquad \Delta V = -\int \mathbf{E} \cdot d\mathbf{r} = -El \tag{2-46}$$

$$\text{part III} \qquad \Delta V = -\int \mathbf{E} \cdot d\mathbf{r} = 0 \qquad \text{because } E \text{ is } \perp \text{ to } d\mathbf{r}$$

$$V_B - V_A = 0 - El + 0 = -El$$

which is exactly what we obtained using a direct, straight-line path. The same answer results regardless of the path taken. Indeed, if a more complicated path is taken, an identical result must be calculated; however, a more complicated path may make the integral in Eq. (2-44) more difficult to evaluate.

From the simple Eq. (2-45) it is seen that the electric field may also be described in units of volts per meter which is also equivalent to a newton per coulomb. The CGS equivalent of a newton per coulomb is the dyne per esu; and it can be shown that 1 dyn/esu $= 3 \times 10^4$ V/m $= 300$ V/cm. This is a handy conversion factor to keep in mind.

If points A and B in Fig. 2-13 are infinitesimally close together, Eq. (2-44) reduces to

$$dV = -E \, dr \qquad \text{or} \qquad E = -dV/dr \qquad (2\text{-}47)$$

where dV/dr is the rate of change of the potential in the direction of the line segment dr. The quantity dV/dr is called the potential gradient. At a specific point in an electric field, the component of $\mathbf{E}$ in a particular direction is equal to the negative of the potential gradient in that direction. If the potential is some arbitrary function of the coordinates x, y, and z, then the magnitude of $\mathbf{E}$ in the x direction is given as*

$$E_x = -\partial V(x, y, z)/\partial x \qquad (2\text{-}48a)$$

and likewise

$$E_y = -\partial V(x, y, z)/\partial y, \qquad E_z = -\partial V(x, y, z)/\partial z \qquad (2\text{-}48b)$$

The negative sign indicates that $\mathbf{E}$ always points in the direction of decreasing potential. If the vector notation introduced in Appendix A is used, we have

$$\mathbf{E} = -\frac{\partial V}{\partial x}\mathbf{i} - \frac{\partial V}{\partial y}\mathbf{j} - \frac{\partial V}{\partial z}\mathbf{k} \qquad (2\text{-}49)$$

where $\mathbf{i}$, $\mathbf{j}$, and $\mathbf{k}$ are unit vectors in the x, y, and z directions, respectively. So, if an expression for the potential is known, then the electric field can be determined via Eq. (2-49).

In the example on p. 43 it was calculated that the electric field 10 Å away from an electron is 4.8×10^4 dyn/esu. Using the appropriate conversion factor, this translates into 14.4×10^6 V/cm which is an enormously large electric field. Since the field is the rate of change of potential with respect to distance, this calculation says that the potential is changing at the rate of over 10^6 V/cm. Again, this is a very large potential change; it reinforces the facts that the electric fields in the vicinity of charged biomolecules can be quite large and that electrical forces between biomolecules can be very significant.

---

* The symbol $\partial/\partial x$ indicates a partial derivative of V with respect to x. The derivative is taken in the normal fashion but assuming the variables y and z are constants.

THE POTENTIAL OF A POINT CHARGE

Let us now develop an expression for the potential of a point charge Q. Referring to Fig. 2-14, we shall calculate $V_B - V_A$ for the field of a positive point charge by moving a test charge in a radial direction from point A to point B. From the definition of potential we have

$$V_B - V_A = - \int_A^B \mathbf{E} \cdot d\mathbf{r} = - \int_A^B E \, dr \qquad (2\text{-}50)$$

The dot product $\mathbf{E} \cdot d\mathbf{r}$ is particularly simple because $\mathbf{E}$ is radial in direction. The field strength E cannot be taken in front of the integral in this case because it is not a constant, but varies with position. Substituting Eq. (2-15) into (2-50), we have

$$V_B - V_A = - \int_A^B \frac{Q}{r^2} \, dr = Q \left[\frac{1}{B} - \frac{1}{A} \right] \qquad (2\text{-}51)$$

Now, if we use the convention that infinity has a potential of zero, and we let point A represent infinity and B (actually r) represent an arbitrary point, the potential in CGS units due to the point charge Q is given by

$$V = Q/r \qquad (2\text{-}52)$$

Equation (2-52) tells us that for a point charge, equipotential surfaces have the geometry of concentric spheres with their centers at the charge. The potential varies as $1/r$ with respect to distance, whereas the electric field varies as $1/r^2$. This makes sense because the relationship between E and V requires $E = -dV/dr$.

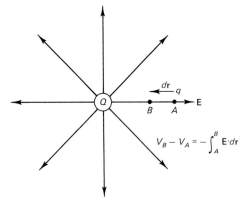

Fig. 2-14   The potential at a point due to a single charge may be found by calculating the work needed to move a charge q from infinity to that point along a radial direction. A radial direction is chosen to make the resulting mathematics simple.

To find the potential at a point in space due to a collection of charges, it is necessary only to algebraically add the contributions from the separate sources, or

$$V_{tot} = \sum_i V_i \tag{2-53}$$

No vector addition is necessary. The fact that potential is a scalar quantity means that you do not have components to worry about, and the entire mathematical process is simpler because of it. In fact, one approach to finding an expression for $\mathbf{E}$ in a particular case is to first calculate V_{tot} and then use expressions (2-48a,b) to get E_x, E_y, and E_z.

As an example illustrating how to calculate the potential when a number of charges are involved, let it be required to find the potential at the spot occupied by charge q_D in Fig. 2-2. The problem is essentially to find the potential due to the three charges q_A, q_B, and q_C. Using the definition of potential, we have

$$V_A = 2/l, \qquad V_B = (1/\sqrt{2})l^{-1}, \qquad V_C = (1/\sqrt{2})l^{-1} \tag{2-54}$$

where V_A, V_B, and V_C are the respective potentials at point D. The total potential is simply $V_{tot} = (2/l)(1 + \frac{1}{2}\sqrt{2})$. If we wanted to find $\mathbf{E}$ at point D, then we would have to set up a cartesian coordinate system and express V in terms of x, y, and z. Then the derivative $-\partial V/\partial x$ could be evaluated to be equal to E_x, $-\partial V/\partial y$ equal to E_y, etc.

ELECTRIC POTENTIAL ENERGY

When an object is raised with respect to the earth, it is said to possess potential energy of magnitude $E_p = mgh$, where m is its mass, g is the acceleration due to gravity, and h is the height. This potential energy exists just because of the spatial relationship between the object and the earth. The energy is called potential because the object has the possibility of releasing that energy when its position is changed. For the time being, the energy is stored. If all constraints on the object were suddenly released, the object would fall to earth converting its potential energy into kinetic energy ($\frac{1}{2}mv^2$). A very similar analogy exists in electrical situations. If two positive charges are constrained to be close to one another, or if a positive and a negative charge are separated by a certain distance, a release of constraints will allow each system to change spontaneously, where the change is governed by the fact that like charges repel and unlike charges attract. If we were to mechanically separate two unlike charges, a certain amount of work, or energy, would be required. Since the conservation of energy principle tells us it is impossible to create or destroy energy, it is said that the work needed to initially separate the two opposite charges is stored as electrical potential energy. The very fact that two oppositely charged particles are separated implies that there is potential energy stored in the system, much as there is when an object is raised to a height with respect to the earth. Electrical potential energy, like gravitational potential energy, is a consequence of the spatial relationship of elementary units: In the case of gravity the units are of mass, whereas in electrical potential energy the units are of charge.

The electrical potential energy of a group of charges is defined as the amount of electrical work needed to assemble the system by bringing the individual charges in from infinity one at a time to assume the positions they have. To illustrate this, consider the simple case of two charged particles that are separated by a distance l (Fig. 2-15). It requires no work to bring the first charge q_1 in from infinity because there are no other forces influencing its motion. The work required to bring charge q_2 in from infinity and place it a distance l away from charge q_1 may be found by remembering the definition of work and Coulomb's law. Refer again to Fig. 2-14. We shall bring charge q_2 in from infinity along a radial direction to its final resting place. We could bring q_2 in by any one of a large number of different routes, but a radial path makes the calculations much simpler since $\mathbf{E}$ due to charge q_1 is also in a radial direction. The required work is given by

$$W = -\int \mathbf{F} \cdot d\mathbf{r} = -q_1 q_2 \int_{\infty}^{l} \frac{dr}{r^2} \tag{2-55}$$

The potential energy U of the system is then

$$U = W = q_1 q_2/l \tag{2-56}$$

The negative sign in front of the integral in Eq. (2-55) indicates that the force $\mathbf{F}$ has a direction opposite to the displacement vector $d\mathbf{r}$. Equation (2-56) is the expression relating the potential energy of two charges separated by

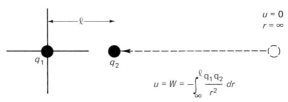

Fig. 2-15 The potential energy of a system can be ascertained by calculating the amount of work necessary to bring the charges in from infinity one at a time and placing them in their desired positions. The energy can be either positive or negative. A negative energy indicates that a system has attractive forces holding it together. The energy is with respect to the reference state where all charges are infinitely separated.

a distance *l*. If one charge is positive while the other is negative, then U is negative; this implies an attractive force between the two charges. A positive value of U indicates a repulsive force. A positive value of U means work must be done to construct the system; whereas, if U is negative, work must be exerted to dismantle it. If U is negative for a particular system, some external energy must be used to separate all the charges to infinity. The above conventions are arbitrary, but are consistent with the idea of giving infinity a potential of zero. The magnitude of the potential energy is important, but so too is its sign, for the sign will tell you whether or not the system under consideration is intrinsically stable or unstable. It should also be known that a single charged particle does not have potential energy all by itself; only with two or more charges, or a system of charges, do we correctly talk about potential energy. As always when we talk about energy, a difference in energy states is implied, rather than absolute energy values. In this case the reference state of zero energy is defined where all charges are separated by an infinite distance. Since potential energy is an energy term, the units are ergs or kilocalories (kcal) in the esu system, and the joule in the MKS system. Also, potential energy is a scalar quantity. The total potential energy of a system can be calculated by considering all pairwise interaction energies, and then summing algebraically. This is illustrated by an example.

Consider the situation depicted in Fig. 2-16 where it is required to find the total potential energy of the system composed of four charges. The

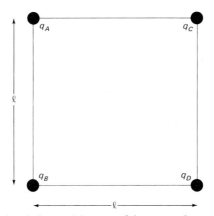

Fig. 2-16   The total electrical potential energy of the system shown above can be calculated by calculating all the pairwise potential energy terms and then summing. If the final answer is negative, it means that an outside force would have to be provided to disassemble the system and remove all the charges to infinity. If the potential energy were positive, this process would occur spontaneously.

individual potential energies for each pairwise interaction are given as:

$$U_{AB} = U_{BA} = q_A q_B/l, \qquad U_{AC} = q_A q_C/l, \qquad U_{AD} = q_A q_D/\sqrt{2}l$$
$$U_{BC} = q_B q_C/\sqrt{2}l, \qquad U_{BD} = q_B q_D/l, \qquad U_{CD} = q_C q_D/l \tag{2-57}$$

The total potential energy of the system then is

$$U_{sys} = \frac{1}{l}\left[q_A q_B + q_A q_C + \frac{\sqrt{2}}{2} q_A q_D + \frac{\sqrt{2}}{2} q_B q_C + q_B q_D + q_C q_D \right] \tag{2-58}$$

The expression for the potential energy of two point charges [Eq. (2-56)] could also have been derived by multiplying the potential due to q_1 at a distance l, $V_1(l)$, by the magnitude of charge q_2, or

$$U = q_2 V_1(l) \tag{2-59}$$

This is because $V_1(l)$ is the work needed to bring one unit of charge in from infinity to a distance l away from charge q_1.

The idea that a group of electrical charges can have potential energy just because of their spatial relationship with respect to one another is the most important concept of electrostatics that we shall use. With it we shall make calculations for the strength of various types of electrical interactions. A chemical bond or a stable electrical interaction takes place because the charge distributions in both cases create situations in which the potential energy of the system is very negative. The more negative the potential energy, the stronger the bonds. As was stated in Chapter 1, the stability of chemical bonds of one sort or another is very important in ensuring the success of a biological molecule. Throughout the rest of this book the strength of electrical interactions will be measured in terms of the potential energy of the system.

It is possible to use the formalism developed in this section to calculate the ionic bond energy between Na^+ and Cl^- in a NaCl crystal. The crystals of NaCl possess one of the simplest and most regular structural geometries of any crystal. Na^+ and Cl^- ions are alternatively located at the corners of a simple cube. This cubic arrangement is found throughout the crystal and is shown in two dimensions in Fig. 2-17. The separation between the closest two ions is 2.81 Å, and to a first approximation the bond energy can be calculated from Eq. (2-56):

$$U = -(4.8 \times 10^{-10})^2/(2.81 \times 10^{-8}) = -8.2 \times 10^{-12} \quad \text{ergs} \tag{2-60}$$

So, the amount of energy stored in the field due to the Na^+ and Cl^- ions being separated as they are is 8.2×10^{-12} ergs. Note that the energy value is negative, meaning that the system is stable as it stands and it will take a force to separate the charges from one another.

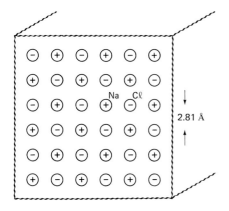

Fig. 2-17 A two-dimensional representation for the crystal structure of NaCl. Each ion is located at the corner of a cube, and this arrangement is propagated throughout the whole crystal. The figure represents one face of a crystal.

2.81 Å

In trying to interpret this number bioscientists generally convert the units of ergs to those of kilocalories and consider how much energy is involved for a mole of such interactions. Remember that a mole is a chemical term, defined as an Avogadro number of particles:

$$1 \quad \text{mole} = 6.023 \times 10^{23} \quad \text{particles}$$
$$1 \quad \text{kcal} = 4.18 \times 10^{10} \quad \text{ergs}$$

The reason for this change of units is due to the fact that biophysicists and biochemists usually work with sample solutions containing very large numbers of biological molecules; hence it is impossible to single out or measure the strength of interaction between any specific molecules. Thus the energy of interaction is measured, or calculated, for a standard number of particles, i.e., the mole. The kilocalorie is used instead of the erg because many times an experimentalist will measure the heat evolved or absorbed in a reaction in order to determine how strong an interaction is, and the kilocalorie is the standard unit of energy that is commonly used when describing heat changes in microsystems. A kilocalorie is 1000 calories; a calorie is defined as the amount of energy needed to raise the temperature of one gram of water one degree Celsius at atmospheric pressure. Units of kilocalories per mole (kcal/mole) will be used almost exclusively throughout this text.

The value of 8.2×10^{-12} ergs when translated to kcal/mole gives 118 kcal/mole. This figure tells us that it will take 118 kcal of energy to separate 6×10^{23} pairs of Na^+ and Cl^- ions that are separated by 2.18 Å. This is a relatively large amount of energy for a bond and indicates that the Na^+ and Cl^- interaction is rather strong. The experimentally measured value for the Na^+–Cl^- bond in the crystal is 183 kcal/mole, which indicates the bond is even stronger than our initial calculation suggests. The discrepancy between the measured value and the calculated value of the bond energy is due to the fact that each ion interacts with more than just one other counter-ion as we assumed in our calculation. A more realistic value for the bond energy may be obtained in the following way. Instead of considering only a single counterion pair in calculating the bond energy, we shall now isolate one ion and then consider its interaction with every other ion in the crystal. This is essentially the same procedure used to calculate the potential energy of the charge arrangement shown in Fig. 2-16.

Let us use as a reference point the ion marked Na in Fig. 2-17, and con-sider the other ions that are in the same horizontal row. The two nearest ions are Cl^- ions, each with a negative charge and distance of separation $l = 2.81$ Å. The energy of interaction for the Na^+ ion with the two Cl^- ions is given by $U = -2e^2/l$.* The next two nearest neighbors in the horizontal row are positive Na^+ ions with each having a distance of separation 5.62 Å from the central Na^+ ion. The energy of interaction between these two pairs is equal to $U = +2e^2/2l$. The next nearest neighbors are again two Cl^- ions. The energy of interaction between our reference Na^+ ion and all the other ions in the horizontal row can be calculated individually and summed to find the total. The total energy is

$$U = \frac{e^2}{l}\left[-\frac{2}{1} + \frac{2}{2} - \frac{2}{3} + \frac{2}{4} + \cdots \right]$$

$$= \frac{-2e^2}{l}\left[1 - \frac{1}{2} + \frac{1}{3} - \frac{1}{4} + \cdots \right] = \frac{-2e^2}{l}\ln 2 = \frac{-1.386e^2}{l} \quad (2\text{-}61)$$

It is seen that the total potential energy is calculated by summing an infinite series, which in the case here just happens to have a very simple closed form, as seen. The fact that $\ln 2$ can be expressed by this infinite sum can be veri-fied by referring to an elementary calculus textbook or to a standard ref-erence of mathematical information. Next, consider the horizontal row immediately above our reference Na^+ ion. The nearest ion is a Cl^- ion with a distance of separation $l = 2.81$ Å, the next two ions are Na^+ with distance of separation $\sqrt{2}l$, the next are Cl^- ions at $\sqrt{5}l$, etc. For this horizontal row, the potential energy of interaction with the reference Na^+ ion is

$$U = \frac{2e^2}{l}\left[-1 + \frac{1}{\sqrt{2}} - \frac{1}{\sqrt{5}} + \frac{1}{\sqrt{10}} - \cdots \right] \quad (2\text{-}62)$$

The reader should also see that there are four rows of ions that interact with our reference ion in exactly this manner. One row is on top of the ref-erence ion, one row is below, one row is in front, and finally one row is in back of the Na^+ ion. If one proceeds in this fashion and considers every

---

* Here e is the charge on an electron or proton.

possible row, then sums all potential energy terms, the final grand total is found to be

$$U_{gt} = -1.75e^2/l \qquad\qquad (2\text{-}63)$$

which is to be compared to Eq. (2-60). Already, we can see that Eq. (2-63) will give us a value of the bond energy that is 1.75 times larger than our original simple calculation. The new calculation gives as $U = -206.5$ kcal/mole which is much closer to the experimental value than our calculation; however, it is still not exact. It turns out that Eq. (2-63) is correct as it stands, but we have not yet taken into consideration several destabilizing forces that will influence our final figure. First, when atoms or ions approach one another very closely, the electrons that orbit the ions tend to limit the distance of closest approach because like charges repel. Therefore, the ions in the crystal have not only attractive electrical forces holding the ions together, they also have repulsive forces due to the electron clouds surrounding the ions. For crystals of Na^+Cl^-, it has been determined experimentally that the repulsive energy of interaction has a value of $U = 21.9$ kcal/mole for each ion pair. The other destabilizing force is due to thermal motion; the higher the temperature, the more disruptive are thermal forces. At $37.5°C$, these forces have an energy of $U = +0.62$ kcal/mole. Since both these effects are destabilizing, their energies are positive (remember the convention); so when combined with the result of Eq. (2-63), we have the final bond energy $U = 184$ kcal/mole, a value that is in good agreement with the measured value. The fact that the argument is so good gives strong support to the ideas that NaCl crystals are held together by ion–ion interactions and that our formalism in describing electrical interactions is essentially correct.

SUMMARY

The basic, most fundamental manner in which point charges interact is described by Coulomb's law. The force of interaction between two charges q_1 and q_2 is proportional to the product of their charge and inversely proportional to the square of the distance separating them. There are two basic systems of units used to describe electrical phenomena, the CGS or esu, and the MKS system. Electrical concepts are described in terms of vectors and scalars. The sum or resultant electrical force acting on a body can be found by vectorially adding all the separate forces. The concept of the electric field was introduced by physicists to help describe the mechanism by which electrical particles interact with one another. The electric field is a vector quantity, and it is through this field that one charged particle influences another. Gauss's law is a relation between electrical flux through a closed surface and the amount of charge contained within that surface. The electrical potential is a scalar quantity that can also be used to describe an electrical field. A potential difference is defined in terms of the amount of work it takes to move a charge between two points. The electric field and the potential are related to one another through an integral or differential equation. The electrical potential energy is a measure of how much work it takes to assemble or dissemble a collection of charges.

REFERENCES

Halliday, D., and Resnick, R. (1963). "Physics for Students of Science and Engineering," Part II. Wiley, New York.
Sears, F. W., and Zemansky, N. W. (1960). "College Physics." Addison-Wesley, Reading, Massachusetts.

3

DIPOLES AND DIELECTRIC CONSTANTS

INTRODUCTION

In the previous chapter emphasis was placed on the characteristics of a single charged entity. This chapter will emphasize the characteristics of another commonly encountered electrical entity, the dipole. A dipole is constructed of one positive and one negatively charged particle, and the whole system acts as a unit.

No study of electrostatics is complete without considering some detailed properties of dipoles. The concept of the electric dipole is important because it acts as a basis for many molecular phenomena and allows fairly simple models to be constructed to explain those phenomena. Molecules that have significant dipole moments interact in a manner different from those that do not. The presence or absence of a dipole moment can, to a large extent, determine the macroscopic, chemical, and physical properties of materials; as we shall see, water owes much of its uniqueness to its large dipole moment. Dipoles are also important in that they help explain the properties of dielectrics. Dielectrics are materials composed of many submicroscopic dipoles that are influenced by electric fields applied to the material. Dielectrics come in gas, liquid, and solid form, and they can grossly affect the manner in which biomolecules interact with one another. Dielectrics differ from conductors in that they do not readily transmit charge and are electrically neutral. From our standpoint, dielectrics are important because the space between biological macromolecules is usually filled by some sort of dielectric material, usually water.

DIPOLES

Let us consider the arrangement shown in Fig. 3-1 where we have two charges of equal and opposite magnitude separated by a distance l. Such a system of charges is an example of simple electric dipole. Just as a single

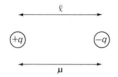

Fig. 3-1   A simple dipole. It consists of a positive and a negative charge of equal magnitude separated by a distance l. The overall charge of a dipole is zero; it has no net charge.

charge can be characterized by its magnitude, a dipole is described in terms of its dipole moment. The dipole moment $\mathbf{\mu}$ is defined as a vector whose magnitude is given by

$$\mu = lq \tag{3-1}$$

and whose direction is represented by the direction from the negative to the positive charge. Notice that the dipole moment consists of two pieces of information about the system, i.e., the charge on each particle and the distance of separation. It should be emphasized that the dipole moment is a product of two independent terms, and if only the dipole moment is known, it does not necessarily give information about the specific value of either. The dipole moment is defined as such because it turns out to be a convenient parameter and allows for an easier description of electrical systems. It is still correct to characterize a dipole system in terms of each individual charge; however, it is also more cumbersome. For the case of an electron and a proton separated by 10 Å (1 Å = 10^{-8} cm), the dipole moment is equal to $\mu = (4.8 \times 10^{-10} \text{ esu}) \times (10 \times 10^{-8} \text{ cm}) = 48 \times 10^{-18}$ esu cm. Because the unit 10^{-18} esu cm appears so frequently when dealing with dipole moments, it has been given a special name called the *debye* (D). It is named after Peter Debye who did much work in physics and physical chemistry. Therefore, a dipole moment of 48×10^{-18} esu cm is represented as 48 D. This notation will be used throughout the rest of this book.

ELECTRIC FIELD OF A DIPOLE

One advantage of using the dipole moment concept is that a system made up of two oppositely charged particles may be thought of as a single unit, instead of two single charges. In this respect it is possible to calculate the electric field of a dipole and express it in terms of the dipole moment. Since this result is important, it will be derived here.

As was described in Chapter 2, the strength of the electric field due to a collection of charges can be found by calculating the individual contributions from each charge and then adding each contribution vectorially to find the total field. However, for this case, it is much more instructive, and mathematically simpler, to first find the potential of the dipole at a point P, and then to use Eqs. (2-47) to find the electric field. This is one situation where it is mathematically easier to calculate the electric potential, which is a scalar, and then to convert this information to find the electric field. Finding the electric field first is possible but more difficult. The problem is depicted in Fig. 3-2. It is then required to find the potential at an arbitrary point P due to the dipole. The variables r_+, r, and r_- are respectively the distances from q_+, the origin, and q_- to the point P. The angles θ_+, θ, and θ_- are the respective angles r_+, r, and r_- make with respect to the x axis. Using Eq. (2-52), we have

$$V(P) = V_+ + V_- = q\left[\frac{1}{r_+} - \frac{1}{r_-}\right] = q\left[\frac{r_- - r_+}{r_+ r_-}\right] \qquad (3\text{-}2)$$

where V_+ and V_- are the separate potentials at P due to the charges q_+ and q_-, respectively, and $V(P)$ represents the potential at point P. Regardless of the values of r_+, r_-, or q, Eq. (3-2) will always give an exact value of the potential. Equation (3-2) is an exact expression for $V(P)$.

For the special case where P is far away, $r \gg l$, it is possible to make several approximations which will simplify Eq. (3-2) and make it somewhat easier to use. The use of simplifying approximations in physics is a recurring theme, and in fact many problems could not be solved as cleanly or neatly without them. The approximations are usually made in such a manner as to make the mathematics a little easier to manipulate. If P is far away, we can make the approximation that $\theta_- \cong \theta$ and that the product $r_+ r_- \cong r^2$. The quantity $r_- - r_+$ is the difference in distance from each respective charge

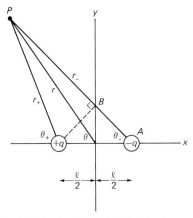

Fig. 3-2 It is required to find the electric field of the dipole at an arbitrary point P that is far away from the dipole. The two charges comprising the dipole are a distance l apart and both lie on the x axis. The angles with respect to the horizontal to P from the two charges are designated θ_+ and θ_-, respectively. The variables r and θ are the respective parameters with respect to the center of the dipole. The distance AB is the difference in length between r_- and r_+.

to P. As the figure is drawn, r_- is longer than r_+ by the distance AB, and this distance can be found by drawing a line from q_+ to intersect r_- at a right angle at B, forming a right triangle (q_+, A, B) in the process. Using straightforward trigonometry, the distance AB is given by

$$AB = l \cos \theta \qquad (3\text{-}3)$$

With these approximations, Eq. (3-2) becomes

$$V(P) = \frac{ql \cos \theta}{r^2} = \frac{\mu \cos \theta}{r^2} \qquad (3\text{-}4)$$

which is the final answer for the potential. It is important to emphasize here that the potential of a dipole at a large distance varies as the inverse *square* of the distance, compared to the $1/r$ dependence for a simple point charge. This is reasonable, however, since the farther away the point P is, the more the dipole looks like a unit of zero charge; hence, the potential decreases more rapidly than for a point charge. Also, the dipole moment μ appears in Eq. (3-3) much as q does in Eq. (2-51) for a single charge.

To find the electric field from Eq. (3-4), it is first necessary to express $\cos \theta$ and r in terms of x and y, where the x and y refer to the coordinates of the point P. These relations are given by

$$\cos \theta = \frac{x}{(x^2 + y^2)^{1/2}}, \qquad r^2 = x^2 + y^2 \qquad (3\text{-}5)$$

Equation (3-4) becomes

$$V(P) = \frac{\mu x}{(x^2 + y^2)^{3/2}} \qquad (3\text{-}6)$$

Using Eqs. (2-47), the x and y components of the electric field are

$$E_x = -\frac{\partial v}{\partial x} = -\frac{\partial}{\partial x}\left[\frac{\mu x}{(x^2 + y^2)^{3/2}}\right] = \frac{\mu(2x^2 - y^2)}{(x^2 + y^2)^{5/2}} \qquad (3\text{-}7)$$

$$E_y = -\frac{\partial v}{\partial y} = \frac{3xy}{(x^2 + y^2)^{5/2}} \qquad (3\text{-}8)$$

Now, since we have both the x and y components of the electric field it is possible to find the value for the total field by summing the squares of the components, (see Appendix A). We then have

$$E^2 = E_x^{\,2} + E_y^{\,2}; \qquad E_{dip} = \frac{\mu}{r^3}\left[1 + \frac{3x^2}{(x^2 + y^2)}\right]^{1/2} = \frac{\mu}{r^3}[1 + 3\cos^2 \theta]^{1/2} \qquad (3\text{-}9)$$

(The details of the algebra are left as an exercise for the reader.) Notice, that E_{dip} varies as $1/r^3$, whereas the field of a simple point charge varies as $1/r^2$.

Equation (3-9) also tells us that E_{dip} is a maximum when $\theta = 0°$, and a minimum when $\theta = 90°$; hence, the field is most influential for points along the axis of the dipole (the x axis). To derive an expression similar to Eq. (3-9) for points close to the dipole, one proceeds in a likewise manner, but without introducing the above approximations—Eq. (3-2) already satisfies this requirement except r_+ and r_- are sometimes difficult to evaluate.

To get a feeling for how far away point P must be for the above approximations to be valid, consider the special cases where P is either on the dipole axis (x axis) or on a line perpendicularly bisecting the dipole axis (y axis). For the latter case, Table 3-1 shows the percentage difference between θ and θ_-, r and r_- for various values for r. For points lying on the x axis, Table 3-1 gives values for the difference between r_- and r for different values of r. The figures in Table 3-1 tell us that if the point P is at least five dipole lengths away, then the approximations are rather reasonable. Table 3-1 also tells us that the approximations are better for points P that are on the y axis compared to equivalent points on the x axis.

TABLE 3-1

Effect of Approximating r by r_- and θ by θ_- [a]

	P on y axis		P on x axis
r	Percent difference between r and r_-	Percent difference between θ and θ_-	Percent difference between r_- and r
$1l$	11.8	30	50
$5l$	0.5	6.3	10
$10l$	0.12	3.2	5
$15l$	0.05	2.1	3.3
$20l$	0.03	1.6	2.5
$25l$	0.02	1.3	2.0
$50l$	0.005	0.6	1.0

[a] The percent differences between r and r_-, and θ and θ_- for a simple dipole when the point P is on the y axis. Also included are the differences between r and r_- when P is on the x axis. l is the dipole length.

DIPOLE IN AN ELECTRIC FIELD

Dipoles can not only set up an electric field and influence other charged particles, but they can also be influenced by external electric fields. In this section we shall consider two examples of how different types of electric fields can influence dipoles. In the first case uniform electric fields will be considered, and in the second case the effect of nonuniform electric fields on dipoles will be examined.

When a dipole is placed in a constant, uniform electric field as shown in
Fig. 3-3, the net effect of the field on the dipole may be derived by considering
the separate effects on each charge of the dipole, and then summing. In the
figure the field **E** is constant over the length of the dipole, and each charge is
influenced by the field. The field creates a force **F**$_+$ on the positive charge and
a force **F**$_-$ on the negative charge, where the direction of each is as shown
and is in accord with the general conventions. The magnitude of each force
is given by Eq. (2-14) as $F = qE$; but since $|q_+| = |q_-|$ and E is constant,
$|F_+| = |F_-|$, and the net force acting on the dipole is zero. This means that
there is no net translational motion of the dipole. However, even though the
net force equals zero, there is still a torque, or tendency for the dipole to
rotate in the field about point O, so that the dipole axis makes an angle θ
with the field. Let us now consider this.

The force **F**$_+$ (or **F**$_-$) in Fig. 3-3 can be separated into two components,
one parallel to the dipole axis and one perpendicular to the dipole axis.
The component of **F**$_+$ (or **F**$_-$) perpendicular to the dipole axis is given by
$F_+ \sin \theta$, and it is this force component that tends to rotate the dipole.
The component of **F**$_+$ (or **F**$_-$) parallel to the dipole axis cannot cause a
torque. Mathematically, torque (Γ) can be defined as a cross product between
the force and the distance vector through which it acts, or

$$\Gamma = L \times F \qquad \text{or} \qquad \Gamma = LF_\perp \qquad (3\text{-}10)$$

where $F_\perp$ is that component of a force causing the torque, and L is the distance
(straight line) from the point of application of $F_\perp$ to the center of rotation.
$F_\perp$ can also be described as the component of the force **F** perpendicular to
L. In Fig. 3-3 it is seen that there are two sources of torque forces tending
to rotate the dipole, i.e., $F_- \sin \theta$ and $F_+ \sin \theta$. Both are tending to rotate
it the same way. The distance from each force to the center of rotation is

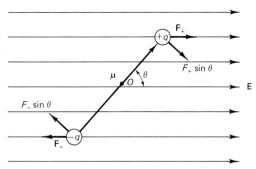

Fig. 3-3   A dipole in a constant, uniform electric field. The net effect on the dipole is to rotate
it to an angle θ with respect to the field. The field is constant over the length of the dipole.

$l/2$, so the total torque is equal to

$$\Gamma = 2F \frac{l}{2} \sin \theta = qEl \sin \theta = \mu E \sin \theta \qquad (3\text{-}11)$$

but this can also be written as

$$\boldsymbol{\Gamma} = \boldsymbol{\mu} \times \mathbf{E} \qquad (3\text{-}12)$$

The torque on the dipole is then given by the cross product between the electric field and the dipole moment. The stronger the field (or $\boldsymbol{\mu}$), the stronger is the torque, and the dipole will become more aligned with the field. For a large field, the alignment of the dipole is essentially perfect, and any further increase in field strength will not result in a greater alignment. This situation is called saturation. At this point the reader should see the advantage of introducing the concept of the dipole moment. By so doing, the mathematics of the situation can be greatly simplified over the case of considering the two charges separately.

Because a constant uniform electric field tends to align a dipole from a random position, work must be performed, or energy must be expended by the field. This energy, however, must be recoverable because of the conservation of energy principle; and it is said that this energy is stored in the field as electrical potential energy. To calculate this energy, we have to calculate the amount of work the field performed in aligning the dipole in the first place. Since this work involves rotational movement, our previous definition of work [Eq. (2-37)] is not quite applicable. From classical mechanics it can be shown that the incremental work dW performed by a torque Γ as the torque moves through the incremental angle $d\theta$, is

$$dW = \Gamma \, d\theta \qquad (3\text{-}13)$$

It should be noticed that Eq. (3-13) is very similar to the case of work involving linear motion where the torque Γ and the force $\mathbf{F}$ are analogous, and where linear displacement dr and angular displacement $d\theta$ are analogous. To find the total work involved in rotating the dipole from some initial angle θ_0 to a final angle θ, it is necessary to add all the incremental values of dW. We then have

$$W = \int dW = \int_{\theta_0}^{\theta} \Gamma \, d\theta \qquad (3\text{-}14)$$

Substituting Eq. (3-12) into (3-14), we have

$$W = \mu E \int_{\theta_0}^{\theta} \sin \theta' \, d\theta' = -\mu E [\cos \theta - \cos \theta_0] \qquad (3\text{-}15)$$

Since we are usually interested in changes of energy rather than absolute energy values, θ_0 is arbitrarily taken to be $\theta_0 = 90°$. This means if the dipole is exactly perpendicular to the field, the potential energy of the system is arbitrarily taken to be zero; this position acts as our reference state. Following this convention, then, Eq. (3-15) reduces to

$$U = W = -\mu E \cos \theta \qquad (3\text{-}16)$$

which can be written in a more general vector form as

$$U = W = -\mathbf{\mu} \cdot \mathbf{E} \qquad (3\text{-}17)$$

The amount of potential energy stored in the field when a dipole is oriented at an angle θ with respect to a constant, uniform electric field is the vector dot product between the field $\mathbf{E}$ and the dipole moment $\mathbf{\mu}$. The better the alignment, the more energy is stored in the field.

DIPOLE IN A NONUNIFORM FIELD

The above derivations are valid only for the interaction between a dipole and a constant, uniform field. Another situation, which will be of special interest to us in Chapter 9, is the interaction of a dipole with an electric field that is not constant over the dipole length. This situation is depicted in Fig. 3-4, where the shaded arrows represent a strong electric field that is getting increasingly larger in the x direction, i.e., $\mathbf{E}$ is a very strong function of x. When $\mathbf{E}$ is very large, the dipole is essentially aligned parallel to the electric field. Because the field in Fig. 3-4 is not constant everywhere, $|\mathbf{F}_-| \neq |\mathbf{F}_+|$. If $E(x)$ is the field strength at the negative charge, then the

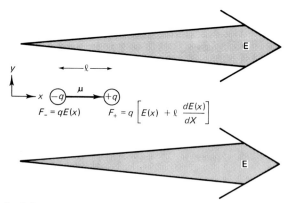

Fig. 3-4 A dipole in an electric field that is not constant over its length. The field is a strong function of position and increases rapidly with the x coordinate. For a large field, the dipole is essentially aligned with the field. The net force on the dipole depends on the rate of change of the field with respect to distance.

electric force acting on q_- is

$$F_- = -qE(x) \tag{3-18}$$

The field at the positive charge can be evaluated from $E(x)$ if it is known how the field changes with distance, i.e., if dE/dx is known. The rate of change of the external field with respect to distance (dE/dx) times the length of the dipole will then be the difference in the field between the charges q_- and q_+. The force acting on q_+ is then

$$F_+ = q[E(x) + l\, dE/dx] \tag{3-19}$$

where $E(x) + l\, dE/dx$ is the magnitude of the field at the charge q_+. The net force acting on the dipole is the sum of Eqs. (3-18) and (3-19)

$$F_{net} = ql\, dE/dx = \mu\, dE/dx \tag{3-20}$$

So, when one considers a nonconstant field, there is a net force acting on the dipole; the net force is proportional to the product of the dipole moment and the electric field gradient. The net force acting on the dipole will then cause the dipole to move in the direction of increasing field. This situation points out the fact that even though a particle may be electrically neutral, its motion can still be influenced by electrical forces if that particle has a dipole moment.

The method of treating the dipole in an electric field also illustrates the point that this two-charge system can be considered as a unit and described through the dipole moment. This is a general technique and future developments will consider μ alone without reference to the individual charges or the dipole length.

GENERAL DIPOLE

So far we have considered the dipole moment to be defined in terms of the simple charge arrangement shown in Fig. 3-1. The general definition of the dipole moment is however not so restrictive in that any group of charges arbitrarily scattered throughout three-dimensional space can be characterized by a dipole moment. Consider, for example, Fig. 3-5 in which we have such a group of positive and negative charges arbitrarily arranged about some local origin O. The vector $\mathbf{r}_i$ is a radius vector and extends from the origin to the charge q_i. The generalized definition of the dipole moment is given by

$$\mathbf{\mu} = \sum_i q_i \mathbf{r}_i \tag{3-21}$$

where the right-hand side of Eq. (3-21) is a vector sum of the individual products $q_i \mathbf{r}_i$. Since $\mathbf{r}$ is a vector, multiplication by the scalar q results in a

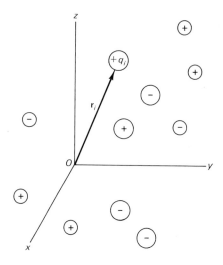

Fig. 3-5   A group of charges arbitrarily scattered in space. The vector **r** is called the radius vector and essentially locates any charge with respect to the origin. A generalized definition can be used to define a dipole moment for any group of charges.

vector. For a continuous distribution of charge, the summation in Eq. (3-21) is replaced by

$$\boldsymbol{\mu} = \int_V \rho(\mathbf{r})\mathbf{r}\, dV \tag{3-22}$$

where $\rho(\mathbf{r})$ is the charge density at a position indicated by the vector **r**. The integral is evaluated over the entire volume V where $\rho(\mathbf{r}) \neq 0$. According to Eq. (3-22), knowledge of $\rho(\mathbf{r})$ is adequate to determine $\boldsymbol{\mu}$; however, the reverse is not true. Knowledge of the value for a particular dipole moment is not adequate to deduce the charge distribution that gives rise to that dipole moment. With this definition of dipole moment it is now possible to see more clearly how a biological macromolecule with charges distributed throughout its whole volume can have a dipole moment, even though there are not necessarily an even number of positive and negative charges. Any system of charges (even if the system has a zero net charge) whose center of positive charge does not coincide with the center of negative charge will have a dipole moment. If the center of negative charges coincides with the center of positive charge, then the dipole moment of the system is zero. In this respect the dipole moment is a measure of the spatial distribution of the charges in a particular system. It is an indication of whether or not the center of positive charge is separated from the center of negative charge.

It should be remarked that the previous definition of $\boldsymbol{\mu}$ [Eq. (3-1)] is just a special case of Eq. (3-21). This can be seen in the following application of Eq. (3-21). Consider Fig. 3-2 where the center of the coordinate system is as shown. Using the generalized definition of dipole moment, we have

$$\boldsymbol{\mu} = -q(l/2)\mathbf{i} - q(l/2)\mathbf{i} = -ql\mathbf{i} \tag{3-23}$$

where the first term is due to the negative charge and the second term is due to the positive charge, and **i** is the unit vector in the positive x direction. The answer states that the dipole moment has magnitude ql and points in the negative x direction.

QUADRUPOLE

Just as a positive and negative charge can form a dipole, two dipoles can form a quadrupole. A quadrupole is a collection of charges which, when considered as a group instead of individuals, can make some of the mathematics easier.

The arrangement of charges shown in Fig. 3-6 is called a quadrupole. However, the mathematics necessary to describe a quadrupole is a little more complicated than for a dipole; and the strength of interaction of a quadrupole with other charges, fields, etc., is generally weak compared to a dipole, so quadrupoles are usually ignored except in the most detailed calculations. It should be noted that the quadrupole is formed by placing two dipoles head to tail so that the net dipole moment is zero, as is the net charge. At large distances the potential due to a quadrupole arrangement varies as $1/r^3$ compared to $1/r^2$ for a dipole and $1/r$ for a point charge (monopole).

Fig. 3-6   An example of a simple quadrupole. Notice that it is made up of two dipoles in such a manner that the dipole moment of the system equals zero, as does the net charge.

Although this book will not consider quadrupoles again, the reader may well wonder why they and the dipole were introduced in the first place, and what possible connection they could have with biomolecules. If one takes an overview, it is seen that as one progresses from a monopole to a dipole, to a quadrupole, etc., one is considering more complicated spatial arrangements of charges. The *etc.* indicates that there are more "poles" that have not, and will not, be mentioned. Biological molecules are real three-dimensional particles, each type having a unique charge distribution throughout its volume. Bioscientists are very much interested in these charge distributions since they help characterize molecular structure and help in understanding biological function. The fact that different biomolecules have different charge distributions also makes it easier for scientists to use special analytical and preparative procedures (Chapter 9) that discriminate against certain types of charge distributions, and make it possible to isolate individual types of molecules in a relatively pure form. By considering the various "poles" of biomolecular charge distributions, it is possible to analytically describe the charge distribution. For instance, the value of the monopole moment tells us the net charge present but nothing about the total positive or negative charges, or their distribution. Knowledge of the dipole and quadrupole moments gives some information on the spatial distribution of the charges; hence, our description of the distribution is more complete than if we knew only the net charge. For instance, if the net charge is zero and the dipole is not, then we immediately know the center of the positive and negative charge distributions are not coincident. By representing a complex charge distribution by its equivalent dipole moment, one is essentially asking whether the centers of positive and negative charge coincide, and if not, how far apart they are. By considering the equivalent quadrupole moment, one is getting yet more information about the charge distribution. So, the answer to the question initially asked is that the dipole moment is introduced so that we may have a quantity to help us better describe a charge distribution. For biomolecules, the monopole is frequently determined either experimentally or theoretically, while the dipole moment is determined less frequently. The quadrupole is almost never determined. In Chapter 4 the experimental methods for determining dipole moments of amino acids, proteins, and nucleic acids will be discussed along with possible explanations of what these dipole moments are telling us about the physical characteristics of biomolecules. It should also be mentioned that knowledge of *all* multipoles is equivalent to knowledge of the spatial distribution of charges. Figure 3-7 briefly summarizes some facts about different charge distributions.

DIELECTRIC CONSTANT

When the mathematical representation of Coulomb's law was introduced in Chapter 2, it was stated that it was applicable only when both charges under consideration were in a vacuum. If the force between two charges

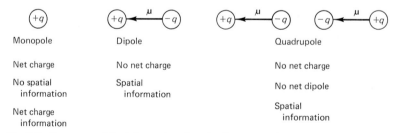

Fig. 3-7 Summary of the information given by the commonly encountered multipoles.

were measured while both were in some medium, say a gas or a liquid, the force would be less than if they were in a vacuum. This is true also for the strength of an electric field. It then seems safe to assume that real-life materials like gases and liquids can somehow alter the magnitude of an electric field by their very presence. Suppose that we have a single point charge in a vacuum, and a similar charge immersed in water (Fig. 3-8), and we are interested in determining the relative strength of the fields generated by each situation. A detailed measurement would show that the magnitude of the electric field in the case where the charge is surrounded by water is reduced compared to the same situation in a vacuum. A series of measurements would show that this reduction of the electric field is dependent upon the exact nature of the medium in which the charge is embedded and that the ratio of the field in vacuum to that in a material can by expressed as

$$E_{vac}/E_{mat} = D \qquad (3\text{-}24)$$

where D is called the dielectric constant. D is defined as one for a vacuum so any other medium will have a dielectric constant that is greater than one. The dielectric constant is a physical characteristic or property of a material and is, in effect, a measure of how much a particular substance will reduce the magnitude of an electric field. The value of a dielectric constant is usually limited to materials that are insulators and do not readily support a free flow of charges. Equation (3-24) will be taken as our definition of a dielectric constant although this is not a rigorous definition. A pure physicist would define it in a different way; but for our purposes, Eq. (3-24) gives a better intuitive feeling of D.

Because electric fields are reduced by a factor $1/D$ in the presence of a dielectric, it is necessary to modify some of the equations we have described so far. The modifications are easy however. For every equation we have derived that describes electric force, field, potential, or electrical energy, it is necessary to multiply the equation by the factor $1/D$. By doing this, and

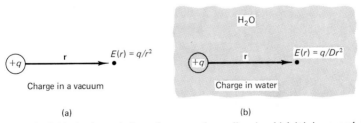

Fig. 3-8 The field of a charge is dependent upon the medium in which it is immersed. (a) The charge is in a vacuum. (b) The charge is in water. The electric field of an identical charge is less in water than it is in a vacuum. This is due to the dielectric nature of water.

then using the appropriate value of D, all our previous equations can be used without further change. In cases where the calculations involve a vacuum, merely set $D = 1$. This correction has the effect of reducing the magnitude of all electrical interactions by a constant amount which depends on the particular dielectric filling the space.

DIELECTRICS AND CAPACITORS

Since dielectrics have such a powerful influence on electrical interactions, it is worthwhile to consider them in a fair amount of detail. One common method of measureing the dielectric constant of a material is through the use of a capacitor. Capacitors are electrical devices that come in many shapes and sizes and are constructed from a variety of materials. A simple capacitor is constructed by placing two conducting plates of area A a distance d apart and connecting the plates electrically through an external circuit. (Fig. 3-9). If a battery is part of this external circuit, no current will flow because the capacitor is essentially a break in the circuit. What will happen is that each side of the capacitor will acquire some charge from the battery and become positively or negatively charged depending on which side of the battery it is connected to. If d is small compared to A, the electrical field between the plates of the capacitor will be uniform and constant.

The capacitance C of the capacitor is defined as

$$C = q/V \qquad (3\text{-}25)$$

where q is the charge on each plate and V is the voltage difference between the plates. If a dielectric is now placed between the plates as in Fig. 3-10,

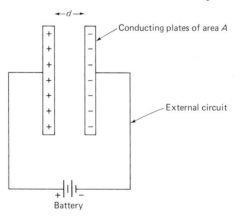

Fig. 3-9   A typical capacitor circuit where charge is built up along the face of each capacitor. The capacitor shown is nothing more than two conducting plates placed a distance d apart.

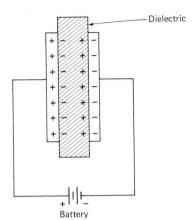

Dielectric

Battery

Fig. 3-10 A capacitor with a dielectric between the plates. In this situation the electric field in the space between the plates is reduced by the dielectric.

it is found that the capacitance increases by a value equal to the dielectric constant of the material:

$$C_{\text{with dielectric}} = Dq/V = DC_{\text{without dielectric}} \qquad (3\text{-}26)$$

Since $C_{\text{with dielectric}}$ is always larger than $C_{\text{without dielectric}}$, the value of D must always be greater than one. By measuring the capacitance with and without the dielectric, a ratio of capacitances will give the dielectric constant.

Let us now look at the mechanism that causes the above behavior and also see why a dielectric has the ability to reduce the strength of electrical interactions. In the examples of the capacitor given in Figs. 3-9 and 3-10 the battery always maintains a constant voltage across the plates, even when there is a dielectric filling the space between. Since the potential difference V does not change when the dielectric is inserted, then the value of q must change to account for the increase in capacitance since, from the definition [Eq. (3-25)], there is nothing else left. This then is a clue as to what is happening.

In order to understand the mechanism of how dielectrics work, we must look at their molecular structure in a general sort of way. Most dielectrics have zero net charge and have no free charges that can migrate to either capacitor plate. If charges did move from one plate through the dielectric to the other plate, the capacitor effect would be lost, and the net result would be a simple closed circuit. If free charges in the dielectric are eliminated, then what other electrical structures can explain the change in capacitance? The answer lies with the dipole. Suppose that our dielectric is constructed of molecules that have permanent dipole moments. We have learned that when a dipole is placed in an electrical field, it tends to align itself with that field. So when a dielectric is placed in the uniform electric field between the plates of the capacitor, the molecular dipoles, which are initially randomly oriented, undergo an orienting effect. In this condition the dielectric is said to be polarized. This is illustrated in Fig. 3-11.

The amount of orientation each dipole receives depends on several factors: the strength of the field, the magnitude of the dipole moment, the extend of any structural constraints placed on the mobility of the dipoles, and the disorienting effects of thermal agitation are among the more obvious. Since there are many molecular dipoles distributed throughout the volume of the dielectric and since each is electrically neutral, the net change in charge of the dielectric under the influence of the capacitor's field is zero. However, if we consider an infinitesimal volume of the dielectric that is one-half a dipole's length thick and adjacent to the positive side of the capacitor plate, we see that the surface of the dielectric acquires a net negative charge (Fig. 3-12). Although the dielectric itself remains electrically neutral, there are surface charges built up on the areas immediately next to the capacitor plates.

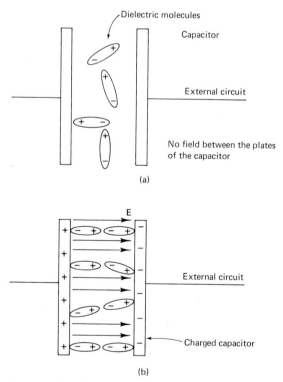

Fig. 3-11   A schematic of how the molecular dipoles within a dielectric are oriented normally (a) and when placed between the plates of a charged capacitor (b).

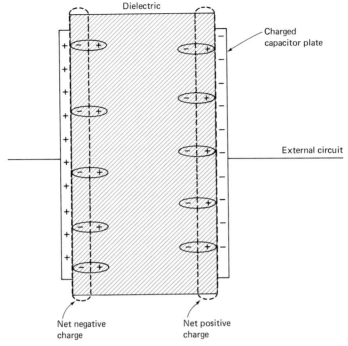

Fig. 3-12 A schematic of a dielectric in the space between the plates of a capacitor. The infinitesimal volume of the dielectric next to the positive plate of the capacitor obtains a net negative charge and that next to the negative plate obtains a net positive charge when the capacitor is charged. The dielectric still maintains a zero net charge throughout.

These charges are in response to the charges on the plates themselves and the electric field of the capacitor in the gap region. These charges then have the effect of neutralizing, or diluting, the charge q on either plate, which in turn reduces the electric field or voltage between the plates. Since the battery wants to keep a constant voltage between the plates, it pumps more charge to both plates to increase the voltage difference back to its original state. Compared to the situation where there was no dielectric in the capacitor, each plate now has an increased amount of charge; hence, by Eq. (3-25), the capacitance has increased.

The orientation of an aligned molecular dipole is always such that its electrical field opposes the field orienting it. If we consider the capacitor as a big dipole of one orientation, then the dipoles within the dielectric will tend to align in exactly the opposite orientation. The molecular dipoles cannot be aligned with the orienting field because if they were, the original field would be strengthened, yet energy has been taken out of the original field to orient the dipoles in the first place, so we have a contradiction.

Also, if the dipoles did orient with the capacitor's field, then the increase in the electric field could be used to produce useful work somewhere else and we would, in effect, be getting something for nothing; and we all know that cannot be true. The net effect of the aligned dipoles in the dielectric is to reduce the applied field, so that the total field in the dielectric is given as

$$\mathbf{E}_{tot} = \mathbf{E}_{appl} + \mathbf{E}_{align} \qquad (3\text{-}27)$$

where $\mathbf{E}_{align}$ is the field due to the aligned dipoles.

Suppose that a particular dielectric has molecules that are constrained so they cannot align themselves when an electric field is applied, or they do not possess permanent dipole moments, or both. Will this general capacitor effect still be seen? The answer is yes, but on a smaller scale than before. When an electric field is applied, most molecules experience a slight deformation in which there is a separation of positive and negative charge centers due to the external field. Positive charges tend to move with the field, negative charges against it. If the distribution of positive and negative charges are initially identical, then the applied field will cause the centers of positive and negative charge to become slightly separated. This is illustrated in Fig. 3-13. The fact that the centers of the positive and negative charges are now separated by a small distance dx causes the formation of a small dipole that exists as long as the field is being applied. The amount of separation dx is very small, typically less than 1 Å. Dipoles formed in this way are called induced dipoles, as opposed to dipoles that are permanent. The induced dipoles act like permanent dipoles in that they are formed in such a way as to oppose the applied field. Since the magnitude of the induced dipole is generally less than a permanent dipole, the opposition to the applied field is less, and hence the dielectric constant is lower for materials without permanent dipoles. When the electric field is turned off, the induced dipoles disappear (except for a group of dielectrics called electrets). In general, the

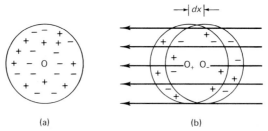

(a)   (b)

Fig. 3-13   The effect of an electric field on a system with coincident centers of positive and negative charges is to separate those centers. (a) A charge system where the centers of positive and negative charge are identical. (b) The effect of an external electric field on this system. The centers of positive O_+ and negative O_- charge are separated by a distance dx.

strength of an induced dipole is proportional to the applied field strength, but this point will not be developed further until Chapter 5.

So now we can see how dielectrics reduce the strength of electrical inter-actions. By orienting permanent dipoles or inducing temporary dipoles in the dielectric medium, or both, the initial field strength is counteracted. When two point charges are interacting in a vacuum, there is no material between them to interfere; however, when these same two charges are sub-merged in a dielectric, the alignment of dipoles in the dielectric sets up a reverse electric field that causes a reduction in the field one charge feels as a result of the other. The dielectric constant D is then a measure of the ability of a material to be polarized either through alignment of permanent dipoles or through the formation of induced dipoles.

When trying to ascertain whether or not a given material will be a good dielectric or a poor one, a number of factors must be considered. To have a high dielectric constant, the basic molecular building blocks of the material must not only possess permanent dipole moments, but the dipoles must be free to move and align in a field. Solids then generally have low dielectric constants because the structural elements are quite constrained and do not have freedom of motion. A good example of this is given by considering water in its liquid and solid states. As a liquid, bulk water possesses a very high dielectric constant of about 78 at 25°C, but when frozen under high pressure, its value drops to as low as 3.0.* In this crystalline state, even the strong dipole moment of the water molecule cannot be aligned by an electric field because the crystal structure rigidly holds each molecule. Since the permanent dipoles cannot be aligned, the main contribution to D comes from induction; but this contribution is relatively small. Based on this principle, then, one would expect gases with strong molecular dipoles to have large dielectric constants; however, this is not necessarily so. True, gas molecules do have great freedom of movement, but the disorienting effects of thermal agitation, or brownian motion, are quite significant. Gas molecules can be easily oriented because they have few constraints, but they can also be easily disoriented; at temperatures commonly found in the laboratory, alignment is usually incomplete because of thermal agitation. Alignment does however increase with increasing external field strength. The hotter the gas, the more randomized are the molecular dipoles; so one would expect the dielectric constant to decrease with increasing temperature. This is usually the actual case.

---

* Ice is known to have as many as eight different crystal structures depending on the tem-perature and pressure. The form of ice that has a dielectric constant of three is rarely en-countered. Actually, normal ice (ice I) has a very high dielectric constant, which implies the molecular dipoles are not as rigidly constrained as one would think. See Chapter 8 for more details.

In considering liquids we have a case where the constraints on the molecular dipoles are intermediate between those on the molecules of a solid and a gas. The degree of association of molecules with one another in a liquid is more than that found in a gas, but they are not held together as rigidly as in a solid. Individual molecules in liquids are capable of alignment in an electrical field, and Brownian motion can also disrupt the alignment; however, viscous effects tend to inhibit both processes. Again, one would normally expect the dielectric constants of liquids to decrease with temperature. The range of dielectric constants typically found can be best illustrated by some examples. Table 3-2 gives values of the dielectric constant for some typical pure liquids commonly found in biophysical work. Table 3-3 gives values for the dipole moment of several molecules in the gas phase. Dielectric constants for amino acids and proteins will be considered in Chapter 4.

In looking over the values of the dielectric constants and dipole moments, a number of trends can be seen. First, there is not necessarily an exact relationship between the strength of the dipole moment in gas phase and the dielectric constant in liquid phase. This is due to the fact that when a gas condenses to a liquid, there is a closer association between molecules with a subsequent change or distortion of electronic structure. This will affect not only the permanent dipole moment, but also the ability of a molecule to have a dipole moment induced by an external electric field. Sometimes two or more molecules that remain isolated in the gas phase will pair up to form a structural unit in a liquid; with the fact that the unit's dipole moment is the vector sum of the individual dipole moments, the resulting dielectric constant can be low.

A trend that can be observed is that the relative strength of the dielectric constant can sometimes be predicted by knowing the structural formula of a compound. C—H and C—C bonds have a relatively weak dipole moment, so one would expect hydrocarbons to have low dielectric constants. Hydrocarbons also have small dipole moments and dielectric constants due to the geometric relationship of individual bonds; i.e., the dipole moment for one bond will vectorially cancel that due to another bond. This can be seen in the case of methane where the structural formula

$$
\begin{array}{c}
\text{H} \\
| \\
\text{H——C——H} \\
| \\
\text{H}
\end{array}
$$

has a high degree of symmetry, and the resulting molecular dipole moment is zero. Another example of this can be seen by comparing methane and carbon tetrachloride. If all the H atoms of methane are replaced by Cl atoms, the dipole moment of the resulting molecule is still zero due to the high degree of symmetry (see Chapter 8). If, however, one chlorine atom and

TABLE 3-2

The Dielectric Constant of Some Typical Liquids

Structural formula	Name	Dielectric constant[a]	Temperature (°C)
H O | // H—C—C | \\ H H	Acetaldehyde	21	10
H O | // H—C—C | \\ H OH	Acetic acid	6.2	20
H O | // H—C—C | \\ H O H | H—C—C | \\\\ H O	Acetic anhydride	22	1
H O H | || | H—C—C—C—H | | H H	Acetone	21	25
H | C // \\ H—C C—H | || H—C C—H \\\\ / C | H	Benzene	2.2	20
Cl | Cl—C—Cl | Cl	Carbon tetrachloride	2.2	20
H H | | H—C—C—OH | | H H	Ethanol	24.3	25

TABLE 3-2   (continued)

Structural formula	Name	Dielectric constant[a]	Temperature (°C)
H—C($\overset{\text{H}}{\underset{\text{H}}{\mid}}$)C($\overset{\text{H}}{\underset{\text{H}}{\mid}}$)—O—C($\overset{\text{H}}{\underset{\text{H}}{\mid}}$)C($\overset{\text{H}}{\underset{\text{H}}{\mid}}$)—H	Ethyl ether	4.3	20
H—C($\overset{\text{H}}{\underset{\text{OH}}{\mid}}$)C($\overset{\text{H}}{\underset{\text{OH}}{\mid}}$)—H	Ethylene Glycol	37	25
H—C$\overset{\text{O}}{\underset{\text{OH}}{}}$	Formic acid	58	16
H—C($\overset{\text{H}}{\underset{\text{OH}}{\mid}}$)C($\overset{\text{H}}{\underset{\text{OH}}{\mid}}$)C($\overset{\text{H}}{\underset{\text{OH}}{\mid}}$)—H	Glycerol	42.5	25
$CH_3(CH_2)_4CH{=}CHCH_2CH{=}CH(CH_2)_7C\overset{\text{O}}{\underset{\text{OH}}{}}$	Linoleic acid	2.6	0
H—C($\overset{\text{H}}{\underset{\text{H}}{\mid}}$)—H	Methane	1.7	−173
H—C($\overset{\text{H}}{\underset{\text{H}}{\mid}}$)—OH	Methanol	32.6	25
$CH_3(CH_2)_7CH{=}CH(CH_2)_7C\overset{\text{O}}{\underset{\text{OH}}{}}$	Oleic acid	2.5	20
H—C($\overset{\text{H}}{\underset{\text{H}}{\mid}}$)C($\overset{\text{H}}{\underset{\text{H}}{\mid}}$)C($\overset{\text{H}}{\underset{\text{H}}{\mid}}$)C($\overset{\text{H}}{\underset{\text{H}}{\mid}}$)C($\overset{\text{H}}{\underset{\text{H}}{\mid}}$)—H	n-Pentane	1.8	20

TABLE 3-2 (continued)

Structural formula	Name	Dielectric constant[a]	Temperature (°C)
	Phenol	9.8	60
	Propane	1.6	0
	1-Propanol	20	25
	Styrene	2.4	25
	Toluene	2.4	0
	Water	(see page 96)	

[a] The value of D is given for the particular temperature stated. Data from "Handbook of Chemistry and Physics," 54th Ed. Chemical Rubber Publ. Co., Cleveland, Ohio, 1973.

TABLE 3-3

The Dipole Moment in Debyes of Some Molecules in the Gas Phase

Structural formula[a]	Name	Dipole moment (debyes)
	Acetaldehyde	2.7
	Acetic acid	1.74
	Acetic anhydride	2.8
	Acetone	2.88
	Benzene	0
$O{=}C{=}O$	Carbon dioxide	0
	Chloroform	1.01
	Ethanol	1.7

TABLE 3-3 (continued)

Structural formula[a]	Name	Dipole moment (debyes)

H   H
  \ /
   C=C
  / \
 H   H
	Ethylene	0

H H | | H—C—C—Cl | | H H	Ethylene chloride	1.34
H H | | H—C——C—H | | OH OH	Ethylene glycol	2.28
H | H—C—H | H	Methane	0
H | H—C—OH | H	Methanol	1.7
H | H—C—Cl | H	Methyl chloride	1.86
H H H H H | | | | | H—C—C—C—C—C—H | | | | | H H H H H	n-Pentane	0
H H H | | | H—C—C—C—H | | | H H H	Propane	0.084
H H H \ | | C=C—C—H / | H H	Propylene	0.35

TABLE 3-3   (continued)

Structural formula[a]	Name	Dipole moment (debyes)
H | C H—C C—H H—C C—H C | OH	Phenol	1.45
CH$_3$ | C H—C C—H H—C C—H C | H	Toluene	0.36
H | H$_2$N—C—NH$_2$ | OH	Urea	4.6
H / O \\ H	Water	1.84

[a] The structural formula is also given so the reader may relate the magnitude of the dipole moment to the chemical composition of the molecule. Data from "Handbook of Chemistry and Physics," 54th Ed.

three hydrogen atoms are bound to a carbon atom (methyl chloride),

$$H—\underset{\underset{H}{|}}{\overset{\overset{H}{|}}{C}}—Cl$$

there is enough asymmetry in the molecule so that the individual bond dipoles do not cancel, but instead yield a relatively large dipole moment.

Another trend in the value of a dipole moment can be seen when oxygen is present in the molecule. Oxygen is a fairly electronegative atom, which

$$\begin{array}{cc} \overset{\delta_+}{C} \overset{\delta_-}{-} \overset{}{O} & \overset{\delta_+}{C} \overset{\delta_-}{=} \overset{}{O} \end{array}$$

Fig. 3-14 Schematic showing the slight negative charge oxygen acquires in a covalent bond with carbon. Concurrently, carbon acquires a slight positive charge. This effect is due to the fact that oxygen is very electronegative and tries to possess a larger share of the common electrons in a covalent bond than does its partner.

means that in a covalent bond it will possess a larger share of the common electrons than will its partner. Because of this, any bond in which oxygen is present will most likely have unequally shared electrons, thus creating a dipole moment. In a C—O or a C=O bond the oxygen atom will tend to acquire a slight negative charge and the carbon atom will acquire a slight positive charge, as is shown in Fig. 3-14. Molecules with these bonds would then possess the potential of having larger dipole moments and also larger dielectric constants than hydrocarbons, i.e., compare methanol to methane, and ethanol to ethane. Notice also that ethanol has one more C atom than does methanol, but a similar dipole moment. Again, this illustrates that the OH group is relatively more important in its contribution to the molecular dipole moment than is the CH_3 group. By placing an OH group onto a benzene ring to make phenol, the dipole moment changes from zero to 1.45 D, whereas the addition of a CH_3 group (toulene) hardly has any effect at all. Molecules or compounds that have a relatively strong dipole moment are called polar; those without a strong dipole are called nonpolar or hydrophobic (water-hating). The initial classification of amino acids in Chapter 1 was made based on the polarity of the side chain groups. The reader might now do well to review these classifications to examine the structure of the side groups in relation to their dipole moments. It is also interesting to note that by cleverly using electronegative elements (F, Cl, O, N), additions can be made to chemical structures that can drastically change their dipole moments and hence their physical and chemical properties to suit a purpose. A good illustration of this point is seen in the case of dithiothreitol (Fig. 3-15).

When a biophysicist or biochemist wants to carry out some research with an enzyme, the enzyme will usually be isolated from the host organism and stored in some sort of aqueous medium at an appropriate pH. One fact that is very well known concerning enzymes is that the part of their structure that is of the highest biological importance can be localized in

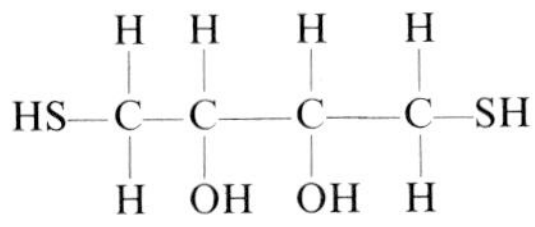

$$\begin{array}{ccccccc} & H & H & & H & & H \\ & | & | & & | & & | \\ HS- & C- & C- & - & C- & - & C-SH \\ & | & | & & | & & | \\ & H & OH & & OH & & H \end{array}$$

Fig. 3-15 A structural formula of dithiothreitol. The OH groups were added to ensure the solubility of the molecule in water.

one area called the active site. Many times this active site will contain a sulfhydryl group (S—H) which plays an important role in the catalytic reaction of the enzyme. Sulfhydryl groups can also be found at other parts of the enzyme. In any event, a disruption of a biologically important S—H group will frequently result in a loss of biological activity. A common example of this is observed when certain heavy metals contaminate the solution. For instance, mercury, copper, and silver combine with the S—H group to form a complex resulting in an inactive enzyme. This is illustrated for the case of mercury:

$$-S-H + Hg^+ \rightarrow -S-Hg + H^+ \tag{3-28}$$

In order to prevent such inactivation, it is usual practice to try to protect the S—H groups. One way that this is accomplished is by adding to the solution a compound that also has S—H groups, so that they will be altered instead of the enzymatic S—H groups. One may wonder how this selective alteration comes about. By making the additive small in size so that it can move about or diffuse in solution much faster than can an enzyme, which is typically 10^2–10^3 times larger, and by placing several S—H groups on it, then the probability of the additive coming into contact with a denaturing agent is much higher than for the corresponding S—H groups of the enzyme. In this respect the biochemist W. W. Cleland synthesized the compound shown in Fig. 3-15 to perform this task. From its structure the molecule is seen to be very small and to possess two S—H groups. In looking at it, though, one would initially wonder about the purpose of the two OH groups. It turns out that without them the compound has a low dipole moment; and because of it, the molecule possesses a low solubility in water. By adding the two OH groups, the dipole moment of the molecule is increased thus making it much more compatible with water, hence the solubility increases. As will become more obvious in Chapter 8, compounds with large dipole moments are easier to dissolve in water than ones that have little or no dipole moment. The OH groups themselves are not of primary biological importance, although the compound is useless without them.

When the dielectric constants for those liquids shown in Table 3-2 are compared for magnitude, it is found that water has the highest value. This is highly significant and is due to water's unique structure and its high dipole moment. In Chapter 2 the question was initially asked as to whether or not electrical forces between biomolecules would be significant. If we were to make a survey of influential forces on a molecular level, would electrical forces be worth considering? In Chapter 2 also a calculation was made showing that the field of a single point charge is enormous, and since biomolecules can have net electronic charges of ± 40 and over, the question now changes to, If the fields surrounding these molecules are so very large, then why do not all similarly charged molecules fly apart from one another, or ones of opposite charge aggregate into one big lump? Since this obviously does not happen in life systems, one must wonder whether nature has not been very clever, and has not somehow been able to dilute the potential strength of electrical forces, whether nature has not figured out some way, or ways, of bringing electrical forces under control, or down to a manageable level. If nature has actually done this, then perhaps a way this has been carried out is by using water as one of the main components in cells. By immersing the molecules necessary to carry out the life processes in a medium of high dielectric constant, electrical forces could be reduced. For water at $20°C$, this reduction in electrical force amounts to almost two orders of magnitude compared to the case where $D = 1$. It is probably no mean accident that life as we know it first started in the primeval oceans. Not only did these waters allow for mobility of molecules and nutrients, they also had dielectric properties that were quite beneficial to the structural framework of early biomolecules. As a simple example of biological composition, consider *Escherichia coli*, which is a bacterium that inhabits the intestinal tract of humans, and which is a very widely studied organism. It has 70% of its total weight made up of water. This is a fantastic amount of a single component in a living organism, and its presence is probably no accident. This is one reason aqueous solvents are routinely used when working with biomolecules that can carry a charge, which includes just about all biomolecules. By using an aqueous soluent it is possible to more closely mimic the natural environment of the molecules. Salts of one sort or another are also dissolved in these solvents. The reason for this varies from case to case; however, the important fact is that water is heavily used as a solvent. Why additional salts are used will be considered in Chapter 7. We shall also see in Chapter 8 that water has other useful properties that make it particularly good as a solvent and host medium for life systems. However, it should be emphasized that a physical property, namely the dielectric constant, by influencing the magnitude of electrical forces, can influence life processes at the molecular level.

So far it has been implied that to correctly use Coulomb's equation (or a subsequent equation), all one has to do is substitute in the value of the dielectric constant for the particular solvent of interest. As in considering many types of physical interactions, this situation is not so simple as it seems. It turns out that the dielectric constant itself depends on the physical conditions of the situation. For instance, the dielectric constant varies with temperature, ionic strength,* distance of separation, and electric field frequency. To determine the exact value of the dielectric constant to be sub-

* The ironic strength is a measure of the amount of salts or ions in a solution.

stituted into an equation takes a fairly detailed knowledge of the problem. Only by considering each case separately can the correct value, or a reasonable approximation, be chosen. In the sections to follow it will be shown how the dielectric constant depends on the physical parameters mentioned above.

DIELECTRIC CONSTANT AND TEMPERATURE

Figure 3-16 shows the experimental values for the dielectric constant of water as a function of temperature. It is to be noticed that there is a steady decrease in the dielectric constant with increasing temperature. This is consistent with the molecular model of dielectrics previously discussed. As the temperature rises, random thermal motion tends to increasingly disorient alignment of dipoles; hence the applied field is not counteracted as much, so D is less. It has been found that the best mathematical fit to the data in Fig. 3-16 is given by

$$D = a - bT + cT^2 - dT^3 \tag{3-29}$$

where

$$a = 87.74, \qquad b = 0.4008,$$
$$c = 7.398 \times 10^{-4}, \qquad d = 1.410 \times 10^{-6}$$

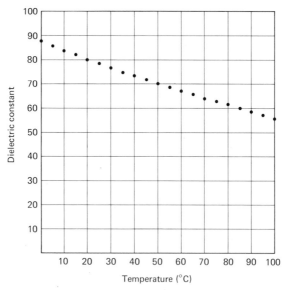

Fig. 3-16 The dielectric constant of water as a function of temperature. (Data from Melmberg and Maryott, 1956.)

For D_2O, the dielectric constant has been reported to vary with temperature according to an equation similar to Eq. (3-29), except that now

$$a = 87.48, \qquad b = 0.4051, \qquad c = 9.638 \times 10^{-4}, \qquad d = 1.33 \times 10^{-6}$$

For other liquids, it has been shown that the dependence of the dielectric constant on temperature can be expressed as

$$D = D_0 e^{-LT} \tag{3-30}$$

where D_0 and L are constants depending on the liquid. Sample values are shown in Table 3-4. In Eq. (3-30) D has no units; therefore L must have units of reciprocal temperature. With formulas similar to Eqs. (3-29) and (3-30) it is possible to calculate values of the dielectric constant for a liquid dielectric for any temperature for which the formulas are valid. It is important to realize that when making a calculation of electrical force, field energy, etc. in which a value of the dielectric constant is needed, it is not enough to realize that a value of D is needed. It must be kept in mind that D is a value that can change its magnitude depending on the physical situation, and that temperature is one factor influencing this change.

TABLE 3-4

Dependence of Dielectric Constant on Temperature[a]

Liquid	D_0	$L (\times 10^3 /^\circ K)$
Hexane	2.33	0.714
Carbon tetrachloride	2.88	0.843
Benzene	2.95	0.876
Chorobenzene	15.52	2.89
Chloroform	12.0	3.33
Methanol	157.6	5.39
Ethanol	146.0	6.02

[a] For a general liquid, the general dependence of the dielectric constant on temperature has the form $D = D_0 e^{-LT}$ where D_0 and L for different liquids are shown above. (Moelwyn-Hughes, 1961.)

DIELECTRIC CONSTANT AND DISTANCE OF SEPARATION

Consider the situation where it is desired to calculate the force of interaction between two ions that are immersed in an aqueous solvent. We have already seen that the field in the immediate vicinity of an ion with one electronic charge is extremely high. Since water molecules possess a permanent dipole moment, the H_2O molecules close to the ion will be aligned with this high field and will form a primary hydration layer around the ion. They will not only be aligned, but because of the intense field, they will be held strongly in place forming a pseudo solid or "crystallike" structure in a region close to the ion. The structure of this "pseudo crystal" is not exactly like a true crystal in the conventional sense; however, this term is used to indicate that there is an ordered water structure in the immediate vicinity of the ion. The structure itself however is dynamic with water molecules moving in and out very quickly. More will be said on this subject in Chapter 8. Suppose that we have two ions where their outer water shells are just touching as shown in Fig. 3-17a. In using Coulomb's law to compute the value of the force between the two ions, what is to be used as a value of the dielectric constant? Since the water molecules in the near vicinity of each ion are rigidly held in place, they are not free to realign in the field produced by the other ion; hence the dielectric constant is going to be less than the dielectric constant of bulk water molecules that could align. Also, the number of water molecules between the two ions is quite small; their

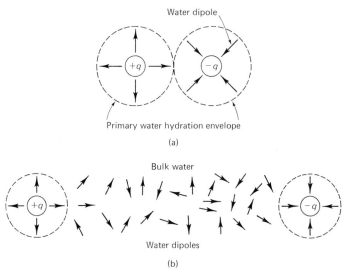

Water dipole

Primary water hydration envelope

(a)

Bulk water

Water dipoles

(b)

Fig. 3-17   The dielectric constant can depend on the distance separating two interacting particles. (a) Two ions are closely spaced, so that only the primary waters of hydration separate the ions. The water dipole moments are indicated by the arrows, and the dotted circle indicates the spherical shell of primary waters. These primary waters are not really as free as bulk water to respond to the field of the other ion. (b) The two ions are separated by a large distance, so that bulk water can gather between the two ions. In this case the bulk water molecules can respond to the electric fields of the two ions, and the appropriate dielectric constant is that of bulk water. In the situation depicted in (a), the appropriate dielectric constant will be less than that for bulk water. In (b) the water dipoles are shown in random orientations.

effect is going to be negligible compared to the situation where bulk water is present as the dielectric. If the ions are now separated as shown in Fig. 3-17b, free water molecules in great numbers will be allowed to enter the space between the two ions. These water molecules will then be able to align with the field produced by the ions; and hence they can effectively reduce the magnitude of the field. The dielectric constant will be equal to that of bulk water in this situation. Remember that alignment of permanent dipoles contributes much more to values of the dielectric constant than does the formation of induced dipoles.

The above situation has been somewhat oversimplified, but it does illustrate the fact that for certain cases, the value of D used in a calculation depends on the distance of separation r. For the specific case of water hydrating an ion, several workers have proposed detailed models of water structure and have made calculations showing an exact relationship between the dielectric constant and the distance of separation (Conway et $al.$, 1951; Grahme, 1950; Ritson, and Hasted, 1948). The various relationships depend heavily on the specific model constructed, although they all show that the effective dielectric constant increases as r does and that for large r, D reaches a saturation value equal to that for bulk water. It has been reported that values of D for bulk water can be used if the distances of separation between ions is greater than about 20 Å. For larger molecules, this distance is probably greater; but little work has been done in this area. A minimum saturation value is also reached as r decreases, where for this case, D is generally believed to take on a value close to 3.0. This value is certainly not absolute and has no real theoretical basis, but is based upon the dielectric value for certain forms of ice.

DIELECTRIC CONSTANT AND SALT CONCENTRATION

Since biological research relies heavily on aqueous salt solutions as solvents for macromolecules, one might wonder how the dielectric constant varies as salts are added to water. The situation unfortunately is not entirely clear, and there are discrepancies among experimental results. However, it is thought that the following trend holds true. When a salt such as potassium chloride is dissolved in water, it dissociates to form two ions which will separately become hydrated as was briefly described in the previous section:

$$KCl \rightarrow K^+ + Cl^- \tag{3-31}$$

Since a certain number of water molecules are fixed in orientation around these ions, they are unavailable to align themselves in an externally applied electric field that would be superimposed on the solution to measure the dielectric constant; hence the value of D will be less in a salt solution than for pure bulk liquid. The effect of these low concentrations of salts is to make a certain number of water molecules unavailable for use as a dielectric medium. For small concentrations, the dielectric constant decreases as the concentration of salt increases. For this situation, Debye has calculated that there should be a decrease of 0.7% for every millimole of KCl per liter in the solvent. This is a rather small effect.

For large concentrations of salt, it seems that the dielectric constant increases and in some instances surpasses the value for the pure liquid. The proposed explanation of this effect is that a large number of positive and negative ion pairs are formed and are aligned by the external field just as molecules with permanent dipoles are, thus increasing the bulk dielectric constant.

DIELECTRIC CONSTANT AND FREQUENCY

So far in our discussion of the dielectric constant, we have restricted our attention to measurements and values of the dielectric constant in a static electric field, i.e., a field that does not change with time. Since much experimental work is performed where molecular solutions are subjected to an alternating or varying electric field, one must consider how the dielectric constant is affected by a field that fluctuates periodically, and how the dielectric constant changes with the speed or frequency of that fluctuation. Consider again Fig. 3-11. In part (b) all molecular dipoles are generally aligned with the field generated by the capacitor. Now suppose that there is connected across the capacitor an ac generator that can reverse the direction of the field, and then reverse it again, etc. What happens to the dipoles? The answer is that it depends on how fast the field changes. If the field changes slowly enough, the dipoles will be able to reorient themselves to the new situation every time the field changes. In order to move, however, the molecules comprising the dielectric have to overcome frictional effects that retard motion. The extent of these frictional forces will depend on the size and shape of the molecules and the amount of association they have with one another. Larger molecules will generally move more slowly, as will ones that are highly asymmetric. Very viscous solvents will also retard the motion of the reorienting dipoles. If frictional forces are easily overcome, the dipoles can easily reorient with the field. However, as the frequency of the field increases, the frictional forces become relatively large compared to the electrical force reorienting the dipoles, and hence the dipole orientation tends to lag the changes of the field; i.e., the dipoles cannot become completely reoriented before the field reverses again. Since the dipoles are not aligning as well in this situation, the dielectric constant must be lower. When the frequency of the external field is very high, the frictional forces are overwhelming compared to the orienting electrical forces; hence the dipoles hardly move at all. When this situation occurs, the contribution of aligned permanent dipoles to the dielectric constant becomes negligible. The only remaining contribution to the dielectric constant comes from the induction of dipoles through distortion of charge centers. The distortion of charge centers to form atomic dipoles can keep up with very fast shifts of the field, and so the dielectric constant takes on a small value that is greater than one. It should be realized that these slight shifts of electronic charge centers are not dependent on shape or size of the molecules or on the solvent's viscosity. They are merely a function of how tightly the electrons are held by the atomic structure of the molecule.

In general then in a static field (zero frequency) the dielectric constant has a maximum value, and the dielectric constant decreases for a field of higher and higher frequency. For water, it is known that the dielectric constant remains relatively constant in fields with frequencies of up to 10^8 Hz (1 Hz = 1 cycle per second). The concept that the dielectric constant can vary with the frequency of the field will be put to use in the next chapter.

What started out as a fairly simple concept has now turned into a very complicated situation. At first the dielectric constant just depended on the material, and now it seems to depend on a host of other parameters. This is generally the way the physics of the situation develops as closer and closer examinations are taken. For most cases, though, the bulk dielectric constant is used as a first approximation in any calculation; however, the reader should at least be aware that the situation can be complex.

SUMMARY

A dipole is defined in terms of a specific charge arrangement, and the dipole moment is a vector quantity whose magnitude is given as a product of the distance of separation and the charge on each particle. The dipole can be treated as a unit when calculating electrical field or potentital without knowing the charge or length. A dipole placed in a constant, uniform electric field will rotate to form an angle with respect to that field. If placed in an inhomogeneous field, the dipole will translate or move in the field. In either situation the response of the dipole depends on the strength of the dipole moment. The dielectric constant is a measure of a medium's ability to have its permanent dipoles oriented or to form induced dipoles under the influence of an external electric field. Dielectric constants are frequently measured with the aid of a capacitor. Liquids that are commonly encountered in biophysical work have dielectric constants that vary from 2 to 80, with water having a high value. The dipole moment of a chemical compound can be frequently deduced qualitatively by knowing its structure, and the dipole moment of a compound will frequently determine whether or not it will dissolve in water. The dielectric constant is a function of temperature, ionic strength, frequency of the applied field, and the distance of separation.

REFERENCES

Conway, B. E., Bockris, J. O., and Ammar, I. A. (1951). *Trans. Faraday Soc.* **47**, 756.

Grahame, D. C. (1950). *J. Chem. Phys.* **18**, 903.

Halliday, D., and Resnick, R. (1963). "Physics for Students of Science and Engineering," Part II. Wiley, New York.

Ritson, D. M., and Hasted, J. B. (1948). *J. Chem. Phys.* **16**, 11.

Sears, F. W., and Zemansky, M. W. (1960). "College Physics." Addison-Wesley, Reading, Massachusetts.

Melmberg, C. G., and Maryott, A. A. (1956). *J. Res. Nat. Bur. Stand.* **56**, 1.

Moelwyn-Hughes, E. A. (1961). "Physical Chemistry." Pergamon Press, New York.

4

DIPOLE MOMENTS OF BIOLOGICAL MACROMOLECULES

INTRODUCTION

In this chapter we shall use the mathematical and physical arguments that have been established in the previous chapters to develop a theory, and a corresponding experiment, whereby the dipole moments of biological molecules can be determined. The value of μ for several typical proteins and amino acids will be presented; and, based on these values, we shall try to see whether any insights or better understanding of molecular structure can be obtained. The reader should be warned that this is a particularly difficult chapter.

The dipole moment of a protein is usually determined by measuring the dielectric constant of an aqueous solution containing the protein at some low concentration and then relating this dielectric constant to the dipole moment of the molecules comprising the solution. In the first part of this chapter a theory will be derived in which the measured macroscopic dielectric constant is related to the permanent dipole moment of the biomolecules in solution. In so doing several new physical concepts must be presented, and they will be explained and discussed in the development. It is important that the reader fully understand these new principles since they will surface at regular intervals throughout the remainder of the text.

POLARIZABILITY

When an electric field is applied to a dielectric, the material is polarized through the alignment of permanent dipoles and/or the induction of atomic or molecular dipoles. The larger the field, the larger is the polarization; permanent dipoles are aligned better, and the positive and negative centers of a molecule are pulled further apart to form larger induced dipoles. The average dipole moment thus formed is then a function of the magnitude of the electric field, and this is usually represented mathematically as

$$\boldsymbol{\mu} = \alpha \mathbf{E} \tag{4-1}$$

where the constant of proportionality α is called the polarizability. Although it is not obvious from Eq. (4-1), α has the dimensions of volume, and its value depends on the properties of the particular material in question. It is a physical constant characteristic of a material, just as the dielectric constant is. Looking at Eq. (4-1), one might suspect that the average dipole moment formed is in the same direction as the applied field; however this is not always true. For materials where the electrical properties show no dependence on direction, the induced dipole moment is parallel to the electric field. For these isotropic materials, α is just a constant depending on the material itself. For anisotropic materials, α takes on the mathematical form of a tensor which has directional properties. For this more complicated case, the x component of the induced dipole will depend not only on the x component of the electric field, but also on the y and z components as well. A similar relationship will hold for the y and z components of the dipole. In general, the tensor form of α has nine components, and α can vary from one spot to another within a dielectric material. The mathematical form α takes depends directly on the atomic and molecular structure of the dielectric. For all cases considered in this book, however, α will be assumed to be a simple constant, depending only on the dielectric.

It should also be realized that α is the sum of separate contributions from induced dipoles and also the alignment of permanent dipoles where we have

$$\alpha = \alpha_{\text{ind}} + \alpha_{\text{perm}} \tag{4-2}$$

If there are no permanent dipoles to be aligned by an external electric field, then $\alpha = \alpha_{\text{ind}}$. The quantity α_{ind} is a measure of how easy it is to displace atomic centers of positive and negative charge in a dielectric. Larger values of α_{ind} indicate that relatively weak forces hold the various charges in place. It is possible to develop physical models so that α_{ind} and α_{perm} may be expressed mathematically in terms of more fundamental quantities or measurable parameters. For our purposes, only α_{perm} is of interest, and rightly so because it is through the alignment of permanent dipoles that the greatest contribution to α is made. For those who are interested in a more mathematical model of α_{ind}, consult the references at the end of this chapter. Our aim then is to derive an expression for α_{perm} in terms of the dipole moment of a biomolecule in solution. We shall proceed to do this after the polarization vector and Boltzmann's distribution are introduced.

POLARIZATION VECTOR

Another quantity that is sometimes used in describing dielectric phenomena is the polarization vector **P**, which is defined as the average dipole
moment per unit volume. As an illustration of this new vector, let us consider a dielectric medium where there are no permanent dipoles. If an
external electric field is applied to this dielectric, the centers of positive
and negative charge of each atom will be displaced from one another forming
tiny dipoles in the process (Fig. 4-1). If the quantity of positive and negative

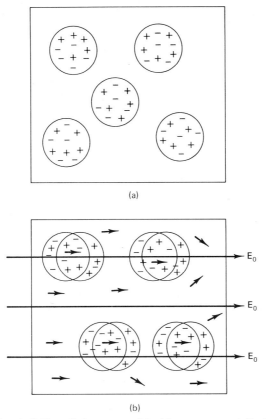

Fig. 4-1   An electric field applied to a material without permanent dipoles will cause the
formation of induced dipoles. (a) The outer boundary represents an infinitesimal volume of
such a dielectric. Each circle represents an atom or molecule with a spherically symmetric
positive and negative charge distribution. (b) The dielectric is placed in a constant, uniform
electric field $\mathbf{E}_0$. Here, the centers of positive and negative charges in each atom or molecule are
separated. The polarization vector **P** is found by vectorially adding all the dipole moments and
dividing by the volume. The small arrows represent induced dipoles.

charge of each atom is q, if the centers of charge O_+ and O_- are displaced by a distance l, and if there are Λ atoms per unit volume, then the polarization vector $\mathbf{P}$ is given by

$$\mathbf{P} = \Lambda q l \tag{4-3}$$

but the last two quantities on the right-hand side of Eq. (4-3) are equal to the induced dipole moments, or

$$\mathbf{P} = \Lambda \boldsymbol{\mu}_{ind} \tag{4-4}$$

where $\mathbf{P}$ can now be seen to be the dipole moment induced per unit volume. If permanent dipoles are present, or the induced dipoles are not all aligned exactly the same way as in Fig. 4-1, or both, then Eq. (4-4) must be modified to read

$$\mathbf{P} = \sum_i \Lambda \boldsymbol{\mu}_i \tag{4-5}$$

where the vector summation is needed. Λ_i is the number of atomic or molecular dipoles per unit volume of the type i.

There are essentially two reasons physicists have introduced the concept of the polarization vector $\mathbf{P}$. First, it presents a link between the microscopic model of atomic and molecular dipoles and a macroscopic description of neutral dielectrics. The polarization vector is a macroscopic quantity, whereas $\boldsymbol{\mu}$ is submicroscopic. The other reason for $\mathbf{P}$'s existence is that it allows other electrical equations to be written more simply. The polarization vector is introduced here because later we shall have to consider an equation with $\mathbf{P}$ in it.

BOLTZMANN'S DISTRIBUTION

The last new concept that must be introduced before finally considering the dipole moment of proteins is Boltzmann's distribution. This is one of the most fundamental and important principles in all of classical physics, and it is involved in many types of theoretical calculations in many different areas. The Boltzmann distribution or principle is essentially concerned with determining how the various components of a system will be statistically distributed throughout the energy states available to them. If a system has a number of energy states that it can occupy, then the Boltzmann distribution can give information on the probability that a particular energy state will be occupied, or that the system will possess that amount of energy. This problem can be illustrated by considering the situation shown in Fig. 4-2. If the horizontal lines represent energy states, and the dark circles

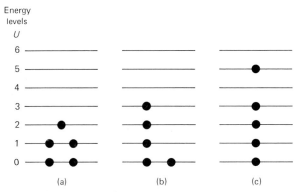

Fig. 4-2 Three possible energy configurations of a system. The horizontal lines represent different energy levels U available to a system, and the dots represent components of the system. Which energy distribution is most likely to occur can be described by the Boltzmann distribution. Case (c) has the highest energy and (a) has the lowest.

represent components within a system, then cases (a)–(c) illustrate three possible energy configurations. Case (c) has the highest energy, and case (a) has the lowest energy per system. The question is then, Which of these three cases is most likely to be found in reality? For a particular temperature T, the Boltzmann distribution will give information as to which case is most probable.

A rigorous derivation of the Boltzmann distribution requires the use of statistical physics which is beyond the scope of this book. A simple example involving the exponential atmosphere will however be considered in which the Boltzmann distribution will be made to seem plausible. This result will then be generalized, and the Boltzmann distribution described along with the criteria for its application. The exponential atmosphere example is used only for the purpose of making the Boltzmann distribution seem reasonable and is not offered as a proof. Otherwise, this example has no bearing on dipole moments of biomolecules.

EXPONENTIAL ATMOSPHERE

Consider a vertical column of a gas having a unit cross-sectional area, and suppose that this gas is in a normal gravitational field and that the column has a constant absolute temperature T throughout the entire volume. This is illustrated in Fig. 4-3. It is now required to find the spatial distribution of the gas molecules as a function of height h and temperature T. The answer will be a special case of the Boltzmann distribution.

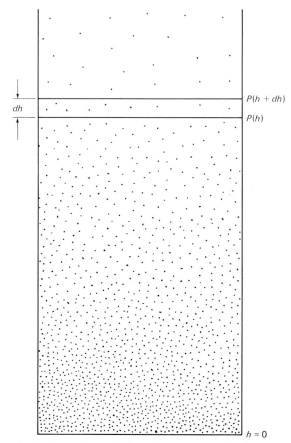

Fig. 4-3   A vertical column of a gas having unit area cross section in a normal gravitational field. The entire column has a uniform temperature T. The pressure at height h is designated $P(h)$ and that at height $h + dh$ as $P(h + dh)$. The density of gas molecules varies with the height of the column and this spatial relationship can be described by a special case of the Boltzmann distribution.

If the total volume of the gas is V, if n is the total number of molecules, and if P is the pressure, then the ideal gas law states that

$$PV = nk_B T \qquad (4\text{-}6)$$

where k_B is called Boltzmann's constant and $k_B = 1.38 \times 10^{-6}$ erg/°K. Equation (4-6) may be rewritten as

$$P = Nk_B T \qquad (4\text{-}7)$$

where $N = n/V$ is the number of molecules per unit volume.

In trying to quantitatively describe the spatial relationship of gas molecules with height, several approaches can be taken. The one used here attacks the problem by deriving first an equation relating the pressure of the column to the height, and then relating this to the number of gas molecules present.

From a purely mechanical viewpoint the pressure of the gas increases as the height from the bottom decreases because each volume of gas must support the weight of gas above it. This is an intuitive fact that the mathematics must describe if the proposed model is correct. If we now consider an infinitesimal slab of gas having a thickness dh (Fig. 4-3), then the difference in pressure between the top and bottom of the slab $P(h + dh) - P(h)$ is equal to the weight of the gas in the slab itself. The weight of each molecule is mg where m is its mass and g is the acceleration due to gravity. The total number of gas molecules in the slab is $N\, dh$ (remember the slab and the column have unit cross-sectional area), so that the total weight is $mgN\, dh$. The difference in the pressure across the infinitesimal slab is then given as

$$P(h + dh) - P(h) = dP = -mgN\, dh \qquad (4\text{-}8)$$

Differentiating Eq. (4-7) with respect to N gives

$$dP = k_B T\, dN \qquad (4\text{-}9)$$

and combining Eqs. (4-8) and (4-9) gives

$$\frac{dN}{N} = \frac{-mg\, dh}{k_B T} \qquad (4\text{-}10)$$

Equation (4-10) is a differential equation relating the change in the number of molecules per unit volume to the change in height. To find a direct relationship between N and h, it is necessary to integrate Eq. (4-10) from $h = 0$ ($N = N_0$) to an arbitrary height h. Doing this yields

$$\ln(N/N_0) = -mgh/k_B T \qquad (4\text{-}11)$$

or

$$N = N_0 e^{-mgh/k_B T} \qquad (4\text{-}12)$$

where Eq. (4-12) is the desired result that describes the number of gas molecules per unit volume as a function of the column height h and N_0, which is the number of gas molecules at height $h = 0$.

It should be noticed that the absolute value of the numerator of the exponential in Eq. (4-12) is equal to the gravitational potential energy of a particle with mass m at height h. Interpreting the physical situation just described in terms of energy levels, it is seen that each gas molecule has available to it different energy states which are given by the gravitational potential energy $U = mgh$. For every different value of mgh, there is a different energy state available to a gas molecule. Equation (4-12) then describes the number of particles N in a particular energy level compared to the number N_0 in the energy ground state ($h = 0$). The factor k_BT has units of energy and is interrupted as being a measure of the amount of thermal agitation or randomizing energy. The higher the temperature, the greater will be the thermal motions of the molecules and the wider will be the distribution of particles throughout the possible energy states. Another way of stating this is that an increased temperature will increase the number of occupied energy states compared to the number occupied at a lower temperature. The increased thermal energy at the higher temperature tends to randomize the system more; thus the system is not as easily described as being in just a few energy states. Referring to Fig. 4-2, the progression from (a) to (c) shows the effect of increasing the temperature of the system. Our vertical column of gas then has its number of molecules per unit volume (density) decreasing exponentially with height. As the temperature goes up, the height distribution of the gas molecules will be greater due to thermal agitation. This model is not strictly analogous to our own atmosphere because in reality the temperature of the atmosphere decreases with height and is not constant, as was initially assumed. The density of the atmosphere does however decrease with increasing height, so our model of an exponential atmosphere is not altogether invalid.

The fact that the numerator of the exponential in Eq. (4-12) is a potential energy term is no small accident. This numerator is actually a difference, the potential energy term $U = mgh$ minus the potential energy term of the ground state $U_0 = mgh_0$ where $h_0 = 0$. Using this notation, Eq. (4-12) can be equivalently written as $N = N_0e^{-(mgh-mgh_0)/k_BT}$. It can be shown that this is a general result and that the population of various energy states can be described as a function of the difference in energies $(U - U_0)$ by an exponential term. This is the Boltzmann principle, and it is written as

$$N_i = N_0e^{-(U_i-U_0)/k_BT} \tag{4-13}$$

where N_i is the number of particles having energy U_i, and N_0 is the number of particles in the ground state U_0. The Boltzmann distribution is a statistical principle, and it states that the probability of finding a system in a particular energy state U_i varies exponentially as the negative ratio of that energy state (relative to the ground state) divided by the thermal energy k_BT. A graphical representation of the Boltzmann is shown in Fig. 4-4. Mathematically speaking, the Boltzmann distribution makes sense. For the higher energy levels, U will be much greater than the ground state U_0, so the difference $U - U_0$ will be relatively large, thus making the right-hand side of Eq. (4-13) relatively small; therefore the ratio N_i/N_0 will be small; the number of particles in the system in energy state U_i is then relatively small

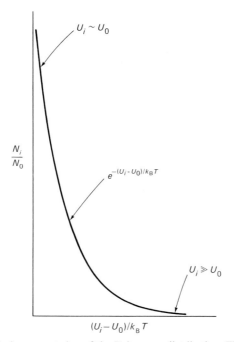

Fig. 4-4 A graphical representation of the Boltzmann distribution. The quantity N_i/N_0 is interpreted as the number of components of a system that are in energy state U_i compared to those in energy state U_0. For the higher energy levels, there are correspondingly fewer components in those states; most components will tend to populate the lower energy levels. As the temperature is raised, the upper energy levels become relatively more populated, i.e., N_i/N_0 is large when $U_i \sim U_0$ and N_i/N_0 is small when $U_i \gg U_0$.

compared to the number in the ground state. One would intuitively feel that the higher energy levels should be more sparsely populated compared to lower energy levels because the higher levels are harder to reach. For higher temperatures, the right-hand side of Eq. (4-13) will be relatively large indicating a higher population in the higher energy levels. Again, one would expect that this is true because at high temperatures the increased thermal energy of the system will allow some components to attain the higher energy levels. So, from a strictly intuitive viewpoint, the Boltzmann distribution seems to predict what might be expected, although why it has an exponential form should not necessarily be obvious.

In order to apply the Boltzmann principle, it is required that the system in question be in macroscopic equilibrium. In the example with the exponential atmosphere it was explicitly stated that all parts of the gas were in thermal equilibrium having the same temperature. Had this not been true, our "derivation" would have failed. If separate fractions of the gas had different temperatures, then each fraction would have a different potential energy population distribution, and no complete overall description would be possible unless the temperature variations of each fraction were known. This kind of information is usually extremely difficult to obtain in practice. Again, let it be emphasized that the Boltzmann distribution is a general principle and may be applied in any number of situations where the assumptions for its derivation are valid. We shall now make such an application to describe the orientation of a freely rotating dipole that is placed in an electric field.

STATISTICAL ALIGNMENT OF FREELY ROTATING PERMANENT DIPOLES

Having introduced the concepts of polarizability and the Boltzmann distribution, we are now ready to proceed to the problem of determining a protein's dipole moment from measurements of the dielectric constant. The approach taken here was first presented by Peter Debye in the early part of the twentieth century, and it has several steps. First, the polarizability of an aqueous macromolecular solution will be related to the dipole moment of the individual macromolecule, i.e., α of the solution will be related to μ of the constituent molecules. Secondly, the polarizability will be related to the macroscopic dielectric constant of the solution; and finally, the relationship between μ and D will be established and modified so that it is in a form such that experimentally measured parameters may be conveniently inserted into the equation.

Let us consider the situation where we have an aqueous solution and every biomolecule in the solution is free to rotate and translate. If it is assumed that the individual biomolecules have permanent dipole moments, then each particle will be partially aligned at some angle θ when a constant, uniform field is externally applied to the solution. Figure 4-5 depicts this situation for a single particle. The orientation of the dipole with respect to the electric field is not fixed because of thermal agitation, and hence θ will vary about some average value. It is now necessary to calculate the average component of the partially aligned dipole in the direction of the electric field. To see why we want to make this particular calculation, it is necessary to remember that the polarizability α relates the dipole moment formed by an electric field to that field ($\mu = \alpha E$), and that for isotropic materials μ is in the same direction as the field. In the situation shown in Fig. 4-5 we have a partially aligned dipole, and we need to know what part of it is in the direction of the field. With this information, we can calculate α, or what is more correct, α_{perm}.

To make this calculation we must know several things: (1) the number of orientations available to the molecular dipole and (2) the relative ease with

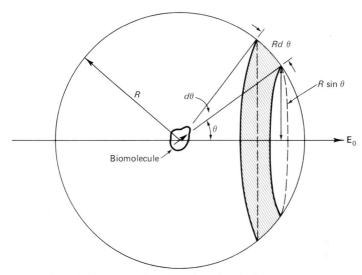

Fig. 4-5 A schematic illustrating the orientation of an individual biomolecule that has a permanent dipole in a uniform, constant electric field E_0. The dipole makes an angle θ with the field. A circle of unit radius is drawn around the biomolecule and the probability of finding a molecular dipole oriented between θ and $\theta + d\theta$ is equal to the area of the shaded annular ring on the surface of the sphere.

which the dipole may attain each orientation; i.e., we have to know for a particular μ and E_0, the relative probability of finding the dipole in an orientation between θ and $\theta + d\theta$. The first of the above requirements is a spatial problem, and the second is an energy problem. If we can calculate the average orientation of a dipole in the situation described, it is then easy to calculate its average component in the direction of the field.

In Fig. 4-5 a spherical surface of radius $R = 1$ has been drawn around our biomolecule. In trying to answer question 1 above, we can see that the probability of finding a dipole oriented between θ and $\theta + d\theta$ is proportional to the shaded annular ring on the surface of the sphere. Remember that the dipole can rotate $360°$ around the E_0 axis and still maintain an orientation angle θ with respect to the E_0 axis. The dipole can be viewed as lying on the outside surface of a cone with half-angle equal to θ and apex at the molecule. The area dA of this annular ring is found by multiplying its width by its length. We have

$$dA = (\text{length})(\text{width}) = (2\pi R \sin \theta)(R \, d\theta) \qquad (4\text{-}14)$$

The answer to question 2 above is given by the Boltzmann distribution. Since the dipole is oriented at angle θ, the interaction energy between dipole and field is given by Eq. (3-16)

$$U = -\mu E \cos \theta \qquad (3\text{-}16)$$

For every different θ, the dipole has a different potential energy level. The expression that best describes the probability a system will be in a particular energy state when a multitude are available (the ease with which the dipole will attain a certain potential energy) is the Boltzmann distribution

$$e^{\mu E \cos \theta / k_B T} \tag{4-15}$$

The number of molecular dipoles oriented between θ and $\theta + d\theta$ in the given situation is equal to the product of the number of orientations available to the dipoles and the relative ease with which these orientations can be achieved. Mathematically, it is equal to the product of the area of the shaded annular ring in Fig. 4-5 [Eq. (4-14)] and the Boltzmann term [Eq. (4-15)]. The number of molecular dipoles between θ and $\theta + d\theta$ is denoted by $N(\theta) \, d\theta$ and is given by

$$N(\theta) \, d\theta = (N_0 e^{\mu E \cos \theta / k_B T})(2\pi R^2 \sin \theta \, d\theta) \tag{4-16}$$

or

$$N(\theta) \, d\theta = \text{const } e^{\mu E \cos \theta / k_B T} \sin \theta \, d\theta \tag{4-17}$$

where the convention introduced in Chapter 3 that a dipole oriented at $\theta = 90°$ to a field is a reference state where $U = U_0 = 0$ has been used to write the exponential term in Eq. (4-15). Now, if the sample dipole is aligned at an angle θ with $\mathbf{E}_0$, then the component of $\boldsymbol{\mu}$ in the direction of the field $\boldsymbol{\mu}_E$ is equal to $\mu \cos \theta$, or

$$\mu_E = \mu \cos \theta \tag{4-18}$$

and it is this value that we want to average. It should also be mentioned that the N_0 factor in Eq. (4-16) can be evaluated by realizing that the ratio $N(\theta)/N_0$ must equal 1 when the right-hand side of Eq. (4-16) is integrated over all angles. This process is called normalization, and we shall not have to do it here.

THE AVERAGE

A brief sidetrack is now necessary in order to discuss the averaging process. In calculating the average of a group of numbers it is typical to total the individual values and then to divide this total by the number of values totaled. Mathematically, this is stated as

$$\bar{A} = \frac{\sum_{i=1}^{n} A_i}{n} = \frac{A_1 + A_2 + A_3 + \cdots + A_n}{n} \tag{4-19}$$

Here $\bar{A}$ is the average, the A_i are the individual values for which an average is desired, and n is the total number of A_i. Equation (4-19) represents a process called number averaging where it is implictly assumed that each individual value A_i is to be treated with equal importance or weight. Suppose now that some values of A are more important than others, or that some values should be given more consideration in calculating the average. In this case the number average process must be replaced by what is called a weighted average, which is given by

$$\bar{A}_w = \frac{\sum_{i=1}^{n} g_i A_i}{\sum_{i=1}^{n} g_i} = \frac{g_1 A_1 + g_2 A_2 + g_3 A_3 + \cdots + g_n A_n}{g_1 + g_2 + g_3 + \cdots + g_n} \qquad (4\text{-}20)$$

where the g_i are the weights, or relative degrees of importance, of each value A_i. As an example using Eq. (4-20), suppose we have a set of numbers $\{1, 2, 3, 4, 5, 6\}$ which represent the values of some quantity and that we want to find the average. However, all odd numbers are twice as important as the even ones. If we arbitrarily weight the odd numbers by 2, and the even numbers by one, we have the weighted average as

$$\bar{A}_w = \frac{2 \cdot 1 + 1 \cdot 2 + 2 \cdot 3 + 1 \cdot 4 + 2 \cdot 5 + 1 \cdot 6}{2 + 1 + 2 + 1 + 2 + 1} = 3.3 \qquad (4\text{-}21)$$

whereas the number average would be 3.5. In a number average all g_i are equal. If the values to be averaged are not discrete values, but are instead continuous functions, then the summations $\sum$ in Eqs. (4-19) and (4-20) are to be replaced by integrations $\int$.

THE AVERAGE COMPONENT OF THE DIPOLE IN THE DIRECTION OF THE FIELD

Having introduced all necessary ingredients, we can now write an equation of the average value of the dipole in the direction of the field. The quantity to be averaged is given by Eq. (4-18) and the weights of each quantity are given by Eq. (4-17). The average value of μ in the direction of the field is then given by

$$\bar{\mu} = \frac{\int_0^{\pi} (\mu \cos \theta)(e^{\mu E \cos \theta / k_B T} \sin \theta)\, d\theta}{\int_0^{\pi} e^{\mu E \cos \theta / k_B T} \sin \theta\, d\theta} \qquad (4\text{-}22)$$

where all constants have been divided out. The integration from 0 to π covers all possible orientations of the dipole. With the writing of Eq. (4-22) the physics of this part of the problem is essentially completed, and it remains only to reduce the mathematics to a simpler expression. To do this, we make the following substitutions. Let

$$x = \cos \theta \qquad \text{so} \qquad dx = -\sin \theta \, d\theta \qquad (4\text{-}23)$$

Equation (4-22) then transforms into

$$\bar{\mu} = \frac{\mu \int_{-1}^{1} x e^{\mu E x / k_B T}}{\int_{-1}^{1} e^{\mu E x / k_B T}} \qquad (4\text{-}24)$$

Notice that the limits of integration have changed from those of θ to those of x, i.e., when $\theta = 0$, $x = 1$ and when $\theta = \pi$, $x = -1$. To evaluate the numerator in Eq. (4-24), use is made of the integral

$$\int x e^{ax} \, dx = \frac{e^{ax}(ax - 1)}{a^2} \qquad (4\text{-}25)$$

where in our case $a = \mu E / k_B T$. The denominator can be integrated directly. Carrying out these processes yields

$$\bar{\mu} = -\frac{\mu}{a} + \mu \left[\frac{e^a + e^{-a}}{e^a - e^{-a}} \right] = \mu \left[-\frac{1}{a} + \coth a \right] \qquad (4\text{-}26)$$

where coth a is called the hyperbolic cotangent and is defined as shown. The bracketed term on the extreme right of Eq. (4-26) is known as the Langevin function. The coth function can now be expanded in an infinite series and an assumption can be made to make Eq. (4-26) simpler still. Consulting any standard mathematical reference, we have

$$\coth a = \frac{1}{a} + \frac{a}{3} - \frac{a^3}{45} - \frac{2a^5}{945} - \cdots \qquad (4\text{-}27)$$

so

$$\bar{\mu} = \mu \left(a - \frac{a^3}{45} - \frac{2a^5}{945} - \cdots \right) \qquad (4\text{-}28)$$

If now we impose the condition that

$$k_B T \gg E\mu \qquad (4\text{-}29)$$

which says the electrical energy of interaction between the dipole and the field is small compared to the thermal energy $k_B T$, then the term $a^3/45$, and higher terms are small compared to $a/3$ and can be ignored, hence

$$\bar{\mu} = E\mu^2 / 3 k_B T \qquad (4\text{-}30)$$

So, for small fields, the average dipole moment in the direction of the field is proportional to the field and inversely related to the absolute temperature.

Both of these results are compatible with the physical reasoning that has been presented so far. The fact that $\bar{\mu}$ is proportional to μ^2, however, was not necessarily anticipated.

For large electric fields, all dipoles will essentially be perfectly aligned, and our treatment here is no longer valid. Figure 4-6 shows the Langevin function plotted as a function of $\mu E/k_B T$. It is seen that as higher and higher field strengths are approached, the alignment of dipoles saturates, and little is gained by a further increase in field strength. For low field strengths, the relationship between $\bar{\mu}/\mu$ and $\mu E/k_B T$ is linear, as was shown in Eq. (4-30).

At 25°C, the value of $k_B T$ is equal to 4×10^{-14} ergs. It will be seen presently that proteins typically have dipole moments on the order of 100 D, so we can calculate an approximate field strength needed in order to satisfy Eq. (4-29). If $E\mu$ is to be 10 times less than $k_B T$, then it can be calculated that E needs to be less than 12×10^3 V/cm.

If we remember the definition of polarizability [Eq. (4-1)], it is seen that the constant of proportionality in Eq. (4-30) between μ and E is such that

$$\alpha_{\text{perm}} = \mu^2/3k_B T \tag{4-31}$$

and the total polarizability is

$$\alpha_{\text{tot}} = \alpha_{\text{ind}} + \mu^2/3k_B T \tag{4-32}$$

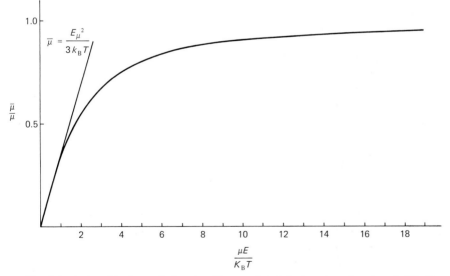

Fig. 4-6 A plot of the Langevin function. The quantity $\bar{\mu}/\mu$ is plotted as a function of $\mu E/k_B T$. Also plotted is the approximation $\bar{\mu} = E\mu^2/k_B T$. It is seen that the two lines initially agree with one another; this is the region where Eq. (4-29) is valid.

This is the desired result, where the polarizability of the solution is related to the dipole moment of the freely rotating molecules contained in the solution. The validity of this result is only as good as the assumptions made to derive it, namely that the solution is in thermal equilibrium and that the electrical field is relatively small. It now remains to relate α to the macroscopic dielectric constant.

POLARIZABILITY AND DIELECTRIC CONSTANT—THE LORENTZ FIELD

In the discussions up to now we have implicitly assumed that the electric field experienced by individual molecules in a dielectric is identical to the externally applied field. This is not aways true. In a gas, where the separations between molecules are large, the field acting on each individual molecule is essentially identical to the external field. However, in liquids and solids where the distance between structural units is small, the polarization of nearby molecules or atoms can alter the value of the applied field "seen" by an individual molecule. This altered field is frequently called the local field, and it can be expressed as

$$\mathbf{E}_{loc} = \mathbf{E}_0 + \mathbf{E}_1 \qquad (4\text{-}33)$$

where $\mathbf{E}_0$ is the externally applied field and $\mathbf{E}_1$ is the field due to the polarization of neighboring molecules. The original field creates a secondary field due to polarization, and this second field in turn alters the influence of the original field on the individual molecules in a dielectric. This effect is most prevalent where the interaction among structural elements is strongest. Since we are interested in finding a relationship between the polarizability of a biomolecule immersed in an aqueous solution to the bulk dielectric constant, it is necessary to know the magnitude of the electric field that actually influences an individual molecule because it is the local field that actually affects individual molecules and determines how much they will be polarized, which in turn affects the value of the dielectric constant. A detailed derivation of the appropriate equations describing this situation will not be given, instead a brief overview will be presented.

The problem of the local field was initially solved by the Nobel prize winning physicist H. A. Lorentz, and the result bears his name. The approach Lorentz took in solving this problem is the following. He considered a general dielectric that was subjected to the action of an external uniform electric field. In deriving the value of the local field he considered a specific point in the dielectric and constructed an imaginary sphere surrounding this point. The size of the sphere he imagined was macroscopically very

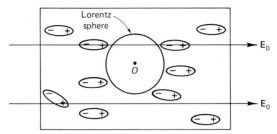

Fig. 4-7 A generalized dielectric in a uniform electric field $\mathbf{E}_0$. An arbitrary point O is surrounded by an imaginary sphere. The local field at the point O is the sum of three distinct contributions.

small, but microscopically large enough to include a large number of structural elements (atoms, molecules, etc.). This situation is illustrated in Fig. 4-7. Lorentz then considered that the local field at a point O was the sum of three separate contributions: (1) the initial field $\mathbf{E}_0$, (2) the polarization of distant parts of the dielectric outside the sphere resulting in the formation of induced charges on the surface of the sphere, and (3) the effect of closely associated molecules within the volume of the sphere. Lorentz showed that the field due to the second contribution was

$$\tfrac{4}{3}\pi\mathbf{P} \tag{4-34}$$

where $\mathbf{P}$ is the polarization vector introduced at the beginning of this chapter. He also showed that the contribution to $\mathbf{E}_{loc}$ from the material within the sphere was zero if this material possessed cubic lattice symmetry (each molecule or atom was located on the corners of a cube). Therefore, the local field could be written as

$$\mathbf{E}_{loc} = \mathbf{E}_0 + \tfrac{4}{3}\pi\mathbf{P} \tag{4-35}$$

Even if the material does not possess cubic symmetry, the assumption is usually made that the field due to the material within the sphere is zero and that Eq. (4-35) holds.

Using Eq. (4-35) for the value of the local field, the classically derived result for the relation between the dielectric constant of a media and the molecular polarizability of the same media has been shown to be

$$(D-1)/(D+2) = \tfrac{4}{3}\pi N\alpha \tag{4-36}$$

where N is the number of molecules per cubic centimeter. Equation (4-36) in a slightly different form is commonly called the Clausius–Mossotti equation; it is most accurate for gases and liquids with low dielectric constants. For liquids like water that have a high dielectric constant and where molecular interactions are strong, Eq. (4-36) is realized to be only approximate. The reason for this lies in the fact that when E_{loc} is calculated for very polar materials via the Lorentz method, the contribution from the material inside the sphere makes a significant contribution. The magnitude of this contribution depends on the exact nature and structure of the dielectric material within the spherical volume, and no general results can be stated. For the specific case where biomolecules are immersed in water, E_{loc} "seen" by an individual molecule will be strongly affected by the water molecules in the immediate vicinity of the biomolecules as illustrated in Fig. 4-8.

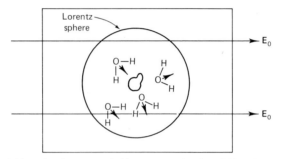

Fig. 4-8   A biomolecule surrounded by water molecules within a Lorentz sphere.

This is due to the fact that water has a relatively large dipole moment which can interact strongly with E_0. Since the polarizability of our biomolecule is dependent on E_{loc}, the relationship between the polarizability and the dielectric constant is expected to be somewhat different than that given by Eq. (4-36). In the middle and late 1930s, L. Onsager (1936) and J. G. Kirkwood (1939) worked on the theoretical aspects of this problem. Utilizing the advances made by Onsager, Kirkwood proposed a specific model for water structure and to relate the dielectric constant and the polarizability derived a new expression that is pertinent to our interests. Kirkwood's result is given by

$$(D - 1)(2D + 1)/9D = \tfrac{4}{3}\pi N\alpha \qquad (4\text{-}37)$$

and this is the equation that we shall use to relate D to α. The exact derivations of Eqs. (4-36) and (4-37) are somewhat beyond the scope of this book and will not be given. Also, Eq. (4-37) is not necessarily the definitive result, for new work in this area may bring more accurate equations; it is however adequate for our purposes.

DIELECTRIC INCREMENT

So far in our development, we have obtained a relation between α and μ [Eq. (4-31)] and between α and D [Eq. (4-37)]. It now remains to combine these two equations and to put the result in a form that can be used in the real world of experimental biophysics.

In making an experimental determination of the dielectric constant, a solution consisting of solvent and molecule (solute) is usually used. In applying Eq. (4-37) to this situation, it is first necessary to modify the right-hand side to account for both solvent and solute in our test solution. The right-hand side then becomes

$$\tfrac{4}{3}\pi(N_1\alpha_1 + N_2\alpha_2) \tag{4-38}$$

where the subscript 1 refers to solvent and subscript 2 refers to the solute molecules. It is implicitly assumed that we are dealing with a monodisperse solute (A single uniform solute), and that any salts that are in the solution are part of the solvent. For dilute solutions, the product $N_1\alpha_1$ can be assumed to be a linear function of N_2 because as N_2 increases, N_1 decreases, i.e., the more solute there is, the less solvent you have. This idea can be expressed mathematically as

$$N_1\alpha_1 = N_1{}^\circ\alpha_1{}^\circ - rN_2 \tag{4-39}$$

where the superscript $^\circ$ indicates values for the pure solvent, and r is a constant. The term $N_1{}^\circ\alpha_1{}^\circ$ is a constant and depends only on the particular solvent used. Equation (4-37) now reads

$$(D-1)(2D+1)/9D = \tfrac{4}{3}\pi(N_1{}^\circ\alpha_1{}^\circ - rN_2 + \alpha_2 N_2) \tag{4-40}$$

It is now useful to rewrite the quantity N (number of molecules per cubic centimeter) in terms of more convenient parameters, namely the concentration and molecular weight of the solute molecule. This is done by means of

$$N = \eta C/M^* \tag{4-41}$$

where η is Avogadro's number (number of molecules in 1 mole $\approx 6 \times 10^{23}$), c is the concentration of the solute in grams per cm^3, and M is the molecular weight of an individual solute molecule (weight in grams of η molecules).

* This equation is derived from the fact that the product of Avogadro's number and the molarity of a solute gives the total number of molecules of that solute. The factor C/M is the molarity of the solute per ml.

Substituting Eq. (4-41) into (4-40), we get

$$\frac{(D-1)(2D+1)}{9D} = \frac{4\pi}{3}\left(N_1\alpha_1{}^\circ - \frac{r\eta c}{M} + \frac{\alpha_2\eta c}{M}\right) \tag{4-42}$$

Differentiating D in Eq. (4-42) with respect to c, we have

$$\frac{1}{4.5}\left(1 + \frac{1}{2D^2}\right)\frac{\partial D}{\partial c} = \frac{4\pi}{3M}\eta(\alpha_2 - r) \tag{4-43}$$

where $\partial D/\partial c$ is called the dielectric increment. Since we are considering aqueous solvents where D is close to 80, the term $1/2D^2$ is going to be very small, and we can neglect it. Substituting Eq. (4-32) for α_2 into Eq. (4-43), yields

$$\frac{\partial D}{\partial c} = \frac{6\pi\eta}{M}\left(\alpha_{\text{ind}} + \frac{\mu^2}{3k_B T} - r\right) \tag{4-44}$$

This equation describes the dielectric increment of a solution subjected to a constant electric field or to a low frequency ac field. Equation (4-44) describes the dielectric increment of a solution that is subjected to an external electric field varying slowly enough with time so that the molecular dipoles have enough time to reorient with it and come to an equilibrium position. A static field of zero frequency is a special case. The reason for relating Eq. (4-44) to the frequency of an oscillating electric field will soon be clear.

One problem that still remains is the constant r in Eq. (4-44). It is generally unknown and hence must be eliminated. This can be done by designing a clever experimental procedure. If an ac electric field is used to measure the dielectric constant, then the molecular dipoles will try to reorient every time the field reverses; however, as the field frequency increases, the molecules in the solution have more difficulty in keeping up with the reversals due to the frictional drag imposed by the viscous solvent. At sufficiently high field frequencies, the permanent molecular dipoles will be completely unable to respond, and the only contribution to the polarizability will be from induction, or molecular distortion, which can keep pace with high frequency changes. For high frequencies, Eq. (4-44) then becomes

$$\left(\frac{\partial D}{\partial c}\right)_{\text{hf}} = \frac{6\pi\eta}{M}(\alpha_{\text{ind}} - r) \tag{4-45}$$

By now subtracting Eq. (4-45) from (4-44), the constant r as well as α_{ind} can be eliminated. We have then

$$\left(\frac{\partial D}{\partial c}\right)_{\text{lf}} - \left(\frac{\partial D}{\partial c}\right)_{\text{hf}} = \frac{2\pi\eta\mu^2}{Mk_B T} \tag{4-46}$$

which relates the dielectric constant to the molecular dipole moment of the particles in solution, and is the desired result.

To get a better understanding of $\partial D/\partial c$ and to see how it is related to a measureable parameter, consider the following. Since we are dealing with dilute solutions, the dielectric constant of the solution can be written as

$$D = D^\circ + \left(\frac{\partial D}{\partial c}\right)c^\dagger \qquad (4\text{-}47)$$

where D° is the dielectric constant of the solvent. The dielectric increment is then given by

$$\frac{\partial D}{\partial c} = \frac{D - D^\circ}{c} \qquad (4\text{-}48)$$

Hence, $\partial D/\partial c$ is determined by measuring the difference between the dielectric constant of the solution and the solvent, and dividing by the concentration of the solute. $(\partial D/\partial c)_{hf}$ is measured in an analogous way at high field frequencies. In order to calculate μ from Eq. (4-46), the molecular weight must be known from a separate and independent measurement.

To illustrate the principles of the previous paragraphs, the reader is referred to Fig. 4-9. This figure shows a schematic plot of the dielectric constant vs. frequency for a two-component system, solvent and solute. The situation depicted assumes that both types of molecules in the solution are polar and that one is substantially larger in size than the other. This situation is representative of the case where a protein, or other biomolecule, is dissolved in water. The variation of the dielectric constant with frequency (v) is divided into five general regions:

(I) At low frequencies of the field reversals, the larger biomolecules have plenty of time to reorient themselves in the alternating field. Here the retarding frictional forces acting on the biomolecule are small compared to the electrical orienting forces. The value of the dielectric constant is essentially identical to the static dielectric constant (D_0), and it is at a maximum value.

(II) As the frequency of the field reversals increases, the frictional forces are no longer negligible, and complete reorientation and equilibrium is

---

†This equation is stating that D of a solution is composed of two contributions: one from the solvent and the other from the solute. The solvent contribution is represented by D° and the solute contribution is dependent on its concentration C. If the solute concentration is zero, then $D = D^\circ$ as it should. If the solute concentration changes the solution's dielectric constant, then $\partial D/\partial C$ times the solute's concentration equals the increase in the dielectric constant of the solution over that of the pure solvent.

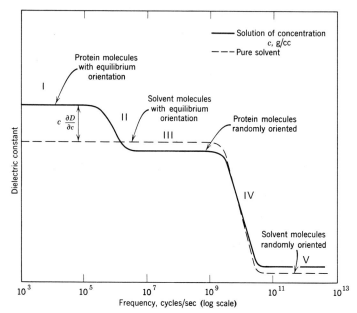

Fig. 4-9   A general representation of how the dielectric constant of a two-component solution varies with the frequency of the applied field. As the frequency of the electric field increases, the effective dielectric constant decreases (Tanford, 1961).

not achieved by the larger molecules. As the frequency of the field increases, the dielectric constant decreases. This region of the curve is particularly interesting because the extent of the reorientation of the biomolecule is dependent upon the hydrodynamic properties of the biomolecules itself; hence this region of the dielectric constant vs. frequency plot can potentially give information regarding the size and shape of molecules. This region where the dielectric constant decreases with increasing field frequency is called the dielectric dispersion region.

(III)   At these frequencies the frictional forces retarding the reorientation of the biomolecules have completely overwhelmed the electrical orienting forces, so the permanent dipoles of the biomolecules make no contribution to the dielectric constant. The solvent molecules, being very small, still achieve complete reorientation of their dipoles, and make the most significant contribution to the dielectric constant. Frequencies in this region are used to measure $(\partial D/\partial c)_{hf}$.

(IV)   As the frequency of the field is further increased, another dispersion region is reached. Here, the solvent molecules are beginning to lag the fluctuations of the field, and complete orientation of solvent dipoles is decreased, so the dielectric constant again decreases.

(V)   In this region the field reversals are so fast ($v = \infty$) that only the atomic distortions of induced dipoles can keep pace; hence the dielectric constant is very small.

DIELECTRIC DISPERSION

Figure 4-10 shows a typical plot of the dielectric constant vs. frequency for the dispersion region. The dielectric constant of the solution decreases here because the permanent dipoles of the biomolecules just cannot keep pace with the field reversals. The frictional retarding forces are larger than the reorienting forces. By examining the shape of the dispersion curve, it is possible to obtain certain information concerning the geometric shape of the biomolecule contained in the solution. In 1929 Debye derived an expression describing the form this curve should take. For the case of a single type of biomolecule in solution (monodisperse) with a spherical shape the dielectric constant is related to the frequency v by

$$D = D_\infty + \frac{D_0 - D_\infty}{1 + (v/v_c)^2} \qquad (4\text{-}49)$$

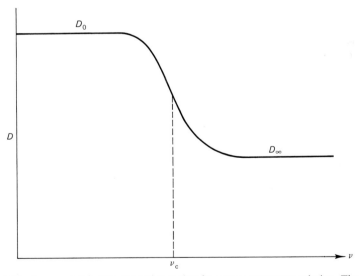

Fig. 4-10   A general dielectric dispersion region for a two-component solution. The region shown illustrates the decrease in the dielectric constant of the solution due to the biomolecule as a function of the electric field frequency. v_c is the critical frequency.

where D_0 and D_∞ are the dielectric constants at low and high frequencies, respectively, and ν_c is called the critical frequency and is defined as the frequency in the dispersion region at which the dielectric constant is midway between the lower and upper values, or where

$$D(\nu_c) = \tfrac{1}{2}(D_0 + D_\infty) \tag{4-50}$$

Since frequency has units of reciprocal time, the inverse of frequency has units of time, and is directly related to the period of the oscillation, or how long it takes for one complete cycle to take place. The period τ_c of the frequency ν_c is given by

$$\tau_c = 1/2\pi\nu_c \tag{4-51}$$

and in the particular case of dielectric dispersion, τ_c is called the relaxation time.

An understanding of the quantity τ_c may be obtained by considering the following. Before an electric field is applied to a group of molecules in solution, they are randomly oriented. When the field is turned on, the molecules become aligned and assume some average orientation angle $\bar\theta$ with respect to the field. This process of alignment is however not instantaneous, and the final aligned configuration is reached as an exponential function of time, or

$$\bar\theta \propto e^{-\gamma t} \tag{4-52}$$

where γ is a constant. The reciprocal of γ is defined as the relaxation time τ_c, so Eq. (4-52) becomes

$$\bar\theta \propto e^{-t/t_c} \tag{4-53}$$

The relaxation time is then the time in which $1/e$ of the change from one equilibrium position to another has occurred. It is a way of measuring how long it takes a molecule to reorient itself in a changing electric field. The relaxation time and also ν_c are explicitly a function of molecular size and shape, and it is through measured values of ν_c that structural information is obtained from the dispersion regions. The relaxation time may be thought of as being a hydrodynamic parameter since the shape of a molecule will determine how it moves in a fluid much in the same way the hull design of a sailboat determines its performance. Intuitively, it can be seen that a larger, or more extended, molecule will take longer to change its orientation, so it will have a larger τ_c (smaller ν_c) than will a small compact molecule. For spherical particles, τ_c has been expressed quantitively as

$$\tau_c = 4\pi\eta_v a^3/k_B T \tag{4-54}$$

where a is the sphere's radius, η_v is the viscosity, k_B is Boltzmann's constant, and T is the absolute temperature. For constant temperature and viscosity, τ_c is very sensitive to changes in the radius of the sphere; hence measurements of τ_c may also be used to determine changes in size.

In Chapter 2 it was explained that globular molecules are very compact; hence they can be, and frequently are, approximated by a spherical shape, so Eq. (4-54) is quite applicable. For molecules that are elongated or have another type of nonspherical shape, Eq. (4-54) does not apply. The appropriate equations for these cases must be derived separately, and in general this is a very complex problem. It will just be mentioned that for molecules that can be approximated in shape by either a prolate or oblate ellipsoid of revolution (Fig. 4-11), equations similar to Eq. (4-54) were derived by F. Perrin in the early part of the 1900s.

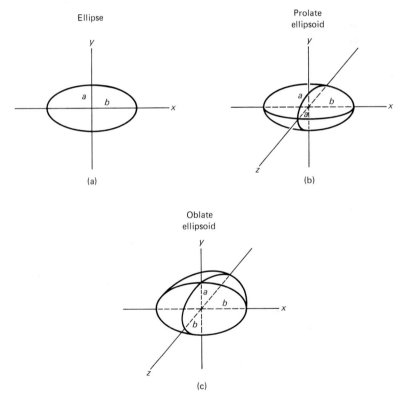

Fig. 4-11 (a) A planar ellipse. (b) A prolate ellipsoid of revolution, which is a three-dimensional figure whose outer surface is defined by rotating an ellipse about its long axis [x axis in part (a)]. A football is an approximate prolate ellipsoid of revolution. (c) An oblate ellipsoid of revolution whose outer surface is formed by rotating an ellipse about its short axis [y axis in part (a)].

For ellipsoids of revolution, there will be two characteristic relaxation times, one for rotation about each axis of symmetry. That is to say, because an ellipsoid of revolution is asymmetric in shape, rotation about one of its axes will be easier for it than rotation about the other. Because of this, Eq. (4-49) will not be exactly correct since it assumes a single relaxation time. Because of its symmetry, a sphere will have only one possible relaxation time, and Eq. (4-49) will be strictly correct. The point to be made here is that the behavior of a biomolecule in an oscillating electric field can also provide information about its size and shape.

Now that the theoretical presentation has been completed, we are ready to consider some actual experiments that have been performed, and to see how different kinds of information about biomolecules can be obtained from dipole moments and dielectric dispersion measurements.

EXPERIMENTAL RESULTS

Historically, measurements of the dipole moments of proteins and amino acids were first performed in the 1920s and 1930s. The most significant early studies were those of Jeffries Wyman of Harvard University, who was concerned with amino acids, and those of J. L. Oncley, who extended the work to a variety of common proteins. In the 1950s and 1960s, one principal worker in the field was Shiro Takashima. It should be mentioned that this field is not an extremely popular one, and this is probably due to several uncertainties in the theory and the interpretation of experimental results. The methods employed do however possess the potential for being a powerful investigative tool, and they already have shed much light on molecular structure. The field of dielectric dispersion today is used mostly in order to gain structural information about biomolecules or polymers.

Amino Acids

When the pioneering work of Wyman and his coworkers was being performed, the theory presented in the earlier part of this chapter was incomplete. Equation (4-36) had been proposed by Debye, but Wyman realized that this equation was inadequate because of the strong polar nature of water in which the amino acids were usually dissolved. Charged amino acids were almost completely nonsoluble in nonaqueous solvents, so the necessary experiments could not be performed even though the theory would fit it. What Wyman did then was to perform empirical experimental studies, realizing that the dielectric increment was directly related to the molecular dipole moment.

If amino acids are truly dipolar in nature as we proposed in Chapter 1, then it should be possible to estimate the strength of their dipole moments. If it is assumed that the positive charge is centered on the N atom, and that the negative charge is centered midway between the two O atoms of the carboxyl group, we have the situation depicted in Fig. 4-12. From X-ray diffraction studies, it is known that the distance between the charges is 2.92 Å, and so the dipole moment can be calculated to be 13.9 D. If the amino acid actually does possess a structure in solution similar to that in Fig. 4-12, then a significant dielectric increment should be measured.

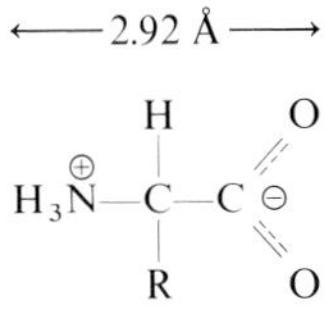

Fig. 4-12   The proposed charge structure of a free amino acid. The carboxyl group is represented as a resonance structure where both bonds to the oxygen atoms are not totally single or double, but are intermediate in character. The positive charge is shown centered on the N atom, and the negative charge is centered midway between the two oxygen atoms.

In order to test the above idea, Wyman reasoned that as more and more carbon atoms were placed between the charges, the dipole moment, and hence the dielectric increment, should increase. Figure 4-13 shows a graph of the dielectric increment plotted vs. the number of carbon atoms between the carboxyl and amino group. For a polypeptide composed of repeat units of glycine, a linear relationship between $\partial D/\partial c$ and n was also demonstrated, except that now the magnitude of $\partial D/\partial c$ was larger (Fig. 4-14). Both of these results are expected because as n increases, so does the distance between the charges. It would also be expected that $\partial D/\partial c$ be larger for the polyglycine case because repeat units of glycine are bigger than repeat units of CH_2. Both these results indicate that there is a regular increase in $\partial D/\partial c$ for every structural unit added. The results also lead one to believe that the individual bonds in the amino acids do not have strong dipole moments, otherwise the linear relationships would not necessarily be expected. The values of $\partial D/\partial c$ for most common amino acids are found to be quite similar. Typical values are shown in Table 4-1.

Wyman also measured the dipole moment of several amino acid esters. The ester form has the advantage of being soluble in nonpolar solvents for which the Debye equation (4-36) was applicable, hence μ could be calculated. The esters were formed by chemically replacing the H of a carboxyl group with a more complicated group. A schematic of this reaction is shown in Fig. 4-15. Since the nonpolar ethyl group would be expected to have a very

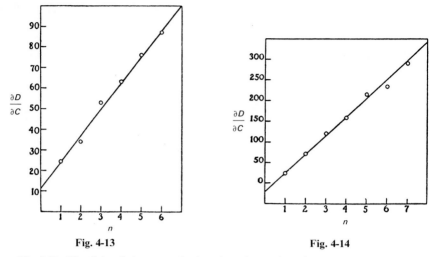

Fig. 4-13   Fig. 4-14

Fig. 4-13 The dielectric increment is plotted vs. the number of carbon atoms (n) between the NH_3 and COOH groups of an amino acid. $H_3N^+ - (CH_2)_n - COO^-$ (Wyman, 1936).

Fig. 4-14 The dielectric increment is plotted vs. the number of glycine groups in the polypeptide.

TABLE 4-1

Values of the Dielectric Increment for Some Typical Amino Acids at 25°C[a]

	Amino acid	$\partial D/\partial C$		Amino acid	$\partial D/\partial C$
	Glycine	22–30	DL	Proline	21
	Alanine	23–28	D	Glutamic acid	26
DL	Valine	25	L	Asparatic acid	28
L	Leucine	25	D	Glutamine	21

[a] The values where a range is given indicate the result of several workers. D and L refer to the stereochemical property of the amino acids. Only L amino acids are usually found in nature (Wyman, 1936).

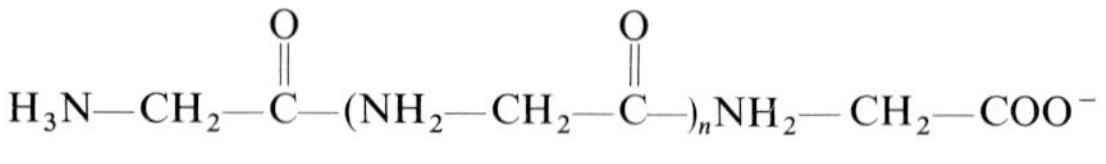

$$H_3N - CH_2 - \overset{O}{\overset{\|}{C}} - (NH_2 - CH_2 - \overset{O}{\overset{\|}{C}} -)_n NH_2 - CH_2 - COO^-$$

The plot shows a linear relationship and the values of dielectric increment are larger than those in Fig. 4-13 (Wyman, 1936).

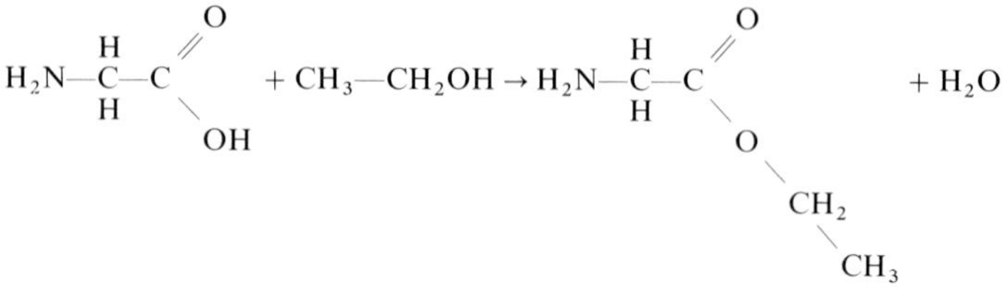

$$H_2N - \overset{H}{\underset{H}{C}} - C \overset{O}{\diagdown} _{OH} + CH_3 - CH_2OH \rightarrow H_2N - \overset{H}{\underset{H}{C}} - C \overset{O}{\diagdown} _{O \diagdown CH_2 \diagdown CH_3} + H_2O$$

Fig. 4-15 A schematic representation for the formation of glycine ethyl ester.

130

small dipole moment itself, its effect on the dipole moment of the amino acid was thought to be minimal, and the amino acid ester's dipole moment was taken to be the same as that of an uncharged amino acid. Typical values obtained are shown in Table 4-2.

In 1938 W. J. Dunning and W. J. Shutt reported on measurements made of $\partial D/\partial c$ as a function of pH for 0.2 M glycine, and their results are shown in Fig. 4-16. The behavior of $\partial D/\partial c$ is thus seen to confirm the model of a dipolar structure for the amino acid. In the pH region 4.5–7.5 $\partial D/\partial c$ is fairly constant where glycine has the structure

$$H_3\overset{\oplus}{N}-CH_2-C\overset{O}{\underset{O}{\diagdown}}{\ominus}$$

TABLE 4-2

Typical Values for the Dipole Moment of Several Different Amino Acid Esters[a]

Ester	μ (D)
Glycine ethyl ester	2.11
α-Alanine ethyl ester	2.09
Valine ethyl ester	2.11
α-Amino butyric ethyl ester	2.13
α-Amino valeric ethyl ester	2.13
β-Alanine ethyl ester	2.14

[a] From Wyman (1936).

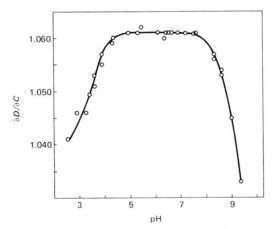

Fig. 4-16 A quantity proportional to the dielectric increment plotted vs. pH for a 0.2 molal solution of the amino acid glycine. $T = 20°C$ (Dunning and Shutt, 1938).

but on either side of this region, where the pK values for the amine and carboxyl are reached $\partial D/\partial c$ decreases sharply, indicating the dipolar nature of the amino acid has been lost. This is entirely in line with the model presented in Chapter 1. The reader should realize that historically it was as a result of these experiments and others like them that the concept of the dipolar structure of the amino acid originated.

For approximate calculations, the formula

$$\mu = (10 \, \partial D/\partial c)^{1/2} \tag{4-55}$$

has been presented by Wyman to relate the dipole moment and the dielectric increment for aqueous solutions. This equation is semiempirical, and the critical assumption in its derivation is that there is a linear relationship between the dielectric constant and the polarization of the liquid per unit volume. It is sometimes useful in making estimates of μ from dielectric increment data.

Proteins and DNA

Molecules the size of proteins present an experimental advantage not possessed by amino acids. Because amino acids are generally not much bigger than water molecules, the dielectric dispersion properties of amino acids do not differ greatly from those of water. In fact the relaxation time of water is $\sim 10^{-11}$ sec, while for amino acids it is $\sim 10^{-10}$ sec. Therefore, when making dielectric measurements of amino acid solutions, it is difficult to get precise data anywhere except in region I of Fig. 4-9. On the other hand, proteins have relaxation times in the range of 10^{-6}–10^{-8} sec and regions II and IV are distinct; hence, values of $(\partial D/\partial c)_\infty$ are obtainable and Eq. (4-46) can be used to calculate μ. For dielectric studies on proteins, frequencies from 10^3 to 10^7 Hz are typically used, whereas for DNA, the range can be extended downward to 10^1–10^2 Hz. This is because DNA is geometrically very asymmetric, giving it a long relaxation time.

The results of some early determinations of the dipole moment for proteins obtained by Oncley are shown in Table 4-3.

The first thing one should notice when looking over the values of μ in Table 4-3 is their magnitude. Compared to the dipole moments shown in Table 3-3 for some typical solvents, the values in Table 4-3 are very large. However, are they as large as one would expect because proteins are significantly larger than solvent molecules? A simple calculation can make all this clearer. Of those biomolecules mentioned in Table 4-3, hemoglobin and serum albumin are about the largest, and it is known that the long dimension of hemoglobin is around 68 Å. The question arises as to what would be the dipole moment of two unit charges separated by 68 Å. A simple calculation

TABLE 4-3

Some Values for the Dipole Moment of Different Proteins[a]

Molecule	Solvent	MW	Temp (°C)	$\left(\frac{\partial D}{\partial C}\right) \times 10^{-3}$	$-\left(\frac{\partial D}{\partial C}\right) \times 10^{-3}$	Difference	μ (D)
Horse carboxyhemoglobin	water	67,000	25	0.33	0.09	0.42	480
Myoglobin	water	17,000	25	0.15	0.06	0.21	170
Insulin	80% Aqueous propylene glycol	40,000	25	0.38		0.38	360
Insulin	propylene glycol		25			0.26	300
β-lactoglobulin	(m/2)-Glycine	40,000	25	1.51	0.07	1.58	730
Ovalbumin	water	44,000	25	0.1	0.07	0.17	250
Horse serum albumin	water	70,000	25	0.17	0.07	0.24	380

[a] From J. L. Oncley, Chapter 22 in "Proteins, Amino Acids and Peptides as Ions and Dipolar Ions" (E. J. Cohn and J. T. Edsall, eds.). © 1971 Litton Educational Publishing, Inc. Reprinted by permission of Van Nostrand Reinhold Company.

yields a value of 326 D, which is of the same magnitude as the dipole moments given in Table 4-3. For myoglobin, the long dimension is about 36 Å, and this corresponds to a dipole moment of 173 D. These figures are very interesting since each of the molecules listed can have over 50 electronic charges (serum albumin can have over 100). What our calculations and experiments are telling us is that all these charges must be pretty much symmetrically spread throughout the molecule's volume or surface, or both. It appears then that the molecules listed in Table 4-3 have their positive and negative charges scattered about in such a manner to result in a relatively small dipole moment compared to what it could be if the separate types of charges were completely polarized. Our simple calculation tells us that the center of positive charge in these molecules almost matches the center of negative charge. If we assume that all charges (both positive and negative) have spherical symmetry, except for one positive and one negative charge which are each on opposite sides of the particle, then it is possible to account for the major part of the dipole moment. This implied symmetry of the charge distribution is not a result we would have necessarily expected, and it gives us an added insight into molecular structure. Only β-lactoglobulin shows a significantly larger dipole moment than the rest, and here it is possible that the charge distribution is a little more asymmetric than the rest.

In conclusion, then, we have learned from dielectric studies a fundamental property about the charge distribution of proteins, namely that the charges present seem to be pretty much symmetrically distributed throughout the protein's volume.

Dielectric studies on DNA are fewer than those on protein, although Takashima has determined some of the essential features. Since the structural details of double helical DNA will be discussed in the next chapter, suffice it to say now that to a rough approximation DNA is akin to a long, thin, somewhat flexible cylinder. One property of DNA that is quite useful in these studies is its ability to be cleaved and still retain its basic structure. Sonication,* mechanical shearing in a blender, and enzymatic degradation are the most commonly used methods for reducing the size of the normally huge DNA strands. In 1963 Takashima investigated the dipole moment of DNA. By measuring the dielectric properties of DNA samples of varying length, Takashima was able to find that there is a linear relationship between its dipole moment and its molecular weight. These results are shown in Fig. 4-17. Since the molecular weight of DNA is proportional to chain length, the dipole moment is also expected to be proportional to length. The results presented in Fig. 4-17 then furnish strong evidence that DNA has a dipole moment in the direction of the long axis rather than in a trans-

---

* Sonification is shearing using sound waves.

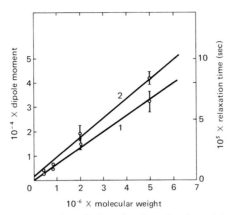

Fig. 4-17   The dipole moment of DNA plotted vs. its molecular weight (curve 1); also plotted (curve 2) is the relaxation time (Takashima, 1963b).

verse direction. It should be noticed also that the magnitude of the dipole moment for DNA is much larger than for globular proteins.

In considering DNA's dispersion behavior, it is to be expected that τ and the dielectric constant should depend on the length of the molecule. Figure 4-18 shows some dispersion curves from salmon sperm DNA of various lengths. As expected, the dielectric constant does depend on the length of the molecule as do the dispersion properties. For the longer lengths of DNA, it is seen that the dispersion is shallower and that the beginning of the dispersion region starts at lower frequencies compared to shorter lengths of DNA. Figure 4-19 is a plot of τ vs. $L^{1.8}$ where L is the DNA's length in angstroms. This graph verifies the fact that longer strands of DNA take longer to reorient and that these relaxation times are fairly low. The DNA from the two different sources seems to behave in a similar manner.

Dielectric dispersion not only can give information relating to the overall shape of a biomolecule, but it can be used to monitor a conformational change, i.e., a change in shape. A good example of this is seen in the case of polyglutamic acid. This synthetic polypeptide exists in a helical conformation below pH 5.0; however, as the pH is raised, the conformation changes to that of a random coil so at pH 8, the polypeptide is completely in the random coil form. This conformational change is shown schematically in Fig. 4-20. The helical form is allowed at low pH because there the carboxyl groups of the glutamic acid side chains are protonated (uncharged), and hence do not repel one another. As the pH increases, the carboxyl groups ionize and mutual repulsion between side chains tends to force the side chains apart, and the helix becomes unstable. Figure 4-21 shows the variation in the dielectric increment and relaxation time as the above transition takes place.

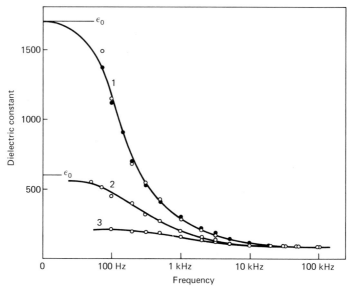

Fig. 4-18 The dielectric constant of Salmon sperm DNA plotted vs. frequency of the alternating electric field. The three curves are for DNAs of different lengths: 7400 Å (curve 1), 5600 Å (curve 2), 1300 Å (curve 3). (Reprinted with permission from S. Takashima, *J. Phys. Chem.* **70**, 1372 (1966). Copyright by the American Chemical Society.)

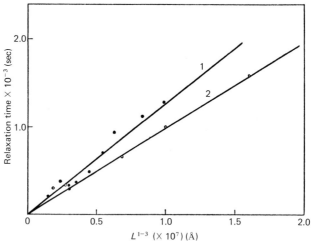

Fig. 4-19 Relaxation time plotted vs. the DNA length in angstroms raised to the 1.8 power. Curve 1 is for salmon sperm DNA, and curve 2 is for calf thymus DNA. (Reprinted with permission from S. Takashima, *J. Phys. Chem.* **70**, 1372 (1966). Copyright by the American Chemical Society.)

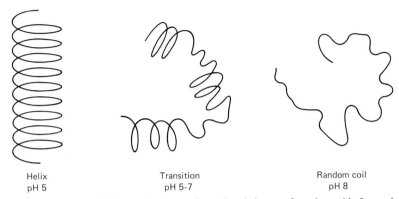

Helix	Transition	Random coil
pH 5	pH 5-7	pH 8

Fig. 4-20 A schematic illustrating the conformational change of a polypeptide from a helix to a random coil. Polyglutamic acid exists as a helix at pH 5.0 and as a random coil at pH 8. Between, it has an intermediate conformation.

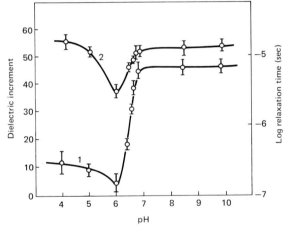

Fig. 4-21 The dielectric increment (1) and relaxation time (2) of polyglutamic acid plotted vs. the pH of the solution (Takashima, 1963a).

Takashima argued that since both τ and $\partial D / \partial c$ underwent a minimum at the same pH value, the existence of an intermediate conformation was plausible. The fact that the dielectric increment went through a minimum was not conclusive proof alone; but since τ had an identical minimum, it could certainly indicate a completely different conformation than either coil or helix. At the time of Takashima's experiment the intermediate state had been postulated and was thought to be composed of a number of short helical regions separated by sections of random coil. The fact that the random coil form of the polymer has a higher dipole moment than the helix implies

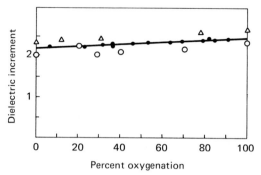

Fig. 4-22   The dielectric increment of horse hemoglobin plotted vs. the percent oxygenation of hemoglobin at 17°C. One hundred percent oxygen means each hemoglobin has four O_2 molecules. The ● symbols are for unpurified hemoglobin and the ○ and △ symbols are for different batches of alcohol-crystallized hemoglobin (Schlecht *et al.*, 1968).

that the charge distribution is more asymmetric or the molecule is more asymmetric in overall shape. It should be pointed out however that a random coil conformation is not a single discrete shape, but that it can be any one of a number of statistically available conformations. Apparently, the lone positive charge of the terminal NH_3^+ group of polyglutamic acid is not symmetrically surrounded by negative charges.

Another conformational change that has been investigated with dielectric dispersion is the oxygenation of hemoglobin. In 1968, Peter Sehlecht *et al.* of Germany published a paper in which the dielectric increment of hemoglobin was measured as a function of the percent oxygenation of hemoglobin (Fig. 4-22). From independent X-ray diffraction data it has been shown that there is a conformation change in the structure of hemoglobin as the molecule complexes with the four oxygens, but the X-ray data could not determine the manner or rate at which the change takes place. It is seen in Fig. 4-22 that there is a linear 10% rise in $\partial D/\partial c$ as hemoglobin progresses from its deoxygenated to its oxygenated form. This indicates that the dipole moment experiences a small increase, and that it appears this change takes place gradually. Whether the change in $\partial D/\partial c$ is due to a redistribution of charges or to a shape change cannot be determined purely from Fig. 4-22.

Takashima has also published data suggesting that oxyhemoglobin and carboxyhemoglobin have different shapes. He has measured τ_c for both species and found at 15°C the values shown in the table. From this, one can

Hemoglobin	τ_c
oxygen	17.4×10^{-8}
carboxyl	15.0×10^{-8}

infer that hemoglobin combined with O_2 has a very slightly more extended conformation than when it is combined with CO. It is not known for sure whether or not this difference is important physiologically, although a change in shape can affect hydrodynamic properties, which in turn can have significant biological consequences.

INTERPRETIVE DIFFICULTIES

So far, the molecular interpretation of the dielectric increment has been solely in terms of the alignment of permanent dipoles under the influence of an electric field. As knowledge of molecular structure and interaction advanced, however, there were suggested several other possibilities which would lead to nonnegligible values of the dielectric increment and the dipole moment. In this section a brief description of these alternative explanations will be presented.

One of the main alternative interpretations to the theory just presented was made by J. Kirkwood and J. Shumaker. They advocated that it was possible to account for the magnitude of most biomolecular dipole moments in terms of the fluctuating charge associated with the chemical equilibria

$$COOH \rightleftarrows COO^- + H^+$$

$$NH_4^+ \rightleftarrows NH_3 + H^+$$

Using this approach, they were able to predict dipole moments as large as 500 D based on this idea alone.

A second contribution to the molecular dipole moment is the presence of permanent dipoles in molecular bonds. Bonds like $C{=}O$, $N{-}H$, and $O{-}H$ do have dipole moments; and if their effects are somehow additive, then this could possibly explain the large μ of proteins. However, it is thought that this is not the case. Due to the contorted conformation of most proteins, the contributions of individual bonds would be expected to mostly cancel in any vector addition.

Another possible contribution to μ could conceivably come from the waters of hydration that accompany biomolecules in solution. Surrounding every macromolecule is a sheath of water so intimately related to the molecule that it moves with the molecule as an intregral part. It is possible that this water is being polarized by the electric field contributing to the measured values of $\partial D/\partial c$. More will be said about water in Chapter 8, but it is generally agreed that water in a hydration layer has somewhat different properties than has bulk water.

It should be realized then that there are other possible contributions to the dipole moment than just the asymmetric distribution of charge and the model that we have developed in detail. In general, it is not possible to tell the importance of each possibility for an arbitrary molecule. It is necessary to study each particular case in detail. The reader should be aware however that alternative explanations are possible.

Arguments favoring the permanent dipole model have been made by Takashima and also by Schlecht for specific cases. According to Eq. (4-54), the relaxation time should be a linear function of the viscosity of the solvent; at least this should be true if the orienting dipole interpretation is correct. In a study with the small proteins myoglobin, ovalbumin, and bovine hemoglobin, Takashima experimentally showed this relationship to hold. His results are presented in Fig. 4-23. For this experiment, glycerine was added to the aqueous solutions to increase the viscosity. For the large protein catalase, however, the change in viscosity seemed to have little effect on τ; this effect remains unexplained.

Because myoglobin has approximately one-quarter the volume of hemoglobin and a similar shape, it is expected that the respective relaxation times for the two molecules should differ by a factor of four since τ is proportional to the molecular volume. This has also been verified experimentally. So, it appears for these cases, that the major contribution to the measured dielectric constant comes from the orientation of molecular dipoles.

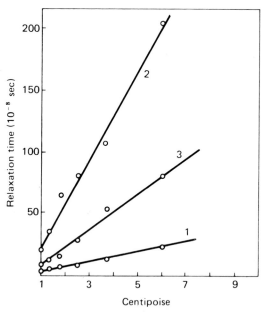

Fig. 4-23 The relaxation time of several small proteins plotted vs. the viscosity of various viscosities. Curve 1 is for myoglobin, curve 2 is for ovalbumin, and curve 3 is for bovine hemoglobin. Theory predicts a linear relationship between the relaxation time and the viscosity (Takashima, 1962).

A most convincing argument for the permanent dipole theory has been shown in the case of myoglobin. Since myoglobin has been studied extensively by X-ray diffraction, the coordinates of each atom are known in detail. Since the coordinates are all known, the exact position of all charged groups can also be found. Using the generalized definition of the dipole moment, it is then possible to calculate the dipole moment based on these coordinates and compare the calculated value with the measured one. Schlecht (1969) calculated for sperm whale myoglobin that $\mu = 170 \pm 10$ D, and his measured value was 167 D. The agreement indicates that by far the major contribution to myoglobin's dipole comes from an asymmetric distribution of discrete charges. To test this further, Schlecht reasoned that by changing the charge distribution in a known way, the calculated dipole moment should change; and it could again be calculated to see whether it still matched the measured value. Chemically, this was carried out by reacting the myoglobin with bromoacetate. In this reaction carboxylmethyl groups are bound to the histidine residues, thus reducing the number of groups that can bear a positive charge. The calculated dipole moment again agreed with measured values, thus strengthening the original postulate.

It should be emphasized that the above results apply only to myoglobin, and generalizations cannot necessarily be made to other molecules. In other cases contributions to the dipole moment other than asymmetric charge distribution may be significant.

MEASUREMENT OF DIELECTRIC CONSTANT

Over the years in which the dielectric constant has been measured many experimental techniques have been used. In this section brief mention will be made of a few so the reader will have an idea of what kind of equipment and principles are used in actually making a measurement in the laboratory. There are three basic classes into which these measurement techniques fall: (1) measuring the electrical force acting through a dielectric medium compared to the same force acting through air or a vacuum, (2) determining the velocity of propagation of electromagnetic waves through a dielectric medium compared to a standard, and (3) measuring the capacity of a capacitor empty compared to one filled with a dielectric. The last method, or modifications thereof, has been the most popular. Only the first and third methods will be discussed here.

Force Methods

Two methods of historical importance exist. With the first, the force between two horizontal capacitor plates was measured. The lower one was fixed, and the upper one was attached to a sensitive balance where the balance measured the force of interaction. The apparatus was standardized and then the medium in question was introduced between the plates. The force attracting the two plates of a capacitor is inversely proportional to the dielectric constant or

$$F = q^2/2DA \qquad (4\text{-}56)$$

where q is the charge on each plate and A is the area of the plates. By measuring the charge and the force, the dielectric constant may be obtained.

In another technique a nickel ellipsoid of revolution was suspended between the plates of a capacitor. In the presence of an electric field, the ellipsoid was deflected. The deflection force exerted on the ellipsoid is dependent on the force between the capacitor plates, which in turn depends on the dielectric constant of the dielectric filling the space. By observing the deflection of the ellipsoid, the dielectric constant could be measured.

Capacitance

A schematic diagram of a simple Wheatstone bridge circuit for capacitors is shown in Fig. 4-24. Four capacitors are arranged as shown where A is an ammeter which measures current flow. Usually one capacitor has an unknown capacitance, whereas the other three have accurately known values. Also, one or several of the three known capacitances is (are) variable. This variable capacitance is then adjusted until the current flowing through the

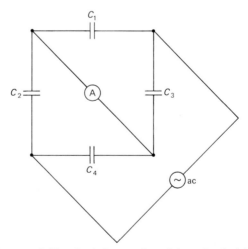

Fig. 4-24 A Wheatstone bridge circuit for capacitors. It is used to find the capacitance of an unknown capacitor.

ammeter reads zero. At this point, simple circuit analysis shows that

$$C_1/C_2 = C_3/C_4 \qquad (4\text{-}57)$$

By measuring the capacitance of the unknown with a reference dielectric between the plates and then again for the unknown dielectric, the dielectric constant can be obtained by a simple ratio. The capacitance of a capacitor depends on the frequency used, and hence the arrangement seen in Fig. 4-24 can be used throughout any frequency range.

Another method using capacitors is called the resonance method. A simple schematic is shown in Fig. 4-25. An inductance L, a resistance R, and a capacitor C are arranged in series. The impedance is defined as the opposition to the flow of alternating current offered by a circuit or a component. The impedance of the coil and the capacitor both depend on frequency, but do so in different ways. For fixed values of L and C, there is one frequency called the resonance frequency at which the current in the circuit is a maximum. This frequency is given by

$$f = (1/2\pi)(1/LC)^{1/2} \qquad (4\text{-}58)$$

By considering the capacitor with a standard and with a dielectric, knowing the inductance L of the coil, a ratio of measured resonance frequencies will give the dielectric constant. This can be expressed as

$$v_{\text{vac}}^2/v_{\text{diel}}^2 = D \qquad (4\text{-}59)$$

where v_{vac} is the resonance frequency when a vacuum exists between the plates. Since coils come with a wide range of inductances, the resonance frequency can be varied over a large range.

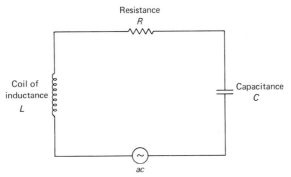

Fig. 4-25 A simple circuit design for measuring the dielectric constant in the resonance method.

SUMMARY

The polarizability α is a measure of how easy it is to polarize a dielectric material through the influence of an external electric field. The polarizability has two main contributions, that due to alignment of permanent dipoles and that due to the formation of induced dipoles. For isotropic materials, the polarization of a dielectric is directly proportional to the strength of the external electric field and the induced dipole moment is in the same direction as the external field.

The polarization vector **P** is defined as the average dipole moment per unit volume of the dielectric. It is a macroscopic quantity, but it can be related to the submicroscopic dipoles composing the dielectric. It is also useful for the development of an electrical theory of real materials.

Boltzmann's distribution is a statistical description of how the components of a system will distribute themselves in the various energy states available to them. It is a very important concept in physics and has many applications.

The polarizability of a molecular solution can be described in terms of the individual molecular dipole moments by using the Boltzmann distribution. This polarizability can then be related to the solution's dielectric constant, which in turn can be related to the dielectric increment. Dielectric increments measured at low and high frequencies can be used to calculate the dipole moment of molecules in solution.

Dielectric dispersion is a technique whereby the dielectric constant of a molecular solution is measured as a function of the fluctuation rate of the external field. This technique can give information on the shape and size of molecules in solution.

The magnitude of the dipole moment for typical proteins indicates that in many cases the individual charges are evenly or symmetrically placed throughout the protein's volume.

REFERENCES

Cohn, E. J., and Edsall, J. T. (1965). "Proteins, Amino Acids and Peptides as Ions and Dipolar Ions." Hafner, New York.
Debye, Peter (1929). "Polar Molecules," Dover, New York.
Dunning, W. J., and Shutt, W. J. (1938). *Trans. Faraday Soc.* **34**, 479.
Kirkwood, J. B. (1939). *J. Chem. Phys.* **7**, 911.
Onsager, L. (1936). *J. Amer. Chem. Soc.* **58**, 1486.
Schlecht, P., Vogel, M., and Mayer, A. (1968). Effect of oxygen binding on the dielectric properties of hemoglobin, *Biopolymers* **6**, 1717.
Schlecht, P. (1969). Dielectric properties of hemoglobin and myoglobin II, dipole moment of sperm whale myoglobin, *Biopolymers* **8**, 757.

Setlow, R. B., and Pollard, E. C. (1962). "Molecular Biophysics." Addison-Wesley, Reading, Massachusetts.

Takashima, S. (1962). Dielectric dispersion of protein solutions in viscous solvent, *J. Polym. Sci.* **56**, 257.

Takashima, S. (1963a). Dielectric dispersion of polyglutamic acid solutions, *Biopolymers* **1**, 171.

Takashima, S. (1963b). Dielectric dispersion of DNA, *J. Mol. Biol.* **7**, 455.

Takashima, S. (1966). *J. Phys. Chem.* **70**, 1372.

Tanford, Charles (1961). "Physical Chemistry of Macromolecules." Wiley, New York.

Wyman, J. (1936). *Chem. Rev.* **19**, 213.

5

TYPES OF MOLECULAR INTERACTIONS

INTRODUCTION

Having laid the basis for electrostatics in Chapters 2 and 3, we are now ready to consider in more detail the various ways in which molecules can interact with one another. In considering this topic the reader should realize that at the lowest, most fundamental level interactions between biomolecules are electrical in nature. The physical behavior of one molecule with respect to another is determined to a very large extent by electrical forces and the laws of electrostatics and electrodynamics. Hydrodynamic influences also play a significant role, but it is essentially the electrical interactions that determine the manner in which proteins and other components associate or dissociate.

From many years of biophysical research, it has been found that complex biological structures are almost always constructed from simple and basic building units, much as the towering pyramids are built from an assemblage of stone blocks. Examples of this principle can be seen by considering the structure of enzymes, viruses, and muscles. Many large enzymes are composed of two or more subunits, all of which must be correctly associated with one another for the enzyme to function properly. Viruses are often built from a collection of protein subunits that encapsulate the genetically important nucleic acid chain. In the case of tobacco mosaic virus (Fig. 1-14) there are approximately 2200 protein subunits interacting with one another to form a cylindrical sheath surrounding a helically wound RNA strand. Each protein subunit here can be likened, from a structural standpoint, to a kernel of corn, and the whole virus to the complete cob. The thin filaments of muscle are composed of countless spherelike proteins called actin which are linked with long thin tropomyosin and short, fat troponin molecules to form an overall helical structure (Fig. 5-1). So, the manner in which these units of structure interact with one another to form a cohesive and well-defined biological entity is a study that has occupied much time in research.

The study of physics, and in particular electrical theory, can contribute greatly to the understanding of biomolecular interactions. It is important

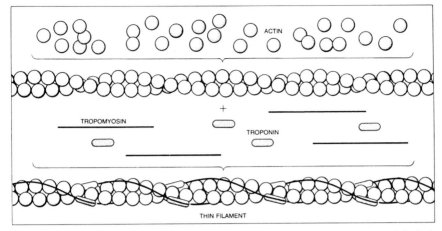

Fig. 5-1 A schematic diagram of the thin filament of muscle. As with most large biological molecules, the thin filament is made up of a number of small subunits that interact and function together as a unit and not individually. (From J. M. Murray and A. Weber, The cooperative action of muscle proteins. Copyright © 1974 by Scientific American, Inc. All rights reserved.)

to realize that in this endeavor it is not necessary to postulate any new physical forces or concepts other than those that have already been presented in previous chapters. The fact that biomolecules are usually considered to be special because they are involved with the process of life does not change the physics of the situation.

In this chapter and the one following, specific types of interactions will be discussed, and calculations will be made concerning their strength and possible importance. It will be noted that individual interactions differ in their directionability and dependence on distance, so when considering the overall effect, the physical behavior of a biomolecule can be dominated by one type of interaction or be influenced by contributions from many types. Finally, it should be emphasized that it is very difficult to determine the details of electrical interactions between biomolecules because their structures are enormously complex and are by no means entirely known, except in a few cases.

CHARGE–CHARGE INTERACTION

In Chapter 2 we saw that Coulomb's law stated that the force acting between two charges q_1 and q_2 is described by

$$F = q_1 q_2 / r^2 \qquad (2\text{-}3)$$

In the same chapter it was also shown that the energy of a system composed of two charges a distance r apart could be given by

$$U = q_1 q_2 / r \qquad (2\text{-}54)$$

This dual description illustrates a general principle that occurs when describing electrical interactions; it is possible to talk about either forces or energies of interaction for a particular situation. If the mathematical expression for the potential energy is known, then it is possible to derive the force dependence. From Chapter 2 it is seen that for the above case, the potential energy is given by the product of a charge q_1 times the value of the electrical potential due to charge q_2 acting on q_1 when the two charges are separated by distance r:

$$U = q_1 V_2 = q_1 (q_2 / r) \qquad (5\text{-}1)$$

If we realize Eq. (5-1) is true, then we can multiply both sides of Eq. (2-47) by q, and it is seen that

$$F_x = -\partial U / \partial X, \qquad F_y = -\partial U / \partial y, \qquad F_z = -\partial U / \partial z \qquad (5\text{-}2)$$

where F_x, F_y, and F_z are the respective components of the electrical forces, and U is the expression for the electrical potential energy. (The reader should remember $qE = F$.) When the potential energy is expressed in polar coordinates, as in Eq. (2-54), the relation between F and U is given by

$$F_r = -\frac{\partial U}{\partial r}, \qquad F_\theta = -\frac{1}{r} \frac{\partial U}{\partial \theta} \qquad (5\text{-}3)$$

where F_r is the component of $\mathbf{F}$ in the radial direction, and F_θ is the component in a direction perpendicular to the radial direction (Fig. 5-2). Equations (5-2) and (5-3) then allow us to derive the magnitude of an electrical force if its corresponding energy of interaction is known. This can be illustrated by a simple example. The energy of interaction between two point charges is $U = -q_1 q_2 / r$, and by use of Eq. (5-3) it is found that $F_r = -\partial U / \partial r = q_1 q_2 / r_2$ and that $F_\theta = 0$. This means the force acting between the two charges is entirely in the radial direction with respect to either charge. This we know to be true from previous considerations. If polar coordinates are used for the energy of interaction equations, and these equations take the form $U = A/r^n$ where A is a constant, then the force dependence on distance can be obtained via Eq. (5-3) to be $F = nA/r^{n+1}$. The force dependence on distance is always weaker than the energy of interaction term.

The relations expressed in Eqs. (5-2) and (5-3) are not restricted to electrical forces and energies alone, but are general in nature where conservative forces are concerned. A conservative force is defined in terms of the type of work that can be performed by the force. The work of a conservative force

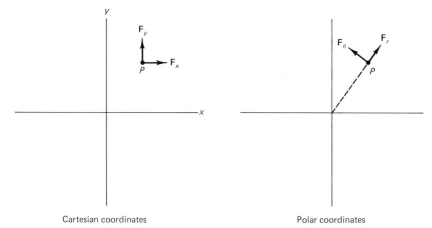

Cartesian coordinates   Polar coordinates

Fig. 5-2 The location of a general point P in a plane can be described in either cartesian or polar coordinates with equal exactness. Likewise, a force acting at a point P can be described in either coordinate system. In polar coordinates the components of the force are in the radial direction (F_r) and perpendicular to it (F_θ), respectively. In cartesian coordinates, the force components are parallel to the x and y axes, respectively. The choice of one coordinate system or the other depends on the convenience of the mathematics.

has the properties: (1) it is independent of the path the force takes; (2) it is equal to the difference between the initial and final value of an energy function; and (3) it is recoverable. Reviewing Chapter 2 should convince the reader that electrical forces are conservative in nature, just as are gravitational forces. A mass raised to a height h above the ground has a potential energy that is completely independent of the manner in which the mass is placed at that height. Likewise, the amount of work necessary to bring a charge q_1 from infinity to a distance r from a second charge q_2 is completely independent of the path that q_1 takes. All that matters in either case is the initial and final resting place. The energy that is stored or used up to place q_1 in its position can be returned to its original state by merely moving q_1 to infinity.

In discussing the charge–charge interaction and those interactions to follow the energy descriptions will be used since it is more convenient to relate to experimental studies. The units to be used for energy will be kilocalories per mole for reasons discussed in Chapter 2. With these restrictions in mind Eq. (5-2) can be written as

$$U = 331.9 n_1 n_2 / Dr \quad \text{kcal/mole} \qquad (5\text{-}4)$$

where n_1 and n_2 are the number of electronic charges on q_1 and q_2, respectively, r is the distance of separation measured in angstroms, and D is the dielectric constant with its value subject to all the complexities discussed in Chapter 3. Written as above, Eq. (5-4) gives the potential energy for a charge–charge interaction directly in kcalories per mole. In general, this type of interaction is quite strong if the two charges are allowed to get close together. The reason for this strength of interaction is due to the $1/r$ dependence of the potential energy. As will be seen soon, other types of interactions have a $1/r^n$ dependence where n is an integer greater than one. By having merely a $1/r$ dependence, the charge–charge interaction is not "diluted" so much by the distance of separation.

The instances where basic charge–charge interactions are biologically important are many; these will be illustrated by the following example. Acetylcholine (Fig. 5-3) is one of a group of chemical compounds, called neurotransmitters, which are discharged from the end of a nerve and travel

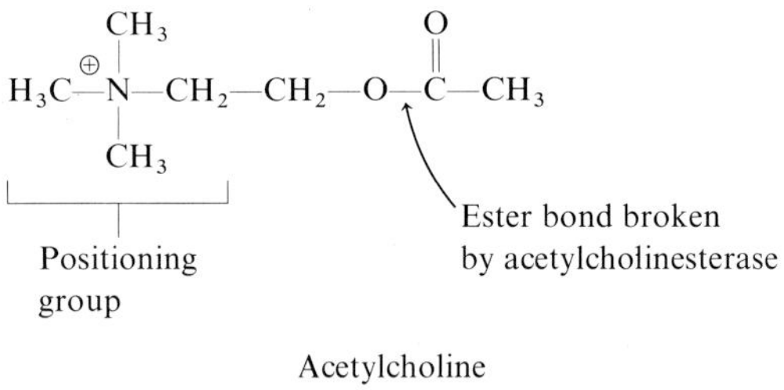

Acetylcholine

Fig. 5-3   The structure of the neurotransmitter acetylcholine.

across the synapse space, to be detected by a receptor on the postsynaptic nerve cell (Fig. 5-4). These neurotransmitters influence such physiological processes as heart beats, muscle movement, and hormone stimulation. Important in the metabolism of acetylcholine is the enzyme acetylcholinesterase, which cleaves the bond indicated by the arrow in Fig. 5-3. Chemical studies of this enzyme have indicated that there is at least one and possibly two negative charges in the active site, where the active site is that part of the enzyme in which the actual catalyzing action takes place. These charges are probably caused by ionized carboxyl groups. In order for the enzyme to work efficiently and repeatedly, it is necessary that acetylcholine be properly aligned with respect to the active site in exactly the correct manner each and every time. One manner of achieving this is for the active site of the enzyme to provide a strong attractive pull on part of the acetylcholine molecule, which will be called the positioning group, thus aligning the ester bond of acetylcholine with the corresponding proper part of the active site. It is thought that part of this attraction in the case of acetylcholinesterase is provided by direct coulombic interaction between the negative charge of the active site and the positive charge of the quaternary ammonium group

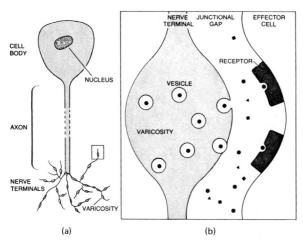

Fig. 5-4 (a) A schematic diagram of a nerve cell. (b) An enlarged view of a nerve cell terminal. The circles, squares, and triangles are neurotransmitters which are released when the nerve cell is excited. They are picked up and detected by a neighboring nerve cell thus transmitting the nerve signal. (From J. Axelrod, Neurotransmitters. Copyright © 1974 by Scientific American, Inc. All rights reserved.)

(Fig. 5-5). Estimates of the distance between the positive and negative charges are on the order of 5.5 Å. The reader should be able to calculate that for this distance of separation, the maximum possible energy of interaction between acetylcholine and the active site is approximately 60 kcal/mole if it is assumed that the dielectric constant equals 1. This is to be compared to a value of about 100 kcal/mole for a C—C covalent bond.

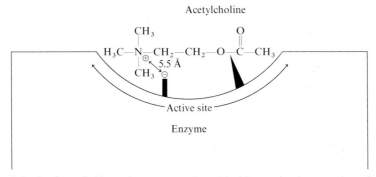

Fig. 5-5 A schematic illustrating a proposed model of interaction between the active site of acetylcholinesterase and its substrate. The spike at the right of the active site indicates clevage of the ester bond.

The above example illustrates not only a case where charge–charge interaction is important, but also a general mechanism whereby enzymes recognize their substrate* by complexing with the positioning group. The reader should realize that the substrate is typically several thousand times smaller than the enzyme, and that finding the one and only alignment where catalysis occurs is not trivial. We shall see in the next chapter that van der Waals forces are also important in positioning acetylcholine in the active site, although their role is not nearly as strong as the charge–charge interaction. Another enzyme where charge–charge interaction is thought to play a major role in positioning the substrate is papain, which comes from the latex of the papaya plant and is the active ingredient in some commercial meat tenderizers.

It is also interesting to note that chemical compounds with a structure similar to that of a substrate can "fool" the enzyme and inhibit activity by competitively binding to the active site, thus blocking the true substrate from access to the enzyme. Figure 5-6 shows a few compounds that are

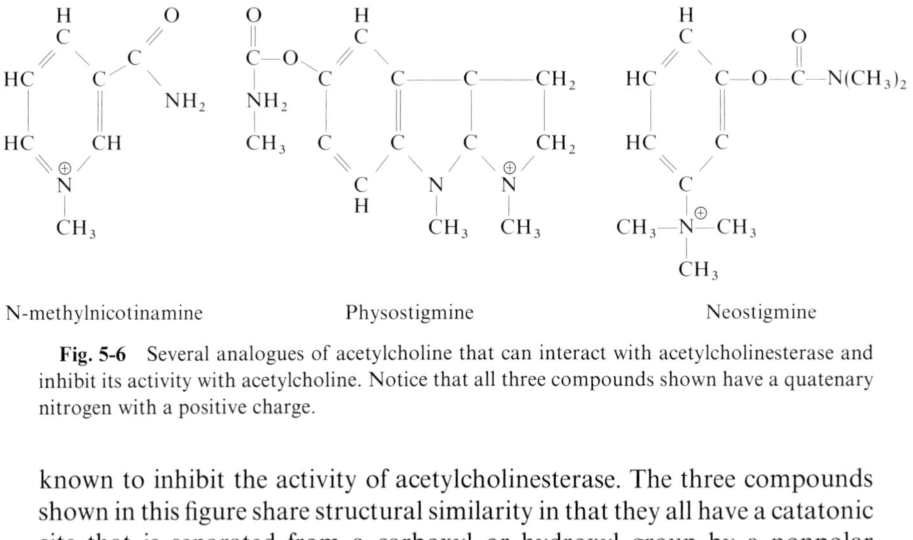

Fig. 5-6   Several analogues of acetylcholine that can interact with acetylcholinesterase and inhibit its activity with acetylcholine. Notice that all three compounds shown have a quatenary nitrogen with a positive charge.

known to inhibit the activity of acetylcholinesterase. The three compounds shown in this figure share structural similarity in that they all have a catatonic site that is separated from a carboxyl or hydroxyl group by a nonpolar region. The proposed model of interaction between enzyme and substrate is further supported by the fact that physostigmine loses its inhibitory effect as the pH of the system is raised and the N atom loses its charge. That the positive charge on the substrate is necessary for binding can also be seen in competitive binding studies that have been performed between isoamyl-

* *Substrate* refers to the molecules on which an enzyme exerts its catalytic activity.

alcohol and dimethyl amino ethanol (Fig. 5-7). Both compounds have essentially the same structure except the dimethylamino ethanol has a positive charge while the isoamyl alcohol does not. It has been found that dimethylamino ethanol is some 30 times stronger as an inhibitor to acetylcholinesterase than isoamyl alcohol.

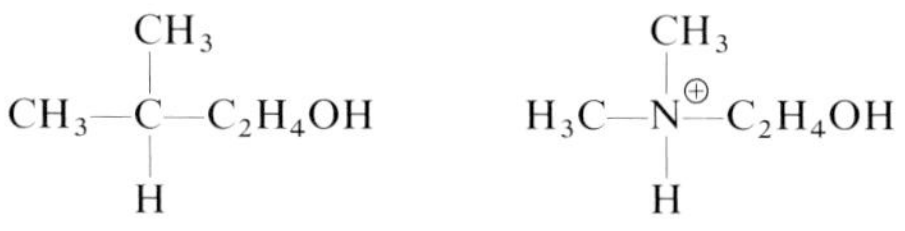

$$CH_3-\underset{\underset{H}{|}}{\overset{\overset{CH_3}{|}}{C}}-C_2H_4OH \qquad\qquad H_3C-\underset{\underset{H}{|}}{\overset{\overset{CH_3}{|}}{N}}^{\oplus}-C_2H_4OH$$

Isoamyl alcohol   Dimethylamino ethanol

Fig. 5-7 The structures of isoamyl alcohol and dimethylamino ethanol. They are similar except for the central atom and their net charge.

CHARGE–DIPOLE INTERACTION

The charge–charge interaction usually evokes the largest energy interactions between two molecules. In general, the next largest type of interaction is between a charge and a dipole.

The interaction between a positive charge and an immobile dipole is shown schematically in Fig. 5-8. For large r, Eq. (3-4) gives the potential

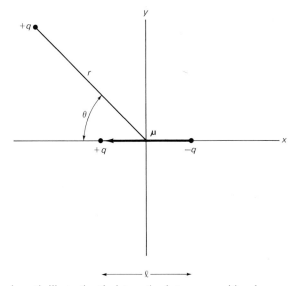

Fig. 5-8 A schematic illustrating the interaction between a positive charge q and a dipole μ.

of the dipole as

$$V = (\mu \cos \theta)/Dr^2 \qquad (3\text{-}4)$$

so the interactional energy of the dipole with the charge q is

$$U = (q\mu \cos \theta)/Dr^2 \qquad (5\text{-}5)$$

(remember that $U = qV$) if the orientation of the dipole is fixed.

According to Eq. (5-5), $U = 0$ for $\theta = 90°$, $U = q\mu/Dr^2$ for $\theta = 0°$ and $U = -q\mu/Dr^2$ for $\theta = 180°$. The maximum energy of interaction thus occurs when the positive charge q is located collinearly with the vector dipole, and the most stable interaction occurs when the dipole points away from the charge ($\theta = 180°$). This makes sense since then the two positive charges are farthest apart.

If the dipole is free to rotate, as would be the case if a protein were immersed in a liquid, then the dipole will assume a statistical orientation about the configuration of lowest energy (Fig. 5-9) with respect to the charge q. The amount of orientation the dipole will assume depends on the strength of interaction between the dipole and the charge. If the interaction is very strong, then the dipole will essentially be aligned collinearly with the charge; however, if the energy of interaction is small compared to the thermal energy $k_B T$, it is then possible to calculate the energy of interaction on a statistical basis.

It can be expected that the energy of interaction for the situation shown in Fig. 5-9 will take on a form similar to that of Eq. (5-5), except that the term $\mu \cos \theta$ will have to be statistically averaged over all possible orientations; i.e., it will be necessary to find the average orientation of the dipole in the direction of the field when the dipole is in the electric field of the charge q. This calculation has however has already been performed in Chapter 4, and the result is

$$\bar{\mu} = \mu^2 E/3k_B T \qquad (4\text{-}30)$$

where now E is given by the field of a point charge, so the component of the dipole in the direction of the field is given by

$$\bar{\mu} = \mu^2 q/3k_B T r^2 D \qquad (5\text{-}6)$$

+q

Fig. 5-9   If a dipole is free to rotate in the field of a charge, the dipole will assume a statistical orientation of lowest energy. The dipole can assume one in a distribution of orientations.

This equation is then the expression for the average component of the dipole moment $\bar{\mu}$ in Fig. 5-9 which points in the direction of q. The potential energy for this situation is then given by

$$U = -q^2\mu^2/3k_B T r^4 D^2 \tag{5-7}$$

where the minus sign indicates an attractive force. The reader should be able to show that Eq. (5-5) can be written as

$$U = -(69.11 n\mu \cos\theta)/Dr^2 \quad \text{kcal/mole} \tag{5-8}$$

and Eq. (5-7) can be written as

$$U = -(80.1 \times 10^4)n^2\mu^2/r^4 k_B T \tag{5-9}$$

where n is the number of electronic charges on q, μ is the dipole moment in debyes, and r is the distance in angstroms from the dipole center to the charge q. It should be pointed out that the fixed dipole situation has a $1/r^2$ dependence and that the statistical dipole has a $1/r^4$ dependence for the energy of interaction. Just from a consideration of the distance dependence, the reader should realize that the charge–dipole interaction is not going to be as strong as the charge–charge interaction and that the statistical dipole will generally be weaker than the fixed dipole case.

To get an insight into the relative energies typically encountered for the two cases of charge–dipole interactions described above, consider a unit positive electronic charge that is a distance of 40 Å away from a myoglobin molecule. Assume that myoglobin has a dipole moment of 170 D at 37.5°C. It will also be assumed that the dipole is centered in the middle of the molecule which can be roughly approximated as a sphere of 20 Å radius. The ion is therefore 20 Å from the molecular surface (see Fig. 5-10). At this

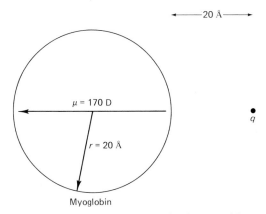

Fig. 5-10 A schematic illustrating the interaction of a charge q with a myoglobin molecule having dipole moment of 170 D. The charge is 40 Å away from the dipole center.

distance the bulk dielectric constant of water can be used. The reader should verify that the maximum energy of interaction available when all charges are fixed is given as $U = -0.10$ kcal/mole, whereas for the case of the fluctuating dipole $U = -0.005$ kcal/mole. So, for the particular case where the dipole is free to fluctuate, a substantial reduction in the energy of interaction results. This is a general result. At 37.5°C the thermal energy $k_B T$ is equal to 0.61 kcal/mole, which is two orders of magnitude larger than the energy for the case of the charge with the fluctuating dipole. For an ion 20 Å away from a myoglobin molecule, the effect of the molecular dipole is minimal compared to the randomizing effect of thermal agitation if the dipole is free to rotate. As the ion moves closer, charge–dipole interaction will exert more influence, the dipole will show more and more orientation, and the energy of interaction will substantially increase.

In the event that a charge–dipole interaction exists where the distance of separation is very small, it is necessary to use a more exact expression in place of Eq. (5-5). For this situation, the potential energy is given by

$$U = \frac{qq_1}{D}\left[\frac{1}{r_+} - \frac{1}{r_-}\right] \tag{5-10}$$

where r_+ and r_- are the distances from the positive and negative charges of the dipole respectively. Here q is the charge on each particle in the dipole and q_1 is the charge interacting with the dipole. The reader should realize that in this situation it is generally necessary to have more information about the dipole than just its dipole moment μ. The situation where the charge is close to the dipole is a more difficult situation to deal with because more exacting information is needed about the system in order to make the calculation. It should be remembered that as a free charge moves closer and closer to a freely rotating dipole, the dipole will reorient itself so that the free charge is collinear with it. This orientation will also have the dipole pointing toward or away from the free charge, depending on whether it is negative or positive in sign.

DIPOLE–DIPOLE INTERACTION

The strength of a dipole–dipole interaction depends not only on the distance of separation of the dipole centers but also on their relative orientations. To derive an expression for the energies involved, it is necessary to calculate only the energy of interaction between all pairs of charges and then to sum the results. This concept is illustrated for the situation depicted in Fig. 5-11 where two dissimilar dipoles (different q and l) have their centers separated by distance L. The exact expression for the potential energy of interaction

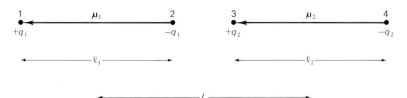

Fig. 5-11 Two dipoles whose centers are separated by distance L. The dipoles are not identical, they have different charges and distances of separation between the charges.

is given by

$$U_{tot} = U_{1,3} + U_{1,4} + U_{2,3} + U_{2,4}$$

$$= \frac{\mu_1\mu_2}{l_1 l_2 D}\left[\frac{1}{L + \frac{1}{2}(l_1 - l_2)} - \frac{1}{L + \frac{1}{2}(l_1 + l_2)} - \frac{1}{L - \frac{1}{2}(l_1 + l_2)} + \frac{1}{L - \frac{1}{2}(l_2 - l_1)}\right]$$

$$(5\text{-}11)$$

and if $l_1 = l_2$, $q_1 = q_2$, Eq. (5-11) reduces to

$$U_{tot} = \frac{\mu^2}{l^2 D}\left[\frac{2}{L} - \frac{1}{L + l} - \frac{1}{L - l}\right] \qquad (5\text{-}12)$$

In deriving Eq. (5-12) the potential energy terms $U_{1,2}$ and $U_{3,4}$ are not included in the sum because they are not part of the interaction between the dipoles. The term $U_{1,2}$ represents how much electrical energy it took to form the dipole, but we are not interested in that. For the above calculation [Eq. (5-11)], it is assumed that the dipoles are already formed, and only the interactional potential energy is desired.

For the case shown in Fig. 5-12, the potential energy is given without approximations as

$$U = \frac{2\mu_1\mu_2}{l_1 l_2 D}\left[\frac{1}{[L^2 + \frac{1}{4}(l_1 + l_2)^2]^{1/2}} - \frac{1}{[L^2 + \frac{1}{4}(l_2 - l_1)^2]^{1/2}}\right] \qquad (5\text{-}13)$$

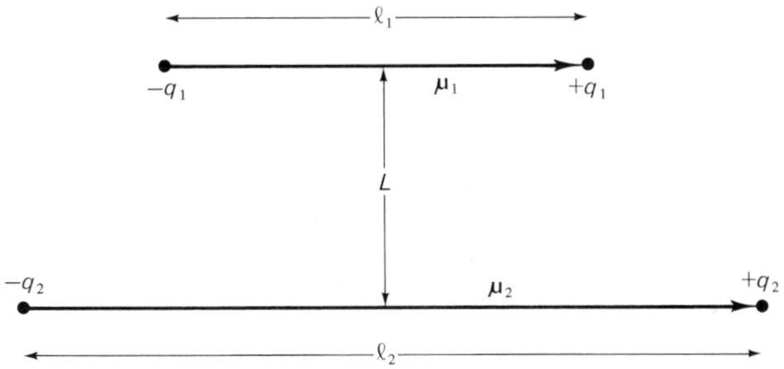

Fig. 5-12 Two parallel dipoles whose centers are separated by a distance L. The respective dipole moments are μ_1 and μ_2.

If both dipoles are identical, Eq. (5-13) reduces to

$$U = \frac{2\mu^2}{l^2 D}\left[\frac{1}{(L^2 + l^2)^{1/2}} - \frac{1}{L}\right] \tag{5-14}$$

where it is still necessary to know the length of the dipole to complete the calculation. Knowledge of the dipole moment alone is not sufficient to calculate the energy of interaction here.

If it is assumed that the two interacting dipoles are far apart (compared to their length), it can be shown that the energy of interaction involves only the magnitude of the separate dipole moments. The equation describing this situation is

$$U = -(\mu_1\mu_2/Dr^3)(2\cos\theta_1\cos\theta_2 - \sin\theta_1\sin\theta_2) \tag{5-15}$$

where the angles θ_1 and θ_2 are defined in Fig. 5-13. Equation (5-15) is derived by using techniques similar to those already employed, and the details are left to the interested reader.

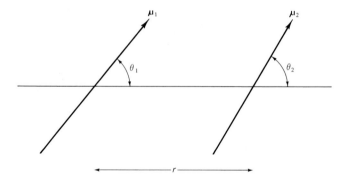

Fig. 5-13   A generalized situation where two dipoles are interacting with one another. The angles θ_1 and θ_2 define the orientation of the two dipoles with respect to a line that passes through the dipole centers. The variable r is the distance between dipole centers.

Five special cases of Eq. (5-15) are shown in Fig. 5-14. Considering the five cases shown in this figure, it is seen that the most stable arrangement between two dipoles occurs when they are collinear and are aligned head to tail. In this case the energy of interaction is negative and the magnitude has its maximum value. Therefore, if both dipoles are without constraints, they will assume this orientation with respect to one another.

The equations shown in Fig. 5-14 can all be written in the form

$$U = 14.3A\mu_1\mu_2/r^3 D \quad \text{kcal/mole} \tag{5-16}$$

where A is either ± 1 or ± 2, μ_1 and μ_2 are in debyes, and r is in angstroms.

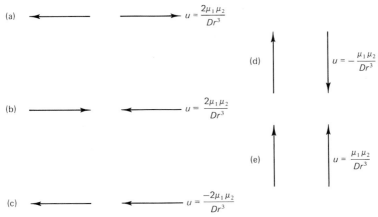

(a) $u = \dfrac{2\mu_1\mu_2}{Dr^3}$

(d) $u = -\dfrac{\mu_1\mu_2}{Dr^3}$

(b) $u = \dfrac{2\mu_1\mu_2}{Dr^3}$

(e) $u = \dfrac{\mu_1\mu_2}{Dr^3}$

(c) $u = \dfrac{-2\mu_1\mu_2}{Dr^3}$

Fig. 5-14 Five separate ways in which two dipoles may orient and interact with one another. The equations shown for each orientation are derived as special cases of Eq. (5–15).

For Eq. (5-16), the appropriate value of A must be substituted in depending on the situation. To get a feeling for the strength of dipole–dipole interactions, consider the example illustrated in Fig. 5-15. In part (a) two myoglobin molecules of dipole moment $\mu = 170$ D are aligned such that their dipoles are collinear, and their dipole centers are separated by 60 Å. Using Eq. (5-12), the energy of the interaction can be calculated to be $U = -0.10$ kcal/mole. In part (b) the dipoles are parallel, and Eq. (5-14) yields $U = -0.02$ kcal/mole. In both cases it was assumed that $T = 37°C$ and that the bulk dielectric constant for water is appropriate. From this example it is seen that dipole–dipole interactions are rather weak, or at least they are when the two dipoles are separated by a relatively large distance. Also, it is seen that a maximum energy of interaction is obtained when the two dipoles are collinear as opposed to being parallel to one another.

The situations considered so far in dipole–dipole interaction have been for fixed dipoles. If one or both dipoles are able to fluctuate, then it is also possible to derive an expression for the potential energy of interaction between two dipoles. Again, this is possible only in the case where the electrical energy of interaction is small compared to the thermal energy. In this respect consider Fig. 5-16 in which dipole 1 is fixed and dipole 2 is free to rotate in the field of dipole 1. The potential energy of a dipole in an electric field has already been given by Eq. (3-17) as

$$U = \mathbf{E} \cdot \boldsymbol{\mu} = E\mu\cos\theta \qquad (5\text{-}17)$$

where $\mu\cos\theta$ is that component of the dipole in the direction of the field. To apply Eq. (5-17) to the situation depicted in Fig. 5-16, it is necessary to

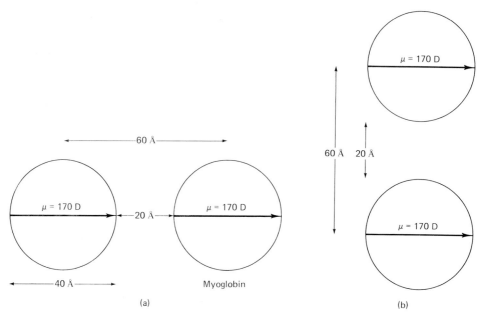

Fig. 5-15   (a) The dipole–dipole interaction between two identical myoglobin molecules where the respective dipole moments are collinear and the centers are separated by 60 Å. The length of each dipole is taken as 40 Å. (b) A similar situation except now the two dipoles are parallel to one another.

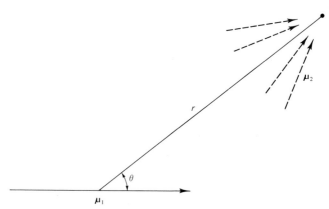

Fig. 5-16   Dipole 1 is fixed and dipole 2 is free to rotate. Dipole 2 can then assume one of a number of statistically accessible positions in the field of dipole 1.

determine the statistical average of the fluctuating dipole in the direction of the electric field generated by dipole 1. This situation is quite similar to that encountered in a previous section, when we considered the energy of interaction between a fixed charge and a freely rotating dipole. The average component of μ_2 in the electric field due to dipole 1 can then be written as

$$\bar{\mu}_2 = \frac{\mu_2{}^2 E_1}{2 k_B T} = \frac{\mu_2{}^2 \mu_1 (1 + 3 \cos^2 \theta)^{1/2}}{3 k_B T \, D r^3} \tag{5-18}$$

and hence the energy of interaction is

$$U = -\frac{\mu_1{}^2 \mu_2{}^2}{3 k_B T r^6 \, D^2} (1 + 3 \cos^2 \theta) \tag{5-19}$$

[Remember that the electric field of a dipole is given by Eq. (3-9).] The minus sign is included because the fluctuating dipole will always be aligned in the most stable state, and hence the energy will be negative indicating attractive forces are occurring between the dipoles. The reader should notice that in this situation there is a $1/r^6$ distance dependence.

In the event that dipole 1 in Fig. 5-16 is also free to fluctuate, the energy of interaction is found by averaging Eq. (5-19) over all possible values of $\cos^2 \theta$. This is equivalent to finding the average value of $\cos^2 \theta$, which is $\frac{1}{3}$. Thus, the energy of interaction between the two dipoles where both are free to rotate is given by

$$U = -2 \mu_1{}^2 \mu_2{}^2 / 3 k_B T r^6 \, D^2 \tag{5-20}$$

This equation would probably be the most appropriate to use for molecules in solution because there individual molecules generally have the freedom to move, i.e., they are not rigidly fixed in space. Again, one should notice the $1/r^6$ distance dependence in Eq. (5-20).

Equation (5-19) can be written as

$$U = -\frac{(3.5 \times 10^4) \mu_1{}^2 \mu_2{}^2 (1 + 3 \cos^2 \theta)}{D^2 T r^6} \quad \text{kcal/mole} \tag{5-21}$$

and (5-20) as

$$U = -\frac{(7.0 \times 10^4) \mu_1{}^2 \mu_2{}^2}{r^6 \, D^2 T} \quad \text{kcal/mole} \tag{5-22}$$

where μ_1 and μ_2 are the respective dipole moments in debyes, and r is in angstroms. Equations (5-21) and (5-22) are valid only for large distances of separation because Eq. (3-9) is restricted by this limitation. However, as the distance of separation between the dipoles decreases, the interaction becomes stronger, and the dipoles tend to become perfectly aligned as in the configuration shown in Fig. 5-11. When this situation develops, Eq. (5-11) is appropriate where it is necessary to know more than just the dipole moments themselves. In general, when interactions occur over large distances and mathematical approximations can be made, it is necessary to know only the dipole moment to calculate an energy of interaction. When identical interactions occur over relatively short distances, these approximations do not hold, and more must be known to calculate energies, e.g., dipole lengths and their respective charges, not just their product.

One would naturally expect that dipole–dipole interactions would be relatively weak when encountered over relatively large distances of separation because of the $1/r^6$ distance dependence, and this is indeed the case. However, over short distances of separation, dipole–dipole interactions can be quite important, and a notable example of this will now be illustrated by considering the hydrogen bond.

THE HYDROGEN BOND

In considering chemical bonds the chemist first encounters two main types: covalent and ionic. An example of an ionic bond is given by NaCl, or any other simple salt. In this type of bond two elements are held in close conjugation by coulombic attraction. Not all elements can form ionic bonds since one atom in the bond must have a low ionization potential, while the other needs a high electron affinity. The ionization potential is a measure of an atom's ability to release an electron, while an atom's electron affinity measures its ability to accommodate an extra electron beyond its normal number. In an ionic bond the atom of low ionization potential donates an electron to the atom with a high affinity. By doing this both atoms become equally but oppositely charged, and hence they are held together by a simple charge–charge interaction. In $CaCl_2$, Ca donates two electrons, one to each Cl atom. In Chapter 2 a calculation was made indicating that the strength of a typical ionic bond in a crystal of NaCl was on the order of 100 kcal/mole. Considering the ionic bond from another viewpoint, it is seen that by completely transferring an electron, each atom involved in an ionic bond achieves a stable, full shell complement of electrons; e.g., by gaining one electron Cl becomes Cl^-, which has 18 electrons, thus completing the 3p subshell of electrons. Atoms with full subshells are known to be extremely stable relative to ones with partially filled subshells. Since atoms on the right-hand side of the periodic table generally have high electron affinities and those on the left have low ionization potentials, these atoms commonly enter into ionic bonds with one another.

The concept of the covalent bond arose because of the inadequacy of the ionic bond in explaining the structure of such molecules as H_2, N_2, O_2, etc. Also, bonds between C atoms, which are quite prevalent in biomolecules, could not be explained by a completely ionic mechanism. The idea of the covalent bond was formulated by the American chemist G. N. Lewis in 1916 in which he postulated that electrons from each atom are shared by both atoms in such a way as to complete the subshells of both atoms. This effort by Lewis was made in part to account for the stability of molecules such as the so-called noble or inert gases (He, Ne, Ar) which have completed subshells of electrons. For molecular hydrogen H_2, two hydrogen atoms are in close vicinity with the two electrons orbiting in such a fashion that each atom "feels" it has completely filled its outer orbit. For atomic chlorine, the outermost unfilled orbit has seven electrons. A chlorine atom needs one more electron to complete a subshell. Thus, in molecular chlorine each atom donates one electron to be shared so each atom has a full subshell of electrons. Two nomenclatures for a covalent bond are shown in Fig. 5-17, where each dot represents an electron. Dots between the atoms represent shared electrons. More frequently, covalent bonds are indicated by a line between atoms where each line represents two electrons.

When Lewis proposed shared electrons, he did not know the shape of the orbits in which the electrons traveled in atoms or molecules, nor did he really know why covalent bonds were formed by electron sharing. After Erwin Schrödinger introduced his famous wave equation in 1926, the techniques of quantum mechanics were available to describe the geometry of orbits and the bond energies to be expected by electron sharing. Using essentially nonclassical physics principles, these calculations soon resulted in a fairly good description of the simple covalent bonds of H_2. More complicated molecules have since been treated using quantum-mechanical principles, but the procedures are generally very complex mathematically. Even for the simple case of H_2, the details of the quantum-mechanical descriptions for the covalent bond are beyond the scope of this textbook. Suffice it to say that a covalent bond is formed when an electron orbit from one atom overlaps an orbit from the other atom. In the simple case where one orbit overlaps another, the bond is termed σ and the electrons involved are called σ electrons. Some general properties of covalent bonds are:

(a)   Atoms in a covalent bond do not equally share common electrons thus giving the bond a slight polar nature.

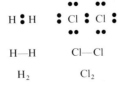

Fig. 5-17   Two ways of representing a covalent bond. In the upper illustration each dot represents an electron in the atom's outer shell. In a single covalent bond each atom donates one electron to the bond. In the lower illustration each line represents a pair of shared electrons.

(b)   Covalent bonds tend to be quite strong with bond energies typically being 100 kcal/mole.

Atoms in covalent bonds are frequently located only 1–2 Å apart.

(d)   Covalent bonds are very directional, meaning that the strongest bond is formed only when the atoms have a specific spatial relationship with respect to one another; i.e., electron orbits of the individual atoms should have a maximum amount of overlapping in space.

The electron density or probability of finding an electron in a covalent bond is large between the two atoms involved.

Although it is not entirely correct, the energy of a covalent bond can be thought of as coming from the increased attraction for a shared electron by two positive nuclei over that of a single nucleus. Obviously, this increased attraction is somehow larger than the mutual repulsion experienced by each nuclei due to the other. The new orbits of shared electrons in the molecule take on the characteristics of both the individual atomic orbits of the participating atoms. Electron orbits and covalent bonds will be further discussed in Chapter 8.

The hydrogen bond is another type of chemical interaction that was initially postulated to explain certain abnormalities in the boiling points of several pure liquids. The hydrogen bond has since been found to play a major role in maintaining the fine structural integrity of many biological macromolecules. The hydrogen bond can be represented schematically as shown in Fig. 5-18, where the hydrogen atom is covalently bonded to the electronegative atom X, R is a general chemical group, and Y is also an electronegative atom. Electronegativity is that property of an atom measuring its ability to attract a shared electron pair. All elements have been assigned an electronegativity number that indicates the relative strength of this property. Fluorine has the highest value of 4.0. Biologically important atoms and their electronegativities are $F(4.0)$, $O(3.45)$, $N(2.98)$, $Br(2.75)$, $C(2.55)$, $S(2.53)$, $I(2.45)$, $H(2.13)$, and $P(2.10)$. Only the first three in this list are generally important in hydrogen bonding in biological systems. Because atoms X and Y are more electronegative with respect to hydrogen and the

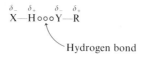

$$\overset{\delta_-}{X}\!\!-\!\!\overset{\delta_+}{H}\,ooo\,\overset{\delta_-}{Y}\!\!-\!\!\overset{\delta_+}{R}$$

Hydrogen bond

Fig. 5-18   An illustration of the hydrogen bond between the molecule X—H and the molecule Y—R. The symbol X represents a general electronegative atom, and Y also represents an electronegative atom. R is an arbitrary chemical group. The symbols δ_- and δ_+ indicate a slight negative and positive charge, respectively; and $\bigcirc\bigcirc\bigcirc$ represents the hydrogen bond interaction.

group R, fractional charges δ result at each atomic site; and hence small dipoles are formed which interact with one another in the head-to-tail arrangement shown in Fig. 5-11. Because electronegative atoms are involved, the covalent bonds X—H and Y—R tend to take on more of a polar nature than they might normally.

One of the initial explanations given for the stability of the hydrogen bond was in terms of the dipole–dipole interaction. This electrostatic explanation was popular because hydrogen, having one valence electron, was able to participate in only one covalent bond (with X); hence the hydrogen bond could not be covalent. The hydrogen bond therefore had to be electrostatic in nature. Since we have already developed an expression covering dipole–dipole interaction, it is now possible to calculate the expected bond energy. To do this, let us take the specific case of hydrogen bonding in water where the situation is as depicted in Fig. 5-19. The distance between dipole

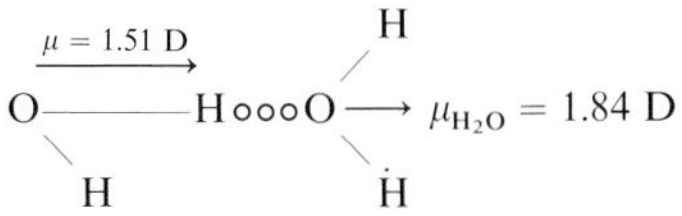

Fig. 5-19 A schematic of a hydrogen bond between two water molecules. The dipole moment of the O—H bond is 1.51 D and that for the entire water molecule is 1.84 D. The O—H bond length is 0.96 Å.

centers is taken as 2.4 Å, the O—H bond distance is 0.96 Å, and the O—H dipole moment is 1.51 D. The dipole moment of the water molecule itself will be taken as 1.84 D. For purposes of this example, the direction of water's dipole moment is taken as midway between the two O—H bonds, and the length of the dipole is taken 0.96 Å. Because of the closeness of the dipoles, the dielectric constant is taken as 1. Substituting these figures into Eq. (5-12) gives $U = -7.0$ kcal/mole, but this value must be corrected for repulsive forces (discussed in a following section). The corrected value is $U = -4.7$ kcal/mole. The experimental values found in the literature range from $U = -4.5$ to $U = -5.8$ kcal/mole, and the agreement is seen to be quite reasonable. This value should be compared to 110 kcal/mole for the O—H bond energy. The close agreement between experiment and calculation supports the use of $D = 1$ in our calculation.

According to the above model for the hydrogen bond, the more electronegative that X and Y are, the stronger the hydrogen bond should be. This model is supported by the facts that when X and Y are fluorine, the strongest hydrogen bonds are formed, a somewhat weaker bond is formed when X and Y are oxygen atoms, and even weaker bonds are formed in the case of two nitrogen atoms.

Other types of dipole–dipole calculations have also been made for various hydrogen bonds, not considering the dipoles as formed from point charges, but instead considering the distribution of charges of all the electrons in their molecular orbits and how these electrons interact with one another and the nuclei. In carrying out this type of calculation a quantum-mechanical description of these orbits is necessary, and the results depend on the particulars of the specific theoretical methods that are used to describe these orbits. These methods also treat the hydrogen bond as an electrostatic interaction, but they tend to be quite a bit more sophisticated than our simple point dipole model. Our model is however satisfying as a first approximation.

Even though there is experimental evidence to support the electrostatic model of the hydrogen bond, there is also a substantial amount of evidence that cannot be adequately explained by the model. The infrared absorption spectra of molecules are sensitive to the distance between atoms that are bonded together and also to the strength of these bonds. The alterations in the IR spectrum due to hydrogen bonds cannot be completely explained by the electrostatic model. Also, the strength of many hydrogen bonds does not vary with the magnitude of the dipole moment of a participating molecule as would be expected. For these and other reasons, there has been a tendency to also treat the hydrogen bond as a partially covalent bond. This approach has the disadvantage of making it necessary to cope with a supersaturated valency for the hydrogen atom, but it has the advantage of being able to explain several phenomena not explained by the electrostatic model. The disagreement over the fundamental nature of the hydrogen bond has not been concluded and is still a topic of research. It is however generally agreed that the hydrogen bond does exist and is an important chemical linkage in determining the structure of biological molecules.

Hydrogen bonds rage in strength between 2 and 10 kcal/mole and as such are relatively weak. Individually, they can be easily broken by thermal agitation, but in large numbers they can give great stability to a structure. Hydrogen bonds can be formed between chemical groups in different molecules or between different groups in the same molecule. It has been found that hydrogen bonds are directional in a manner similar to that of covalent bonds and that individual hydrogen bonds have characteristic bond lengths. In general the strongest bonds are formed when the atoms X, Y, and H are collinear, and the bond strength usually lessens as the bond is bent or distorted (Fig. 5-20). Because of the electrostatic nature of the bond, they are affected by the pH and the amount of free ions in solution.

The importance of hydrogen bonds in maintaining biological structure will be illustrated by three well-known examples: α-helix, β-structure, and the double helix of DNA. These illustrations should also give the reader a better idea of the spatial arrangement of biomolecules at the secondary structure level.

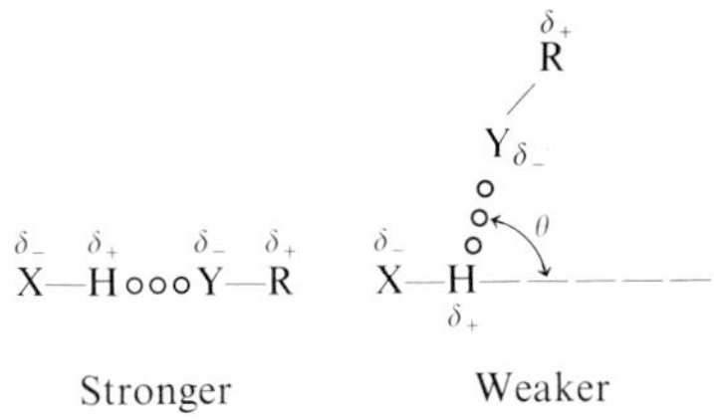

$$\begin{array}{ccc} & & \overset{\delta_+}{R} \\ & & \diagup \\ & & Y_{\delta_-} \\ & & \| \\ & & O \\ \underset{\delta_-}{X}\!\!-\!\!\underset{\delta_+}{H}\overset{\delta_-}{ooo}\underset{\delta_+}{Y}\!\!-\!\!R & & \underset{\delta_-}{X}\!\!-\!\!\underset{\delta_+}{H}\overset{\theta}{\diagdown}\!\cdots \\ & & \end{array}$$

Stronger   Weaker

Fig. 5-20 The relative stability for two orientations of a hydrogen bond. The hydrogen bond is strongest when the atoms X, Y, and H are collinear.

Alpha Helix

The α-helix is one possible structure that a polypeptide or protein chain can assume. This model was first proposed by Linus Pauling and R. Corey in 1951 and has since gained wide acceptance. Basically, the polypeptide backbone of the protein assumes the shape of a helix with the side chains extending generally outward in space. The helical shape of the backbone is maintained by an array of intrachain hydrogen bonds. Each hydrogen bond is formed between the hydrogen attached to the nitrogen in one peptide linkage, and the oxygen of the carboxyl group which is in the third peptide bond down the chain. Each hydrogen bond therefore spans three residues. The α-helix is shown in Figs. 5-21a and 5-21b. From these figures it is seen that the hydrogen bonds are roughly parallel to the helical axis. The reader might think of them as being analogous to the cross bracing on a bridge or large building. Essentially, every peptide linkage in the α-helix is involved in a hydrogen bond, thus giving the whole structure a rigid cylindrical appearance. At least three successive hydrogen bonds have to be broken before any flexibility becomes evident. There is a small hole running the length of the helix, but it is too small to accommodate water molecules. The α-helix can be either right-handed or left-handed, like a screw; although in native proteins, only right-handed helicies have been found. Not all amino acids can participate in an α-helix, and some can form helical arrangements other than the α-helix. If all polypeptide chains did form α-helical structures exclusively, every protein would be expected to have a rodlike shape, which is clearly in conflict with experimental evidence. What actually happens in most proteins is that the polypeptide chain forms α-helical structures for only small stretches of its length. Where there is no α-helix, the chain is flexible enough to double back on itself, and hence a globular or compact looking protein can result. Such is the case for hemoglobin and myoglobin (Fig. 1-10) where the sausagelike portions are actually α-helical.

It should be remaked that the details of the α-helix were proposed with regard to theoretical and experimental considerations. X-ray diffraction studies on a protein called keratin (a constituent of hair) had been performed, and the α-helical model was in part an attempt to interpret the results.

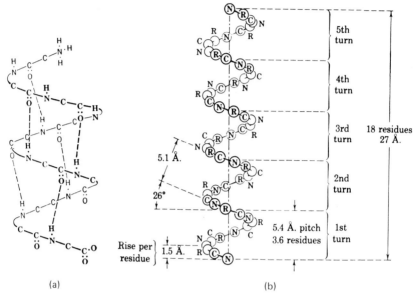

Fig. 5-21   Two schematics of the α-helix found in protein secondary structure. (a) The hydrogen bonding within the chain. (b) Some of the physical parameters of the α-helix (From R. F. Steiner, "The Chemical Foundations of Molecular Biology." Copyright 1965 by D. Van Nostrand Company, Inc. Reprinted by permission of D. Van Nostrand Company.)

Also, the α-helix is a structure in which the number of hydrogen bonds present is maximized, thus giving the total structure a potential energy that is a minimum. In this way the most stable configuration is assured.

Beta Conformation

The β-sheet (also proposed by Pauling and Corey in 1958) is another possible structure for polypeptide chains where again the number of hydrogen bonds present is maximized. Here, instead of the bonds being intramolecular, they are intermolecular. Two parallel, polypeptide chains are in an extended conformation so that the α-helix cannot be formed, but hydrogen bonds can be formed between two separate chains. This is illustrated in Fig. 5-22. The C=O and N—H groups are oriented roughly perpendicular to the backbone and come in close contact, thus permitting the formation of hydrogen bonds. Here, the hydrogen bonds are perpendicular to the backbone direction.

Double Helix

The last example to be considered in this section illustrating the biological importance of H bonds is the double helix of DNA. This structure was

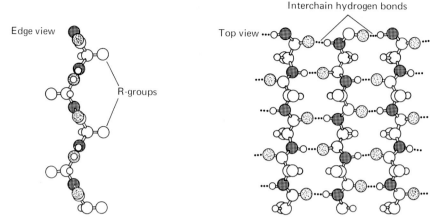

Fig. 5-22   Two views of the β-sheet structure found in proteins (Springall, 1954).

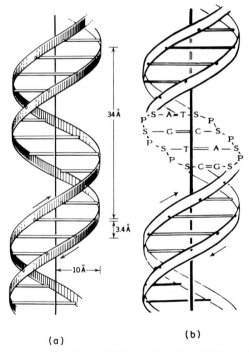

(a)   **(b)**

Fig. 5-23   Schematic diagram of the DNA double helix. (a) The helical shape of both nucleotide chains. The two chains are antiparallel to one another. In part (b) the P stands for the phosphate group, S for the sugar moiety, and G, C, A, and T stand for the nucleotide bases (From R. F. Steiner, "The Chemical Foundations of Molecular Biology." Copyright 1965 by D. Van Nostrand Company, Inc. Reprinted by permission of D. Van Nostrand Company.)

first proposed by James Watson and Francis Crick in 1953; it is partially stabilized by hydrogen bonds. In this structure two nucleotide backbone chains are wound around one another with both chains assuming the shape of a right-handed helix (Fig. 5-23). The chains are antiparallel and the bases are located between the two chains pointing perpendicular to the helical axis. Geometrically, these bases are almost planar and the planes of these bases are roughly parallel to one another. This arrangement is called base stacking. Bases from one chain are hydrogen bonded to bases on the other chain in a specific manner. The allowed base pairs are adenine–thymine and cytosine–guanine, and the hydrogen bonds of each are shown in Fig. 5-24. The pairing is such that each pair can exactly fit between the two backbone chains. The sequence of bases on one chain does not necessarily match the other, although the antiparallel sequences are complimentary

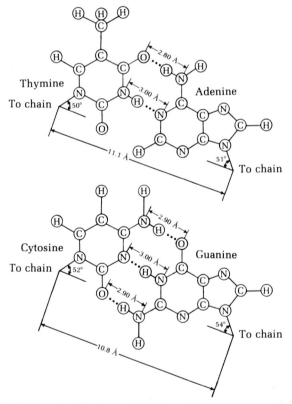

Fig. 5-24   The hydrogen bonding base pairing scheme of adenine with thymine and cytosine with guanine. The C–G base pairs have three hydrogen bonds (Pauling and Corey, 1956).

because of the base pairing. The base pairs of the DNA are generally shielded from the outside environment, whereas the backbone sugars and phosphate groups are in close contact with the solvent. The diameter of the double helix is about 20 Å, the perpendicular distance between the planes of two successive base pairs is 3.4 Å, and there are exactly 10 base pairs per turn of the helix. In the α-helix of proteins, there are 3.6 amino acids per complete turn. The GC base pair of DNA has three hydrogen bonds, whereas the AT has only two. One would then expect that DNAs rich in GC base pairs should be more stable than ones rich in AT pairs. This is indeed the case, as is shown in Fig. 5-25 where the percentage GC content of different DNAs is plotted vs. the melting temperature of those DNAs. The melting temperature is roughly defined as that temperature at which half the base pair interactions are broken as the DNA is heated. A higher melting temperature implies a more stable molecule and a higher percentage of GC base pairs. DNA melting temperatures are usually independent of molecular weight. As with the α-helix, the DNA double helix conformation gives the appearance of a long thin rod. It takes on the order of 10 base pairs to form a stable structure, and DNAs can exist in long chains with some having molecular weights over 10^9 daltons.

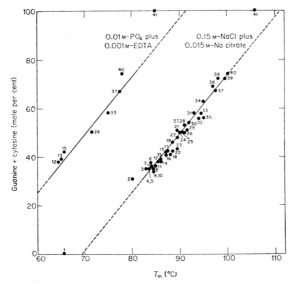

Fig. 5-25   The mole percent of G–C base pairs in a particular DNA plotted vs. the respective melting temperatures. The different numbers represent DNAs from different sources. The two lines are for solvents of different ionic strengths (Marmur and Doty, 1962).

ALPHA OR BETA

Another example of the importance of dipole–dipole interactions can be seen in a polypeptide's choice between the alpha or beta structure. From what has been stated so far it would seem reasonable that all proteins can attain or form the α-helix. However, some prefer the β-structure, or have very unstable α-helicies. One of the reasons for this has been elucidated by two Russian scientists, and again it has to do with a physical property of the amino acids.

In working with synthetic polypeptide systems it had been found that polyserine, polycysteine, and polythreonine each preferred the β-structure over the α-helix. This must mean that for these particular polypeptides, the β-structure is energetically more favorable than the α-helix or that there is some steric hinderance that will not allow comfortable formation of the α-helix. The clue to the problem can be found by examining the structure of the basic amino acids that make up the above polypeptides. In each case there is either an O or an S atom in the γ position of the side chain. The fact that the γ position is occupied by an atom more electronegative than C means that the C_β—M_γ (atom M represents O or S) bond will have a dipole moment larger than normal compared to the case when atom M is a carbon. Birshtein and Ptitsyn (1967) reported the results of some calculations that showed that the dipole–dipole interaction between the dipole of the C_β—M_γ bond and the dipole of the adjacent peptide bond

$$\begin{array}{c} \quad\quad\;\; O \\ H \quad\;\; \diagup\!\!\diagup \\ -N-C \end{array}$$

was repulsive in nature and could destabilize the tight formation of an α-helix, but that it did not affect the formation of the β-structure. Where such repulsive dipole–dipole interactions took place, the β-structure would be formed as being energetically more favorable (U is more negative).

Because serine, cysteine, and threonine do have this increased dipole moment in the C_β—M_γ bond which interacts unfavorably with the dipole of the peptide bond, these three amino acids acquire the reputation of being α-helix breakers. These three amino acids can be incorporated into α-helicies, and this has been shown experimentally; but as their percentages increase, the stability of the helix decreases. Proline is also an alpha helix breaker, but for a different reason. It has no free amino hydrogen to donate to a hydrogen bond.

INDUCTION ENERGY

In considering the charge–charge, charge–dipole, and dipole–dipole interactions, we have been cataloging the common ways in which energy can be stored in a field due to a specific arrangement of electrical charge. Up to now all interactions have involved an asymmetric distribution of charge, either as a monopole or as a dipole. By storing energy in the field, these particular electrical arrangements have been shown to possess certain amounts of stability, thus giving meaning and strength to molecular structures. Another way of storing energy in an electrical field is through polarization or induction. When a dipole is induced by an electrical force, energy is needed to keep the dipole formed and aligned with the field; hence energy is said to be stored in the electric field as potential energy. It is now desired to calculate the energy of interaction stored in a medium of polarization α when it is subjected to an electric field.

When a dipole is induced by an external field $\mathbf{E}$, it is positioned such that it inhibits the efforts of the original field. According to Eq. (3-17), the electric potential energy between a dipole and an electric field is given as

$$U = -\mathbf{\mu} \cdot \mathbf{E} \qquad (3\text{-}17)$$

but the induced dipole is given by

$$\mathbf{\mu} = \alpha \mathbf{E} \qquad (4\text{-}1)$$

so

$$U = -\alpha E^2 \qquad (5\text{-}23)$$

But Eq. (5-23) is not strictly correct in this case because it does not take into account the amount of work that was necessary to separate the charges in forming the dipole in the first place. This amount of work must be subtracted from Eq. (5-23). It can be calculated in the following manner: In displacing a charge q an infinitesimal distance dx to form the dipole, the infinitesimal work dW done is

$$dW = \mathbf{F} \cdot d\mathbf{x} \qquad (5\text{-}24)$$

The force $\mathbf{F}$ however is given by $\mathbf{F} = q\mathbf{E}$, so we have

$$dW = Eq\,dx = E\,d\mu \qquad (5\text{-}25)$$

where $d\mu$ is the infinitesimal dipole formed. The total work performed in inducing the dipole μ can then be calculated by summing all incremental elements of work. This is given by

$$W = \int_0^\mu E\,d\mu' = \int_0^\mu \frac{\mu'\,d\mu'}{\alpha} = \frac{\mu^2}{2\alpha} = \frac{\alpha E^2}{2} \qquad (5\text{-}26)$$

where μ' is used as a dummy variable. The limits on the integral represents the beginning and the end of the dipole formation. Subtracting Eq. (5-26) from (5-23) then gives us the net energy stored in the field when a material of polarizability α is polarized by an electric field $\mathbf{E}$:

$$U = \tfrac{1}{2}\alpha E^2 \qquad (5\text{-}27)$$

CHARGE-INDUCED DIPOLE INTERACTION

Using Eq. (5-27), the energy stored in the field due to the induction of a dipole by a point charge is given by

$$U = -\tfrac{1}{2}\alpha q^2/r^4 D^2 \qquad (5\text{-}28)$$

where q is the magnitude of the charge and r is the distance from charge to dipole center. If α is given in units of $10^{-24}\,cm^3$/molecule, r in angstroms, and q is given in the number n of unit electronic charges, then Eq. (5-28) can be written as

$$U = -165.2n^2\alpha/r^4 D^2 \quad \text{kcal/mole} \qquad (5\text{-}29)$$

To get an idea of the magnitude of energy involved, consider a unit charge 40 Å away from a water molecule where $\alpha = 1.46 \times 10^{-24}\,cm^3$/molecule. In this case, $U = -1.8 \times 10^{-8}$ kcal/mole. For a separation of 2 Å, $U = -15.1$ kcal/mole where at this distance of separation the dielectric constant can be safely assumed to equal 1.

Compared to a charge–dipole interaction, induction effects are usually relatively minor. Where there is a permanent dipole, this dipole will usually interact with point charges in such a manner as to render the induced dipole interaction insignificant. Also, induction effects are generally significant only for small distances of separation, as the above example illustrates. Here they can be fairly strong. One case where a charge-induced dipole interaction would be expected to play a role would be in enzyme substrate binding. In this situation the distance of separation is low, and the small forces involved would help in the fine alignment while the charge–permanent dipole or charge–charge interaction would determine the gross alignment. In considering the charge–induced dipole effect between an ion and a complex molecule, it is typical to consider only those parts of the molecule that are in the immediate neighborhood of the ion. To consider the whole molecule would result in a hopelessly complex problem. From a practical standpoint, only simple chemical groupings, like CH_3, $COOH$, NH_2, etc., that are attached to the molecule should be considered, and then the total interaction is taken as the sum of the individual interactions.

The main contribution to the polarizability α is from electron displacement, and different chemical groupings will differ in their polarizabilities depending on how the electrons are held in the various bonds. Usually the electrons in the outer molecular orbits are most important in the value of a molecule's polarizability since they can be influenced most by an external field. Electrons closer to the nucleus are under more of its influence and are not as subject to external electric fields. The inner electrons also have a tendency to be "shielded" by the electrons in the outer orbits. These principles should be kept in mind since they will be used in describing one theory of van der Waals forces in the next chapter.

Polarizabilities are usually determined experimentally by measuring the index of refraction and then making a subsequent calculation to obtain the polarizability. The mathematical relationship between the index of refraction and the polarizability is beyond the scope of this book and will not be discussed. Suffice it to say that the propagation of light through a medium depends on the ability of molecular electrons to be displaced from their equilibrium positions. This is the physical basis underlying the relationship between the polarizability and the index of refraction. The index of refraction of a medium depends on the density of electrons and also on their environment, i.e., how tightly they are held and whether or not they are participating in a bond. It has been argued that from a theoretical standpoint the total index of refraction of a medium can be interpreted in terms of the individual atoms composing the material where each atom will contribute a characteristic amount to the total index of refraction. Over the past 100 years the various atoms have had their incremental contributions determined, so that if the chemical composition of a substance is known, then a theoretical calculation of the index of refraction could be made merely by adding up all the incremental contributions. An alternative explanation for describing the refractive index, and hence also the polarizability, is in terms of the chemical bonds of the material. Since the bonds determine the position and environment of the electrons more so than do the atoms involved, it is natural to think that certain types of bonds will have a higher (or lower) refractive index or polarizability. This concept has led to the assignment of atomic polarizabilities for commonly encountered chemical bonds. Typical values of bond polarizabilities are shown in Table 5-1.

To obtain the polarizability of a simple chemical group, it has been found that a good value can be obtained by totaling the polarizabilities of each bond present in the particular chemical group in question. For example, methane (CH_3) has three C—H bonds, and its polarizability is given by $3 \times 0.655 \times 10^{-24} = 1.97 \times 10^{-24} \, cm^3/molecule$. As another example, consider a carboxyl group

TABLE 5-1

Values of the Polarizability for Different Types of Covalent Bonds[a]

Bond	Length (Å)	Bond energy (kcal/mole)	Polarizability (10^{-24} cm^3/molecule)	Bond	Length (Å)	Bond energy (kcal/mole)	Polarizability (10^{-24} cm^3/molecule)
C—C	1.54	81	0.475	N—H	1.01	92	0.721
C—H	1.09	99	0.655	O—O	1.32	34	0.641
C—O	1.42	80	0.559	O—H	0.96	105	0.733
C—N	1.46	66	0.598	S—S	2.08	66	2.93
C—S	1.81	61	1.75	S—H	1.34	87	1.83
C—F	1.38	102	0.705	C=C	1.35	140	1.59
C—U	1.76	76	2.53	C≡C	1.20	193	2.31
C—Br	1.93	63	3.64	C=O	1.22	178	1.31
C—I	2.13	47	5.57	C≡N	1.15	211	1.86
N—N	1.44	37	0.602	N=O	1.16	104	1.32

[a] From Webb (1963).

Here, there are one C=O bond, one C—O, and one O—H bond. Using the appropriate values from Table 5-1 yields a polarizability of 1.6×10^{-24} cm^3/molecule. Values of α for some common biochemically important groups are given in Table 5-2.

The simple additivity rule for finding the total polarizability of a chemical group if all the individual bond polarizabilities are tabulated is known to be systematically incorrect in several cases, and the reader should be aware of this. The most important case involves chemical groups that have conjugated bonds. Compounds with conjugated multiple bonds have polarizabilities higher than expected based on the sum of contributions from individual bonds.

When examining Tables 5-1 and 5-2, the reader should notice several things. First, the bonds involving carbon and a halide have substantially higher polarizabilities than the other bonds shown. This means that the electrons of these bonds are particularly easy to distort from an equilibrium position with an external electric field. Multiple bonds also tend to have higher polarizability than simple bonds involving the same atoms. This is due to the arrangement of electrons in multiple bonds and in particular, the delocalized nature of the so-called π electrons.* The polarizabilities of

TABLE 5-2

Polarizability of Some Common Biochemical Groups

Group	(10^{-24} ml/molecule)
CH_2CH_3	3.75
CH_3	1.97
NH_2	1.44
SH	1.83
COOH	2.6
COO^-	1.87
OH	0.733
OCH_3	2.5
NH_3	2.2
$COCH_3$	3.75
CHO	1.97
CN	1.86
H_2O	1.46

* π electrons are a general class of electrons that are also shared in a covalent bond by the two participating atoms. These electrons have the property of being more loosely associated with any one atom or pair of atoms in a compound than do the σ electrons which have been previously discussed in the sections on covalent bonds.

the bonds found in saturated hydrocarbons (C—C and C—H) are quite low; however, the polarizability of a hydrocarbon can become large as the chain length grows. A simple rule of thumb to remember is that the polarizability is directly proportional to the volume of the atom or molecule in question. Another aspect of the polarizability is that it is not always isotropic. The polarizability of a compound or a bond can be higher in one direction compared to another. This is merely relating the fact that in a chemical bond the arrangement of electrons is such that it is quite frequently easier to displace an electron from its equilibrium position in one particular direction. Although it is not obvious, the polarizability is usually largest in the direction of the bond; however, for purposes of this text, this anisotropy will be neglected.

DIPOLE-INDUCED DIPOLE INTERACTION

The potential energy stored in the field when a permanent dipole of magnitude μ induces a dipole in a polarizable material can be shown to be

$$U = -\alpha\mu(1 + 3\cos^2\theta)/2r^6 D^2 \tag{5-30}$$

where θ is as previously defined. Equation (5-30) can also be written as

$$U = -7.17\alpha\mu^2(1 + 3\cos^2\theta)/r^6 D^2 \quad \text{kcal/mole} \tag{5-31}$$

where α is in units of $10^{-24}\,\text{cm}^3/\text{molecule}$, μ is in debyes, and r is in angstroms. When both dipoles are free to fluctuate, the energy of interaction is given by

$$U = -\alpha\mu^2/r^6 D^2 \tag{5-32}$$

or

$$U = -14.34\alpha\mu^2/r^6 D^2 \quad \text{kcal/mole} \tag{5-33}$$

As one might expect, the energies involved in dipole-induced dipole interactions are extremely low unless the distance of separation is quite small.

SHORT-RANGE REPULSIVE INTERACTIONS

So far in this chapter we have considered a number of different types of electrical interactions that can possibly affect macromolecular behavior and structure. In each specific case the interactions have the potential of being stabilizing or destabilizing depending on the exact arrangement of charges. In this section one additional interaction will be considered which is always repulsive or destabilizing in character. This interaction comes into play only when atoms or molecules are very close together, and hence its name.

When one considers the structure of an atom, a picture is usually visualized in which a central positive nucleus is surrounded by a cloud of orbiting electrons. When these atoms are combined to form a molecule, no matter how large or complicated, there is still an envelope of negative electrons surrounding or lying on the outermost limits of the molecule. Therefore, the first thing an external probe would encounter in approaching a molecule would be an electron. If two separate molecules were to approach each other, then the electron clouds surrounding each molecule would be expected to interact and create a repulsive force tending to push them apart. This repulsive force would arise from simple coulombic repulsion of like charges. Thus, in considering the net result of all the electrical forces acting between two molecules, the repulsive interactions must also be included. Because of the differences in shape and size of molecules, no overal theoretical treatment of the short-range repulsive force has been possible; however, a few general characteristics are possible.

For all the different interactions considered so far, it has been possible to write the energy of interaction dependence on distance of separation mathematically as

$$U_A = -A/r^a \tag{5-34}$$

where A and a are specific constants depending on the particular interaction in question. For situations depicted in this chapter, a ranges between 1 and 6. It is implicitly assumed that the constant A includes the value of the dielectric constant, polarizability, etc. The minus sign again represents an attractive electrical force. In a similar manner it is possible to describe the short-range repulsive interactions in terms of its energy of interaction by the equation

$$U_B = +B/r^b \tag{5-35}$$

where B and b are constants, and the sign is positive. The exponent b is thought to lie somewhere between 9 and 15, indicating that the repulsive force is extremely strong if r is small and that the distance dependence decreases rapidly with increasing r. The uncertainty in the value of b is due to a deficiency in the theoretical understanding of repulsive forces, and b will be taken as equal to 12 here. Therefore, the total energy of interaction when two molecules are very close is

$$U_{AB} = U_{tot} = \frac{B}{r^b} - \frac{A}{r^a} \tag{5-36}$$

Equation (5-36) implicitly assumes that there is only one attractive interaction. If two or more attractive forces are significant, then more terms of the form $-A/r^a$ have to be included in Eq. (5-36) where the separate a, and As have different numerical values characteristic of the type of interaction they represent. Equations (5-34)–(5-36) are shown schematically in Fig. 5-26.

It will be noticed in Fig. 5-26 that Eq. (5-36) has a minimum at $r = r_e$ which is interpreted as the equilibrium distance of separation. This is the distance of separation between two molecules in which the repulsive forces equal the attractive ones. At shorter distances the repulsive forces increase dramatically, and the system eventually acquires a positive potential energy, which is an unstable situation. At distances longer than the equilibrium value the electrical potential energy also increases (becomes less negative), indicating that the system is not as stable as it was when the distance of separation was equal to r_e. The equilibrium separation, then, is that value of r that makes the system the most stable from an energy viewpoint. Equation (5-36) can be used to calculate a value of r_e. To evaluate this

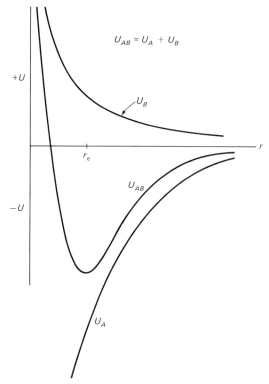

Fig. 5-26 A schematic illustrating the distance of separation dependence for a generalized attractive (U_A) and a generalized repulsive electrical force (U_B). The line labeled U_{AB} is a plot showing the generalized form the total energy takes as a function of distance of separation.

equilibrium distance, it is necessary to differentiate Eq. (5-36) with respect to r, set the result equal to zero, and then solve the equation for r. This procedure is the common one used in calculus to find the minimum or maximum points of a mathematical function. This procedure thus yields

$$r_e = (bB/aA)^{1/(b-a)} \tag{5-37}$$

Using Eq. (5-37), it is now possible to rewrite Eq. (5-36) as

$$U = -\frac{A}{r_e{}^a}\left[1 - \frac{a}{b}\right] \tag{5-38}$$

This equation should now be compared to Eq. (5-36). It is then seen that when the short-range repulsive forces are considered, the original energy of interaction represented by U_A has to be multiplied by a correction factor $1 - (a/b)$. This is strictly true only at $r = r_e$. Another way of stating this is the following: If it is agreed that Eqs. (5-34) and (5-35) are appropriate ways of representing the attractive energy term and the repulsive energy term, respectively, then the repulsive term can be taken into consideration by merely multiplying U_A by the correction factor $1 - (a/b)$. Looking at Fig. 5-26, we see that at distances less than r_e, repulsive interactions destabilize the system and make the potential energy more positive. At distances greater than r_e the correction to U_A for repulsive forces decreases rapidly. If we use $b = 12$, then the correction factors for the various types of interactions considered in this chapter are given in Table 5-3.

In addition to the correction for short-range repulsive forces there is also one additional correction term that must be considered when calculations of electrical interactions are made. This is due to the effect of thermal energy. Referring to Fig. 5-27, it is seen that the energy of interaction at an equilibrium distance of separation is given by the minimum in the curve. This value is arrived at by calculating the energy due to one or several types of interactions, and then correcting for repulsive forces. However, this does not take into consideration the destabilization effect of thermal energy. For a specific temperature, a value equal to $k_B T$ must be added to the electrical potential terms as a correction factor. This is illustrated in Fig. 5-27 by the short vertical line XY where at physiological temperature $k_B T = 0.62$ kcal/mole. This has the added implication that the equilibrium distance no longer has a single value but instead varies between r_i and r_m. Since the energy curve is asymmetric around the line XY, this also means that the average distance of separation at equilibrium is larger than r_e, which makes sense from an intuitive standpoint.

In the section on dipole–dipole interactions a calculation was made to determine the strength of the hydrogen bond energy of interaction between

TABLE 5-3

Short-Range Repulsive Force Correction Needed for Various Types of Electrical Interactions

Interaction	Original equation	Numerical correction factor	Corrected equation
Charge–charge	$-\dfrac{331.9 n_1 n_2}{r D}$	$\left(1 - \tfrac{1}{12}\right) = 0.92$	$-\dfrac{304.2 n_1 n_2}{r D}$
Charge–dipole	$-\dfrac{69.1 n \mu \cos\theta}{r^2 D}$	0.83	$-\dfrac{57.6 n \mu \cos\theta}{r^2 D}$
	$-\dfrac{80.1 \times 10^4 n^2 \mu^2}{r^4 D^2 T}$	0.67	$-\dfrac{53.4 \times 10^4 n^2 \mu^2}{r^4 D^2 T}$
	$-\dfrac{14.3 A \mu_1 \mu_2}{r^3 D}$	0.75	$-\dfrac{10.7 A \mu_1 \mu_2}{r^3 D}$
Dipole–dipole	$-\dfrac{3.5 \times 10^4 \mu_1{}^2 \mu_2{}^2 (1 + 3\cos^2\theta)^{1/2}}{r^6 D^2 T}$	0.5	$-\dfrac{1.75 \times 10^4 \mu_1{}^2 \mu_2{}^2 (1 + \cos^2\theta)^{1/2}}{r^6 D^2 T}$
	$-\dfrac{7.0 \times 10^4 \mu_1{}^2 \mu_2{}^2}{r^6 D^2 T}$	0.5	$-\dfrac{3.5 \times 10^4 \mu_1{}^2 \mu_2{}^2}{r^6 D^2 T}$
Charge-induced dipole	$-\dfrac{165.2 n^2 \alpha}{r^4 D^2}$	0.67	$-\dfrac{110.1 n^2 \alpha}{r^4 D^2}$
Dipole-induced dipole	$-\dfrac{7.17 \alpha \mu^2 (1 + 3\cos^2\theta)}{r^6 D^2}$	0.5	$-\dfrac{3.59 \alpha \mu^2 (1 + 3\cos^2\theta)}{r^6 D^2}$
	$-\dfrac{14.34 \alpha \mu^2}{r^6 D^2}$	0.5	$-\dfrac{7.17 \alpha \mu^2}{r^6 D^2}$

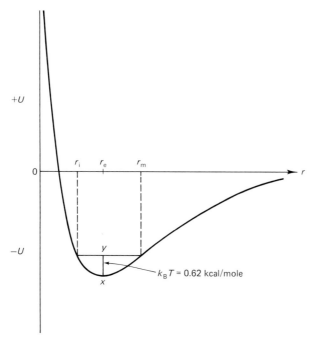

Fig. 5-27 A schematic illustrating the concept that the equilibrium distance between two molecules can take on a range of values.

two water molecules. Subsequent to the calculation it was stated that a correction for repulsion forces was necessary. Since the dipole–dipole interaction assumed has essentially a $1/r^3$ dependence, it can now be seen that the calculated answer should be multiplied by the correction factor 0.75 to give $U = -5.3$ kcal/mole, and that $+0.62$ kcal/mole should be added to this to yield the final answer of -4.7 kcal/mole. It should be emphasized that the repulsive force correction is made because it was assumed that the two water molecules are at an equilibrium distance with respect to one another and that $T = 37°C$.

The reader should now realize that there is a wide variety of ways in which two compounds can interact electrically. It should be emphasized once again that no new physical laws have been proposed in describing these interactions and that biological systems can be understood in terms of classical physics. The difficulty of applying physical principles to biological systems lies in the complexity of biological macromolecules and the general lack of detailed knowledge concerning them.

FURTHER EXAMPLES

In order to get a better feel for making electrical calculations, two other examples will be considered. These are somewhat artificial situations, but they do illustrate the application of principles and equations derived in this chapter.

For the first case, let us consider a cylindrical molecule 3000 Å long with a uniform negative charge distribution γ. It is desired to know what charge distribution is necessary to fully orient a water molecule, i.e., what charge per unit length γ will align a water molecule in the fashion shown in Fig. 5-28. In order to solve this problem, it is first necessary to find the strength of the field due to the biomolecule that is influencing the water molecule. An exact method would consider each charge on the rod and sum their effect to find the total field as a function of distance away from the rod. This however would be a very laborious and difficult job. An alternative approach makes use of a result previously derived. Consider that the water molecule is close enough to the biomolecule so that it appears to be an infinite rod of uniform charge. The field of the rod is then given by Eq. (2-34):

$$E = 2\gamma/r$$

The question now is, How close does the water molecule have to be for the biomolecule to appear as an infinite rod? If the water's dipole center is 7.5 Å away from the rod, then the distance from the dipole center to either end of

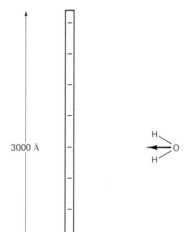

3000 Å

Fig. 5-28 Situation depicting a water molecule next to a biomolecule 3000Å long having a uniform negative charge. It is desired to find what charge per unit length γ will orient the water molecule perpendicularly to the biomolecule.

the rod is 200 times the perpendicular distance to the rod. This means that the angle subtended by the dipole center to either end of the rod is given by $\theta = \tan^{-1} 200 = 89.7°$; this is very close to $90°$, which would occur in the case of a true infinite rod. The conclusion then is that if $r = 7.5$ Å, the rod appears as an infinite line of charge to the water dipole.

If the water dipole is to be perfectly aligned, then the energy of interaction is given by the equation

$$U = -\mathbf{E} \cdot \boldsymbol{\mu} \qquad (5\text{-}39)$$

or

$$U = -2\mu\gamma/r \qquad (5\text{-}40)$$

Also, the electrical energy of interaction has to be much larger than the thermal energy in order to have the perfect alignment maintained. Let us assume that $T = 37°C$ and that the electrical energy must be 10 times that of the thermal energy. With these assumptions the reader should be able to show that the necessary charge per unit length is given by

$$\gamma = 10k_B T r/2\mu \qquad (5\text{-}41)$$

Substituting the appropriate values yields $\gamma = 8.4 \times 10^{-3}$ esu/cm, or about 0.018 electronic charges per angstrom length of the rod. This translates into 54 electronic charges for the entire biomolecule, a value that is quite within the realm of reality. It should also be obvious that any other water molecules located closer than 7.5 Å to the rod will also be aligned, and those located farther out will experience more and more thermal disorientation. Water molecules that are essentially held in a rigid orientation by a charged bio-molecule are termed primary waters of hydration.

As a second example illustrating the use of the concepts presented in this chapter, let us consider the electrical interaction between two water molecules with a charge, say a NH_4^+ group. This situation is diagramed in Fig. 5-29. Suppose it is desired to calculate the energy of the system. For this case, the total energy is the sun of several terms. The interactions of

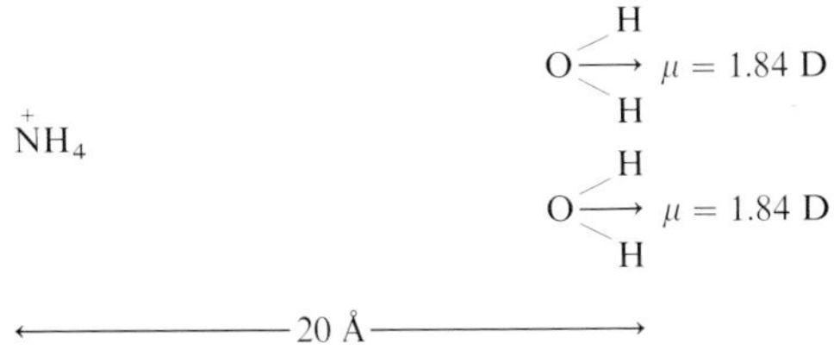

Fig. 5-29 Two water molecules interacting with a charged group; in this case the charge is a NH_4^+ group. The calculated electrical energy of interaction for the system is desired.

interest are charge–dipole between the $NH_4{}^+$ group and the water dipoles, charge-induced dipoles, and dipole–dipole repulsion between the two waters. Assuming the values

$$\mu_{H_2O} = 1.84 \quad D, \qquad \alpha_{H_2O} = 1.44 \times 10^{-24} \quad cm^3$$

the equation for the energy of electrical interaction is

$$U = -\frac{2(69.11)n\mu_{H_2O}}{r^2} - \frac{2(165.2)n^2\alpha_{H_2O}}{r^4} + \frac{14.3\mu_{H_2O}^2}{r^3} \qquad (5\text{-}42)$$

Substituting the above values, the total electrical energy is $U = -0.6$ kcal/ mole. This means the system is stable as described, and it will take energy to remove all components to infinity. It should also be remarked that the answer is an approximation since several assumptions were made in writing Eq. (5-42); e.g., the $NH_4{}^+$ charge is collinear with both dipoles.

SUMMARY

   Biological systems are constructed of countless numbers of molecule–molecule interactions and a great deal of these may be explained by simple electrical interactions. Common types of electrical interactions include charge–charge, charge–dipole, dipole–dipole, charge-induced dipole, and dipole-induced dipole. These interactions can be developed from the basic laws of electrostatics. Interactions among molecules can contain all or just a few of the different types. The strength of these interactions depends primarily on the asymmetry of the charge distribution on each molecule, the distance of separation, and the value for the dielectric constant. Generally speaking, interactions involving net charges are stronger than ones having just permanent dipoles, which are in turn stronger than those whose dipoles are induced. Two correction factors must be applied to electrical calculations. If the molecules are at an equilibrium distance apart, short-range electron repulsion will destabilize the interaction, as will the effect of thermal energy.

REFERENCES

Axelrod, J. (1974). *Scientific American* June, p. 61.
Birshtein, T. M., and Ptitsyn, O. B. (1969). *Biopolymers* 5, 785.
Marmur, J., and Doty, P. (1962). *J. Mol. Biol.* 5, 109.
Murray, J. M., and Weber, A. (1974). *Scientific American* Feb., p. 61.
Pauling, L. and Corey, R. (1956). *Arch. Biochem. Biophys.* 65, 164.
Pimental, G. C., and McClellan, A. L. (1960). "The Hydrogen Bond." Freeman, San Francisco, California.

Rich, A., and Davidson, N., eds. (1968). "Structural Chemistry and Molecular Biology." Freeman, San Francisco, California.

Settow, R. B., and Pollard, E. C. (1962). "Molecular Biophysics." Addison-Wesley, Reading, Massachusetts.

Springall, H. D. (1954). "The Structural Chemistry of Proteins." Academic Press, New York.

Steiner, R. F. (1965). "The Chemical Foundations of Molecular Biology." Van Nostrand Reinhold, Princeton, New Jersey.

Webb, J. L. (1963). "Enzyme and Metabolic Inhibitors," Vol. I. Academic Press, New York.

6

VAN DER WAALS FORCES

INTRODUCTION

In the previous chapter we considered the various ways in which molecules can interact with one another via different types of electrical forces. The various forces considered were shown to depend on the distance of separation and the type of charge distributions present in the molecules of interest, with the more asymmetric charge distributions giving rise to potentially larger forces. Molecules with a net electronic charge are capable of producing the most powerful interactions, while those with permanent dipoles, or structures in which a dipole could be induced, are capable of correspondingly weaker interactions. In all these previous situations there had to be some type of real or potential asymmetric charge distribution in order to cause a substantial force or electrical field. This chapter will consider interactions among molecules that have neither a net charge nor a permanent dipole moment. That this type of force exists has been realized for over a hundred years, but it has been only relatively recently that its basic nature has been verified experimentally and its significance in biological systems considered. This last type of electrical interaction that we shall consider is called a van der Waals (VDW) or London dispersion force, and it is yet another way in which matter holds itself together in an orderly array.

Van der Waals forces are among the least understood of all electrical interactions although their influence is quite widespread. They have one outstanding characteristic that makes them particularly significant; they are universally present, just like gravitational forces. In this respect van der Waals forces are quite different from covalent, ionic, or dipolar interactions. Although they are of prime importance with interactions among nonpolar molecules, polar molecules can also exhibit van der Waals forces. In strength van der Waals forces are generally considered weak; however, there are a number of common situations where they are the only appreciable force present. Van der Waals forces play a role in a myriad of physical, chemical, and biological phenomena: Surface tension, gas properties, stability of biological membranes, condensation properties, adhesion properties, friction, and enzyme–substrate recognition are just a few specific examples where VDW forces are important.

In most situations VDW forces are considered to be always attractive, and this is especially true with interactions among identical molecules. They can also be considered to have a long range, and the basic nature of their influence can undergo a fundamental change depending on the distance over which the interactions occur. This aspect is quite different from those forces that we studied in Chapter 5.

At this point the reader may get the feeling that VDW forces are really fundamentally different from the classical models of electrical interactions presented in the previous chapter, and this feeling is quite correct. Van der Waals forces cannot be fully explained in classical terms since a quantum-mechanical treatment is necessary for their understanding. However, classical or intuitive parallels can make them understandable and give insights into their mechanism of operation.

The concept of VDW forces came out of work investigating the nature of real gases. In this area of study it was realized that even for gases consisting of nonpolar molecules, there were forces responsible for holding the gas molecules together; i.e., gas molecules would show an attractive force for one another even when there was no apparent reason that they should. The Dutch scientist J. C. van der Waals postulated a quantitative equation for the behavior of real gases in which these forces were taken into consideration, although he did not really understand the basis for their action. In the 1930s F. London proposed the first successful model for the mechanism responsible, and further he derived a mathematical formalism describing the situation. London's treatment was the first explanation of VDW forces in terms of fundamental principles, and it stands as a landmark in the field. One drawback to London's approach however was that his theory was rather cumbersome to handle when applying it to macroscopic bodies. In the 1950s the Russian physicist C. M. Liftshitz and his co-workers proposed a completely different treatment in which they considered matter as a continuum and not as a collection of numerous individual atoms as London had done. Lifshitz's theory is very complicated, but it is significant in that only macroscopic information concerning the materials of interest is needed in order to ascertain the importance and influence of VDW forces in a particular situation. Even though it has been 40 years since London first proposed his original theory explaining the basic nature of these forces, it has been only rather recently that experiments have verified the basic correctness of his work. The importance of VDW forces are just beginning to be appreciated in biological systems, and in them lies the possible explanation of a great many biological processes that occur on a molecular level.

In this chapter we shall discuss van der Waals' work, consider the evolution of London's model and theory, briefly describe Lifshitz's theory, and consider several experiments that investigate the physical nature of VDW forces. Finally, we shall take a look at the chemical and biological importance of VDW forces and see that these forces are responsible for a wide variety of phenomena in biological systems.

REAL GASES

In 1873 van der Waals proposed an equation to describe the behavior of real gases in contrast to the actions of an idealized gas. Previously, the ideal gas law read

$$PV = nRT \qquad (6\text{-}1)$$

where P is the gas pressure, V is the gas volume, T is the absolute temperature, R is the gas constant ($R = 1.98$ cal/mole), and n is the number of moles of gas present. Equation (6-1) then describes the pressure, volume, and temperature relationships of an ideal gas. Although it is not explicitly seen, Eq. (6-1) is most applicable when the pressure is low or the amount of interaction between gas molecules is minimal. This has been discovered by experiment. In other words Eq. (6-1) is not the best detailed description for the behavior of gases found in the real world.

To account for the behavior of real gases, van der Waals modified Eq. (6-1) to read

$$\left(P + a\frac{n^2}{V^2} \right)(V - nb) = nRT \qquad (6\text{-}2)$$

where a and b are constants depending on the specific gas, and the other variables are as previously defined. As can be seen, van der Waals made essentially two corrections or additions to the classical theory. The term nb is called the excluded volume correction and represents the volume of space occupied by the gas molecules themselves. If one visualizes a gas as a group of billiard ball type molecules moving randomly within the confines of some container, then the volume of space any one ball can have access to is equal to the volume of the container minus the volume of the rest of the billiard balls. No two objects can occupy the same space at the same time, hence the volume accessible to the gas is less than just plain V. The constant b is the excluded volume per mole of gas, and it will differ for each gas depending on the size of individual molecules.

The other correction that van der Waals introduced is related to the attractive forces gas molecules exert on one another. The fact that these forces exist, even for nonpolar gases, was realized by the fact that they would condense to liquids or even solids if the temperature was lowered enough. Since gas molecules attract one another, the pressure a gas can exert on the outside world will be somewhat reduced; hence van der Waals added the an^2/V^2 term to compensate for this effect. The force helping to hold the gas together is going to be proportional to the number of molecules per unit volume or n/V; and since each molecule can attract and be attracted by its neighbor, the total force is proportional to $(n/V)^2$. The proportional factor is the constant a. When van der Waals proposed his modifications to the ideal gas law, the mechanism of the attraction between gas molecules was unknown; yet van der Waals' name has since been associated with those forces. It should be emphasized that for polar molecules, the different types of interactions discussed in Chapter 5 could account for an attractive force; however, in this chapter, van der Waals forces are considered to be those forces arising from an entirely different mechanism.

MECHANISM OF VDW FORCES

The nature of van der Waals forces between nonpolar molecules can be visualized by examining the following. Consider an interaction between a dipolar molecule and a molecule without a permanent dipole. Via a dipole-induced dipole interaction, this situation can result in an attraction between the two molecules. This type of interaction was originally proposed as the mechanism for VDW forces, but it failed to explain how two molecules without permanent dipole moments can attract one another. Also, the classical theory would predict that a dipole–dipole type of interaction would decrease in strength as the temperature increased, but early workers investigating gas phase interactions found that attractive forces did not vanish as fast as, say, Eq. (5-20) would predict. Another mechanism must then be at work. This mechanism has to do with instantaneous dipole moments.

Even if a molecule has no permanent dipole moment, it can have an *instantaneous* nonzero dipole moment. If a molecule is said to have a zero permanent dipole moment, this means that over the period of time needed to measure the dipole moment, the time average is zero. This does not mean that at any specific point in time there is no dipole moment $\mu(t)$, where the symbol $\mu(t)$ represents an instantaneous value of the molecule's dipole moment. If a molecule has no net permanent dipole, then the time average of $\mu(t)$ is zero. However, at any specific time, $\mu(t)$ may have a nonzero value; it is then possible for this *transient* dipole to interact with a neighboring molecule by inducing a momentary dipole in it, thus generating an attractive force. The force is attractive, remember, because dipoles are induced in such a fashion as to cause attractive forces. When the transient dipole in the first molecule changes as a function of time, the induced dipole will respond by having its value change also. In this manner two nonpolar molecules can then continuously attract one another if they are sufficiently close enough to feel the influence of the other.

The reader may gain some insight into the origin of these instantaneous or transient dipoles by remembering that electrons are in constant cyclic motion throughout the volume of a molecule thus changing the charge distribution and hence the molecule's instantaneous dipole moment. Because the magnitude of these transient dipoles is small compared to that of a permanent dipole, van der Waals forces are generally considered weak. However, as in the case of hydrogen bonds, a multitude of weak forces can add to give a very substantial effect.

A quantitative example illustrating the above model can now be shown.* Consider two hydrogen atoms, where each atom is composed of a central proton that is orbited by a lone electron. According to the model of the hydrogen atom proposed by the Danish physicist Niels Bohr, the electron can circulate around the proton only in well-defined circular orbits. The smallest orbit has a radius of r_0 which is about 5.3×10^{-9} cm (Fig. 6-1).

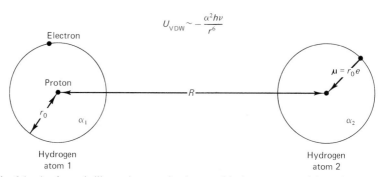

$$U_{VDW} \sim -\frac{\alpha^2 h\nu}{r^6}$$

Fig. 6-1   A schematic illustrating two closely spaced hydrogen atoms. A lone electron orbits the nucleus consisting of a single proton at a distance of r_0. At any time the magnitude of the instantaneous dipole of either atom is equal to r_0e where e is the electron charge.

If we consider the electron to be in this orbit of radius r_0, then the dipole moment formed by the proton–electron system is zero when averaged over one complete orbit of the electron, i.e., the time average of $\mu(t)$ is zero. Remember that $\mu(t)$ is a vector quantity and that the average of such a quantity must take into account the direction as well as the magnitude. At any one specific instant, however, the dipole is equal to the magnitude of

* First derived by David Tabor (Israelachvili, 1974)

the quantity

$$\mu_1(t) = r_0 e \tag{6-3}$$

which is nonzero. The symbol e stands for the electronic charge. The subscript one indicates the first hydrogen atom. The instantaneous electric field of this dipole can be found by using Eq. (3-9). Paying attention only to the distance of interaction dependence, we have

$$E_1(t) \sim \mu_1(t)/R^3 \tag{6-4}$$

If the second hydrogen atom has polarizability α_2 and is close enough, an induced dipole will be formed in it. The magnitude of this induced dipole is given by

$$\mu_2 = \alpha_2 E_1(t) \tag{6-5}$$

or

$$\mu_2 \sim \alpha_2 \mu_1(t)/R^3 \tag{6-6}$$

The energy of interaction U between the two dipole separated by a distance R in a vacuum is given by Eq. (5-15) and can be written

$$U(t) \sim \mu_1(t)\mu_2(t)/R^3 \tag{6-7}$$

Substituting into Eq. (6-7), the van der Waals energy of interaction is then

$$U(t) \sim -\alpha_2 \mu_1^2(t)/R^6 = -\alpha_2 r_0^2 e^2/R^6 \tag{6-8}$$

This equation then tells us that the energy of interaction has a $1/R^6$ dependence and that the force has a $1/R^7$ distance dependence. It is important to realize that the energy of interaction described in Eq. (6-8) is negative, indicating an attractive force. This is due to the manner in which $\mu_2(t)$ is induced by $\mu_1(t)$.

From Bohr's model of the atom it can be found that

$$r_0 \sim e^2/hv \tag{6-9}$$

where h is Planck's constant (6.6×10^{-27} erg sec) and v is the frequency of the electron in its orbit around the proton. Remembering that the polarizability α is proportional to the volume of the hydrogen atom

$$\alpha \sim r_0^3 \tag{6-10}$$

then Eq. (6-8) can be written as

$$U \sim -\alpha^2 hv/R^6 \tag{6-11}$$

The reason for manipulating Eq. (6-8) into the form of (6-11) is to facilitate its comparison to another equation to be seen presently.

In 1930 London used a quantum-mechanical argument to derive an equation for the energy of interaction involved in van der Waals forces. For two identical molecules, London's equation has the form

$$U = -\tfrac{3}{4}hv\alpha^2/r^6 \qquad (6\text{-}12)$$

where the symbols are similar to those defined previously. It is seen that the results of London's rigorous treatment is identical to our simply derived equation except for a constant numerical factor. It should be mentioned that the α^2 term arises due to the fact that molecule 1 can just as likely polarize molecule 2 as be polarized by molecule 2. In arriving at this result London did not use the model of two hydrogen atoms, but instead considered the case of two neutral harmonic oscillators. These oscillators can be thought of as having a positive and negative charge attached respectively to either end of a spring with the positively charged end being held stationary while the negative end is free to oscillate at frequency v. In their rest position the two charges of the oscillator are superimposed on one another; but when oscillating, there is a dipole moment formed due to the separation of the charges (Fig. 6-2). Since the springs are assumed to have three-dimensional freedom, the dipole moment averaged over all time is zero, as it should be for a symmetrical charge distribution. By examining the energy of interaction between two oscillators from a quantum-mechanical standpoint, London was able to see that there were forces involved beyond those that were described by classical physics. The reasons London used the harmonic oscillator model were that it was well understood from a quantum-mechanical viewpoint and that it was a fairly simple model which could represent the motion of electrons about a nucleus of positive charge.

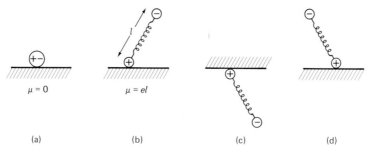

$\mu = 0$ $\mu = el$

(a)   (b)   (c)   (d)

Fig. 6-2   Schematic illustrating a harmonic oscillator. (a) The oscillator is in its equilibrium position with the positive and negative charges superimposed on one another. Here $\mu = 0$. In (b)–(d) the oscillator is extended, resulting in an instantaneous nonzero dipole moment. The negative charge has the ability to oscillate in three dimensions; therefore, the charge distribution over time is symmetrical and the time-averaged dipole moment is also zero.

The reader may well wonder, What is quantum mechanics, and why is it needed to describe VDW forces which have previously been described so nicely in terms of transient dipole moments? Without going into great detail, let it be said that quantum mechanics is a branch of theoretical physics that was first developed over the last years of the nineteenth century and the first part of the twentieth century. It is used almost exclusively for describing nuclear, atomic, and molecular systems; in many ways it is completely different from classical physics. In classical physics matter and energy are considered continuous, whereas in quantum mechanics the energy states of matter are considered discrete or quantized. Energy levels are well defined and are not continuous; i.e., energy levels can take on only certain values as opposed to any value. This is the main difference between classical and quantum theory. In his model of the hydrogen atom Bohr imposed quantum-mechanical restrictions by allowing electrons to circulate the nucleus only in certain descrete orbits. At times quantum mechanics predicts things that do not really have any counterpart in classical physics or in common physical intuition. It was one such concept that London used in deriving the mathematical description of VDW forces. The model of a transient dipole in one molecule inducing a dipole in another molecule in such a way as to have the two dipoles attract one another was London's effort to describe this phenomenon in terms of classical physics. It should also be mentioned that in the derivation of Eq. (6-11) quantum-mechanical assumptions were indirectly introduced through Eq. (6-9), so that Eq. (6-11) is not derived solely from principles of classical physics. Without further studying some simple elements of the Bohr atom and principles of quantum mechanics, the reader will just have to accept many statements presented here. It should also be noticed that Eq. (6-12) has no temperature dependence as does the case of unhindered dipole–dipole interaction derived in Chapter 5.

In the situation where two nonidentical molecules are interacting, London showed that Eq. (6-12) becomes

$$U = -\frac{3}{2}\frac{(h\nu_1)(h\nu_2)\alpha_1\alpha_2}{h(\nu_1 + \nu_2)r^6} \tag{6-13}$$

where the subscripts 1 and 2 refer to the respective molecules. The frequencies ν_1 and ν_2 are those frequencies of the oscillating electrical dipoles responsible for the formation of induced dipoles. Since the nuclei of atoms are relatively stable, these frequencies describe the motion of the mobile electrons since it is this motion that is responsible for the transient dipoles. For the most part these electrons are those that lie on the outermost parts of a molecule, for these electrons can be influenced by an external electric field more readily than can electrons that are buried in the interior of the molecule. The outer electrons tend to shield the inner electrons from external fields. Therefore, when considering induction forces of any kind, the outer or valence electrons are the most important.

London also realized that the quantity $h\nu$ has units of energy, and he equated this term to the ionization energy I of the molecule, where the ionization energy is the energy needed to pull one of the outer electrons off the molecule. This relationship was motivated by quantum-mechanical considerations in interpreting the physical meaning of the quantity $h\nu$. Equation (6-13) then becomes

$$U = -\frac{3}{2}\frac{I_1 I_2 \alpha_1 \alpha_2}{(I_1 + I_2)r^6} \qquad (6\text{-}14)$$

where the Is are measured in kilocalories per mole. The reason for equating $h\nu$ to I is an effort to simplify the situation somewhat since the ionization potential is an experimentally obtainable quantity whereas the factor $h\nu$ is harder to obtain.

Knowing that the ionization potential for a hydrogen atom is 13.6 ev, or 312.5 kcal/mole, and that $\alpha_H = 0.4 \times 10^{-24}$ cm^3, the reader should be able to verify that $U = -0.002$ kcal/mole for the VDW interaction between two hydrogen atoms separated by 5 Å in a vacuum. For a separation of 2 Å, $U = 0.6$ kcal/mole. Compared to charge–charge and charge–dipole interactions, this is a relatively weak energy of interaction; but this is a deceptive point as we shall see later.

In considering VDW forces the reader should realize that any molecule is not limited to a single fluctuating dipole, but is capable of having a multitude of transient dipoles, each of which can induce a dipole in another molecule. The resultant force is then a cumulative effect. The oscillating frequencies of the transient dipoles responsible for VDW forces depend, of course, on the electrons in the outer layers of their particular orbits; but from Bohr's simple model of the hydrogen atom, these frequencies can be estimated to be on the order of 10^{15} Hz. Induced dipoles then will also be oscillating at these frequencies. Two identical molecules will tend to experience a greater VDW force compared to a similar situation between two dissimilar molecules. This can be seen by realizing that a maximum force will result when the two transient dipoles can stay exactly in phase with one another, and this situation is easiest to obtain when both molecules have the same natural frequencies for the oscillating dipoles. If two oscillating dipoles are interacting with one another, the strength of the interaction depends on how well one can respond to the other. If the response is instantaneous, the interaction will tend to be maximized, whereas it will be progressively weaker if one dipole lags the other. If the two dipoles are completely out of phase, a minimum interaction occurs. This aspect of VDW forces tends to draw together similar molecules, as opposed to dissimilar ones, and helps stabilize macromolecules that are composed of identical subunits.

From a biochemical standpoint VDW forces are most important in those situations where the chemical components are nonpolar and are in close proximity to one another. In these cases VDW forces are quite significant and can play a major role in holding the system together. Where net charges or permanent dipoles are present, VDW forces are relatively small; here they play only a secondary role, but one that can still be important or crucial. An excellent example illustrating the importance of VDW forces in a biological system can be seen in the biological membrane and lipid–lipid interaction.

LIPIDS, FATTY ACIDS, AND MEMBRANES

By definition, lipids are organic compounds that are insoluble in aqueous systems, but which can be dissolved in such liquids as chloroform, ether, or benzene. There are several subclasses of lipids with each having a unique type of structure. In this section we shall be mainly concerned with those called fats and derivatives of fats. One of the basic units of a fat is called a fatty acid, where a fatty acid is an organic acid that has a long hydrocarbon tail that can be either saturated or unsaturated. Fatty acids are rarely found free in nature since they are usually combined with other compounds. They serve as both structural elements in biological membranes and as intracellular storage sites of metabolic energy. Several samples of fatty acids are given in Fig. 6-3. By looking at these examples it can be seen that fatty acids have a long nonpolar chain that is terminated by a polar COOH group, which can be charged or not depending on the environment. The

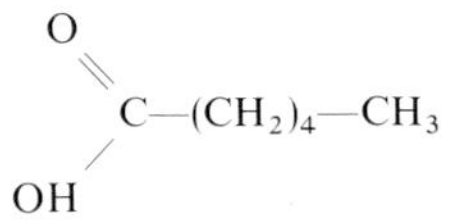

$$O$$
$$C-(CH_2)_4-CH_3$$
$$OH$$

Palmitic acid

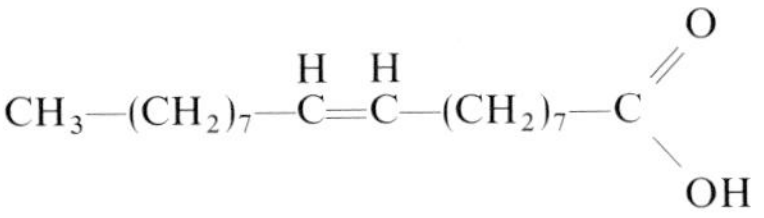

$$\qquad\qquad H \quad H$$
$$CH_3-(CH_2)_7-C=C-(CH_2)_7-C$$
$$\qquad\qquad\qquad\qquad\qquad\qquad O$$
$$\qquad\qquad\qquad\qquad\qquad\qquad OH$$

Oleic acid

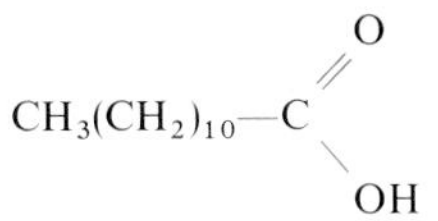

$$CH_3(CH_2)_{10}-C$$
$$\qquad\qquad O$$
$$\qquad\qquad OH$$

Dodecanoic acid

$$\qquad H \ H \quad H \ H \quad H \ H$$
$$CH_3-CH_2-C=CCH_2C=CCH_2C=C(CH_2)_7-C$$
$$\qquad\qquad\qquad\qquad\qquad\qquad\qquad\qquad\qquad O$$
$$\qquad\qquad\qquad\qquad\qquad\qquad\qquad\qquad\qquad OH$$

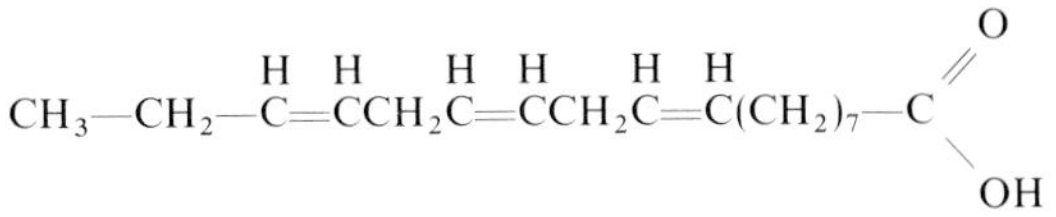

Linolenic acid

Fig. 6-3 Several examples of some fatty acids. The hydrocarbon portion can be either saturated or unsaturated.

TABLE 6-1

The Name, Structure, and Melting Point of Some Saturated and Unsaturated Fatty Acids[a]

Carbon	Structure	Name	Melting point (°C)
Saturated fatty acids			
11	$CH_3(CH_2)_9COOH$	Hendecanoic acid	28.6
12	$CH_3(CH_2)_{10}COOH$	Lauric acid	44
14	$CH_3(CH_2)_{12}COOH$	Myristic acid	58
15	$CH_3(CH_2)_{13}COOH$	Pentadecanoic acid	53
16	$CH_3(CH_2)_{14}COOH$	Palmitic acid	63
18	$CH_3(CH_2)_{16}COOH$	Stearic acid	71.5
19	$CH_3(CH_2)_{17}COOH$	Nonadecanoic acid	69.4
20	$CH_3(CH_2)_{18}COOH$	Arachidic acid	77
24	$CH_3(CH_2)_{22}COOH$	Lignoceric acid	86.0
Unsaturated fatty acids			
16	$CH_3(CH_2)_5\overset{H}{C}{=}\overset{H}{C}(CH_2)_7COOH$	Palmitoleic acid	−0.5
18	$CH_3(CH_2)_7\overset{H}{C}{=}\overset{H}{C}(CH_2)_7COOH$	Oleic acid	13.4
18	$CH_3(CH_2)_7\overset{H}{C}{=}\overset{H}{C}(CH_2)_7COOH$ (trans)	Elaidic acid	45
18	$CH_3(CH_2)_4\overset{H}{C}{=}\overset{H}{C}CH_2\overset{H}{C}{=}\overset{H}{C}(CH_2)_7COOH$	Linoleic acid	−5
18	$CH_3(CH_2)_5\overset{H}{C}{=}\overset{H}{C}(CH_2)_9COOH$ (trans)	Vaccenic acid	44
18	$CH_3(CH_2)_5CHOHCH_2CH{=}CH(CH_2)_7COOH$	Recinoleic acid	77

[a] Unless otherwise indicated, all double bonds are in the cis configuration. The cis configuration introduces a kink in the hydrocarbon's structure thus keeping the separate hydrocarbons chains apart. The trans structure of double bonds is similar to a straight-chained hydrocarbon with only single bonds. Data from "Handbook of Chemistry and Physics," 54th Ed.

polar end part of a fatty acid is commonly called the head group and the hydrocarbon portion is called the tail; each part has distinctly different electrical properties. The tail can interact with other molecules only through VDW type forces, whereas the head with its ionizable group can interact via a number of different mechanisms. Thus, fatty acids show a diversity among themselves in chemical structure and in electrical properties.

In pure form the fatty acids behave very much like the hydrocarbon tail of the specific compound. Saturated fatty acids with chains containing 10 or more carbons are solid at room temperature, and the melting temperature of these solids increases as the hydrocarbon chain length increases (see Table 6-1). This phenomenon can be understood by realizing that as the hydrocarbon chain length increases, the total VDW force between separate molecules increases also; thus the molecules become more closely associated with one another, which is a property of the solid state. If the hydrocarbon chain is short, the total VDW forces are relatively small and less molecular association is seen, hence a liquid or gas state is formed. As a simple example of this phenomenon consider that the short-chained hydrocarbons like methane, ethane, propane, and butane are gases at room temperature, but that longer chained hydrocarbons like those found in kerosene or gasoline are liquid. This is again due to the VDW forces of attraction between neighboring molecules. As the carbon chain length increases, the total VDW forces increase, and a more condensed phase of matter is formed. For the short-chained hydrocarbons, the VDW forces are weak and easily overcome by random thermal motion; hence, each molecule is more independent, which is characteristic of a gas phase.

Let us now take a more detailed look at the VDW interaction between the hydrocarbon tails of two fatty acids. As a model, consider Fig. 6-4 in which segments of two separate hydrocarbon chains are in close proximity. From London's treatment we know that the VDW energy of interaction between two separate atoms or molecules is equal to

$$U = -A/r^6 \qquad (6\text{-}15)$$

where A is a constant depending on the nature of the interacting groups. We now wish to find an expression for the total energy of interaction between the two chains shown in Fig. 6-4 as a function of their length L and their distance of separation R. To do this, it is assumed that the total energy of interaction can be found by summing all pairwise interactions between the two separate chains. This assumption is not strictly correct, but it will be good enough for our purposes. What will be done then is to calculate the VDW energy between the CH_2 group labeled 1' and the one marked 1, between 1' and 2, between 1' and 3, etc. for all possible pairs, and then assume that the total energy is additive. Next, interactions between the CH_2 group

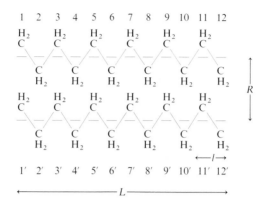

Fig. 6-4 Schematic illustrating two closely spaced saturated hydrocarbon chains. The carbon atoms in the backbone are staggered to represent the tetrahedral structure of their bonds. The axis of each chain is separated by a distance R. The length of the chain is L, and each unit has length l.

marked 2′ and all those of the other chain are also calculated and summed, being careful not to duplicate previous calculations. When the interaction between all possible pairs is summed, the total interacting energy will result. To do this, a specific structural form is needed for each hydrocarbon chain so that the position of each CH_2 group is known in detail. A pictorial representation of this model is shown in Fig. 6-4. It will be immediately noticed that each hydrocarbon backbone is not straight but zigzagged. This is due to the particular directionability of the covalent bonds associated with carbon. More will be said about this in Chaper 8.

With the above model, Salem (1962) derived an expression for the total VDW energy of interaction between the two hydrocarbon chains. It was found that

$$U_{tot} = -\frac{AL}{4l^2 R^5}\left[3\tan^{-1}\left(\frac{L}{R}\right) + \frac{L/R}{1+(L/R)^2}\right] \tag{6-16}$$

which reduces to

$$U_{tot} = -\frac{3\pi A N}{8lR^5} \tag{6-17}$$

if the distance of separation R is much smaller than the total length L. In Eqs. (6-16) and (6-17), N is the number of CH_2 groups in the hydrocarbon backbone, and is equal to L/l. Equation (6-17) shows that the total VDW energy of interaction between two hydrocarbon chains is directly proportional to the number of carbon atoms present, which verifies earlier predictions that as the hydrocarbon chain length increases, the greater is the association between molecules and the more condensed is the state. The other significant fact coming from Eq. (6-17) is that the energy is extremely sensitive to the distance of separation R. As Salem pointed out, two chains initially 5 Å apart will have their VDW energy increased by a factor of three if they move only 1 Å closer together. Salem also calculated from Eq. (6-17) that the VDW energy per CH_2 group was of the order 0.4 kcal/mole when $R = 5$ Å. It is easy to see then that the VDW interaction can be quite significant, being on the order of 5–20 kcal/mole for long chain hydrocarbons. This is further emphasized by calculations that have been made that describe the total energy of interaction between the hydrocarbon tails of neighboring fatty acid molecules in a monolayer.* Based on experimentally arrived at values of R, Salem calculated energies for interactions involving stearic acid, hexatriacontanoic acid, and isostearic acid, respectively. The results are shown in Table 6-2. In the monolayers it is presumed that the individual molecules are all aligned with the long axis of the molecules roughly parallel.

TABLE 6-2

Calculated VDW Energy of Interaction between Individual Molecules in a Monolayer Composed of Identical Molecules[a]

Name	Structure	R (Å)	VDW energy (kcal/mole)
Stearic acid	$CH_3-(CH_2)_{16}COOH$	4.8	-8.4
Hexatriacontanoic acid	$CH_3(CH_2)_{34}COOH$	4.8	-16.8
Isostearic acid	$CH_3-CH-(CH_2)_{14}COOH$ $\quad\quad\; \mid$ $\quad\quad CH_3$	6.0	-2.8

[a] Reproduced by permission of the National Research Council of Canada from L. Salem, Can. J. Biochem. Physiol. **40**, 1287–1298 (1962).

It is interesting to note the difference in energies between the cases of stearic and isostearic acid. The individual molecules of isostearic acid cannot approach one another as closely as can those of stearic acid because of the side chain methyl group; hence, because of the strong distance dependence, molecules of isostearic acid in a monolayer attract one another three times less strongly than do stearic acid molecules in a similar situation even though both have the same number of carbon atoms.

Another example illustrating the sensitivity of distance separation in VDW interactions is seen by considering cis and trans double bond formation in fatty acids. To understand what a cis or trans double bond is, refer to Fig. 6-5. Both compounds shown make up 2-butene, but the geometric

* The monolayer referred to here is essentially an array of hydrocarbons one layer thick, whose axes are all parallel to one another, e.g., matches in a match box.

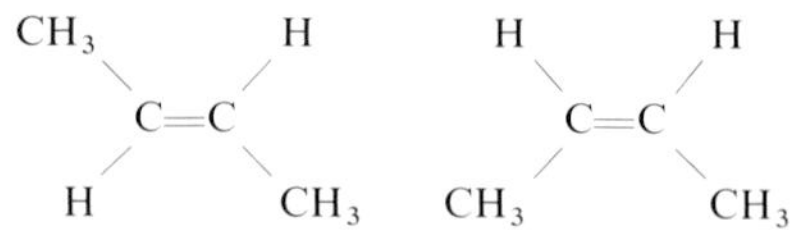

$$CH_3 \quad H$$
$$C=C$$
$$H \quad CH_3$$

$$H \quad H$$
$$C=C$$
$$CH_3 \quad CH_3$$

Trans-2-butene   Cis-2-butene

Fig. 6-5   Illustrations of cis and trans structures for the same chemical compounds participating in a double bond. Cis and trans structures are essentially stereoisomers, and one structure cannot be converted to the other without the breaking of bonds.

arrangement of the methyl groups is different in each case. These two possible arrangements of the same molecule are called geometric or stereoisomers. Cis-2-butene cannot be coverted to trans-2-butene except by breaking and reforming bonds, much as a right hand cannot be made into a left hand without pulling some fingers off one, or both, hands. Cis and trans then refer to the geometric arrangement of a double bond. Two groups are cis at a double bond if they are on the same side and they are trans if on opposite sides. The importance of configuration in a double bond is that it can easily affect the physical properties of a substance. Consider Fig. 6-6 where we have three fatty acids with equal numbers of carbons, except they all have different configurations. Because the cis configuration creates a kink in the hydrocarbon's backbone, whereas a saturated hydrocarbon or a trans configuration has little or no disfigurement, it is harder for unsaturated cis compounds to approach one another. Since Eq. (6-17) predicts a dramatic decrease in energy of interaction between two chains as R increases, it is expected that unsaturated fatty acids in the cis form should have lower melting temperatures than the corresponding saturated or trans structures. The higher the melting temperature, the stronger is the association among the molecules comprising the material. Table 6-3 shows the melting temperature for oleic acid, stearic acid, elaidic acid, and linoleic acid, all with

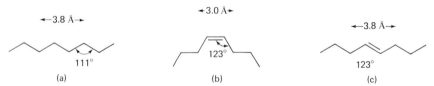

(a)   (b)   (c)

Fig. 6-6   Schematic diagram of three hydrocarbon structures with each having an equal number of carbon atoms. (a) A normal saturated hydrocarbon. (b) A cis double bonded structure. (c) A trans double bonded structure. Notice that the cis structure exhibits a bulge, preventing close approach of several cis structures if they are to be aligned parallel to one another. In practice with common chemical convention, each corner in the drawings represents a carbon atom with the correct number of associated hydrogen atoms.

TABLE 6-3a

Comparison of Melting Points for Fatty Acids Having the Same Number of Carbons but with Different Double Bonded Configuration

Name	Melting temperature (°C)	Configuration	Molecular weight (D)
Oleic acid	16.3	$CH_3(CH_2)_7CH{\underset{cis}{=\!=}}CH(CH_2)_7COOH$	282.5
Elaidic acid	45	$CH_3(CH_2)_7CH{\underset{trans}{=\!=}}CH(CH_2)_7COOH$	282.5
Stearic acid	71.5	$CH_3(CH_2)_{16}\underset{saturated}{COOH}$	284.5
Linoleic acid	−5	$CH_3(CH_2)_4CH{=\!=}CHCH_2CH{\underset{cis}{=\!=}}CH(CH_2)_7COOH$	280.5
Linolenic	−11.3	$CH_3(CH_2CH{\underset{cis}{=\!=}}CH)_3(CH_2)_7COOH$	278.4

TABLE 6-3b

Comparison of the Melting Points for Several Pairs of Fatty Acids, Identical Except for the Cis or Trans Configuration of Their Double Bonds[a]

Name	Melting temperature (°C)	Configuration	Molecular weight (D)
α-Elastearic	49	$CH_3(CH_2)_3[CH\!\!=\!\!CH]_3(CH_2)_7COOH$ cis	278.4
β Elastearic	71	$CH_3(CH_2)_3[CH\!\!=\!\!CH]_3(CH_2)_7COOH$ trans	278.4
Oleic	16.3	$CH_3(CH_2)_7CH\!\!=\!\!CH(CH_2)_7COOH$ cis	282.5
Elaidic	45	$CH_3(CH_2)_7CH\!\!=\!\!CH(CH_2)_7COOH$ trans	282.5
Erucic (cis)	33	$CH_3(CH_2)_7CH\!\!=\!\!CH(CH_2)_{11}COOH$ cis	338.6
Erucic (trans)	61.5	$CH_3(CH_2)_7CH\!\!=\!\!CH(CH_2)_{11}COOH$ trans	338.6

[a] Data from "Handbook of Chemistry and Physics," 54th Ed.

18 carbons. The table shows quite clearly that as the number of cis double bonds increases, the melting temperature drops, and that even a trans double bond arrangement has a lower melting point than a corresponding saturated fatty acid. It should also be mentioned that of the naturally occurring unsaturated fatty acids, almost all have their double bonds in the cis configuration.

MICELLES

Fatty acids as such are almost insoluble in water; however, their K^+ or Na^+ salts (a soap) are soluble in aqueous solutions although the solvated molecules form a special type of structure known as a micelle. Schematic representations of a spherical and an ellipsoidal micelle are shown in Fig. 6-7. The head group of the individual fatty acids is represented by a dot and the hydrocarbon tail is indicated by a zag line. As can be seen, the micelle is formed in such a way that the hydrocarbon portion of the fatty acids are pretty much excluded from contact with the aqueous surroundings. The reason this particular structure exists is due to the fact that the system has a minimum potential energy, and hence it is most stable this way. One contribution to the stability of the micelle is from the VDW forces existing between hydrocarbon chains within the micelle. A lower potential energy can be achieved by this structure than by having the hydrocarbon chains interact

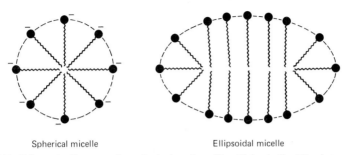

Spherical micelle   Ellipsoidal micelle

Fig. 6-7 Schematic diagram of a spherical and a ellipsoidal micelle. The dots represent
the charged head groups of fatty acids and the zigzag lines represent the hydrocarbon tail. In
fact the number of individual fatty acids needed to make a micelle is much larger than those
shown. In this structure the hydrocarbon portion of the fatty is not exposed to the surrounding
aqueous media.

with the aqueous solvent. Because of their location, the charge head groups
of the micelle can interact via charge–dipole or charge-induced dipole type
interactions with the polar water molecules surrounding the micelle. Since
each head group of the fatty acids comprising the micelle supports a negative
charge, one would expect that repulsive coulombic forces would tend to
destabilize the structure and cause its destruction. On the other hand, the
VDW forces between the hydrocarbon tails are attractive and tend to help
keep the structure intact. They are not the *sole* reason for the micelle's
stability, but without them the micelle would have a difficult time.* Because
the VDW forces are weak, it is required that each micelle have a certain
minimum number of fatty acids for the total attractive forces to be strong
enough to stabilize the structure. Two or three fatty acids cannot form a
micelle, it takes on the order of a hundred to a thousand. The micelle then
is a stable molecular structure that is partially stabilized by VDW forces.
But what does this have to do with biological structure? The answer lies
in the membrane.

* In discussing the situation of how micelles are formed and stabilized, there are several
approaches one can consider. In order to get a complete picture of the overall situation however, it is necessary to use a thermodynamic treatment in which the electrical interactions
considered here are just a partial description. The more complex description also takes into
account the concept of entropy. This concept essentially states that some physicochemical
reactions take place because there is a natural tendency for systems to move toward a more
random state, away from order. This tendency also contributes to the stability of micelles
although in a rather indirect way concerning the structure of water, which should not be
obvious to the reader. The emphasis of this chapter is to show that VDW forces can also contribute to the stability of biological systems.

BIOLOGICAL MEMBRANES

Membranes are thin structures composed of lipids and proteins that surround a cell and separate it from the external world. They prevent the contents within from mixing with the surroundings, and also help regulate the flow of selected metabolites into and out of the cell. In the case of cells comprising more complex organisms (eukaryotic cells), many subcellular structures are surrounded by membranes to compartmentalize the cell, much as walls separate the various functional rooms of a house. The nucleus, mitochondria, Golgi apparatus, and lysosomes are typical subcellular structures that have their own membranes. Bacterial membranes that surround the cell are themselves usually surrounded by the cell wall. Membrane structure and functions vary in detail from one source to another, although their general purposes are similar. In recent years there has been much research done in order to determine the structure and working mechanisms of these ubiquitous structures. One concept that has emerged from this work is that VDW forces play a role in holding membranes together, and for this reason it is worthwhile to consider membranes as an example of how VDW forces influence biological structure. In order to do this we must consider another type of lipid, the phosphoglyceride.

Phosphoglycerides are lipids that are found in the structure of biological membranes and whose general structure can be schematically represented as in Fig. 6-8. R_1 and R_2 represent hydrocarbon chains, where phosphoglycerides commonly have one saturated and one unsaturated hydrocarbon. The X component usually has a net charge that can be either positive or negative. Typical X components are shown in Table 6-4. A phosphoglyceride then is analogous to a fatty acid in structures with both a head and a tail, although to be strictly correct the phosphoglycerides have a double tail represented by the two separate fatty acids. In aqueous solution the phosphoglycerides also form micelles, although of a different nature than those shown in Fig. 6-7.

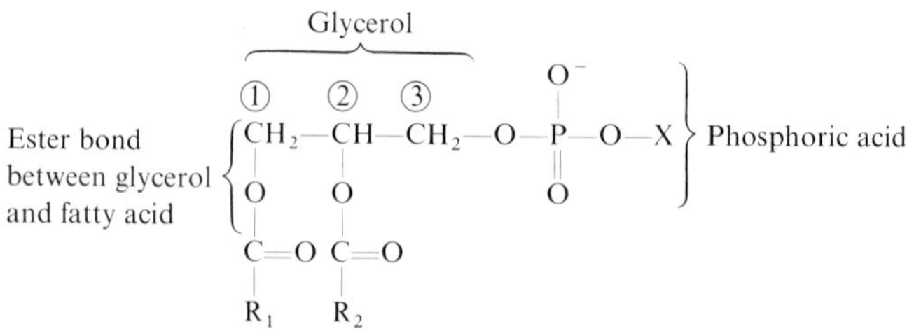

Fig. 6-8   The general structure of a phosphoglyceride. R_1 and R_2 represent hydrocarbon chains that serve to distinguish one phosphoglyceride from another.

TABLE 6-4

**Typical Substitutions on the Phosphate Group in a
Phosphoglyceride**

X Component	Name of phosphoglyceride
—H	Phosphatidic acid
—$CH_2CH(NH_3^+)COO^-$	Phosphatidylserine
—$C_6H_6(OH)_5$	Phosphatidylinositol
—$CH_2CH_2NH_3^+$	Phosphatidylethanolamine
—$CH_2CH_2N(CH_3)_3^+$	Phosphatidylcholine

When placed in an aqueous solution, phosphoglycerides tend to form what have come to be known as lipid bilayers. This structure is shown in Fig. 6-9. In the lipid bilayer, as in the spherical micelle, the hydrocarbon portion of the individual molecules extend toward the interior of the structure, while the polar head groups comprise the perimeter and point toward either the outside or the inside aqueous environment of the cell. By forming a completely closed loop, this lipid bilayer can isolate an internal aqueous environment from an exterior one. Real biological membranes, however, are composed not only of lipids, but contain proteins as well. The exact structural relationship between lipid and protein in membranes is not known, although there is evidence that there is a diversity of proteins associated with different membranes. Several proposed models of membranes relating protein to lipid are shown in Fig. 6-10.

By looking at these proposed structures, the reader can see that a membrane is a special form of a micelle and that VDW forces between the closely spaced hydrocarbon chains will help stabilize the structure. The importance of VDW forces is also emphasized by the lack of evidence for any covalent bonds between adjacent lipid and protein molecules. The

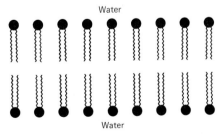

Fig. 6-9 A schematic of the liped bilayer formed when phosphoglycerides are placed in water. Again, as with the micelle, the hydrocarbon chains are buried in the interior of the structure with the polar head group located on the outside in contact with water.

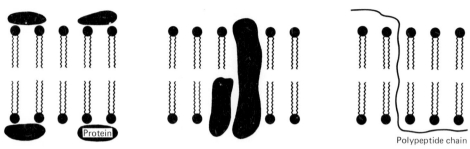

Fig. 6-10 Three proposed models of the physical relationship between proteins and lipids in a biological membrane.

whole structure seems to be stabilized by VDW and polar forces. Exactly how the protein molecules interact or affect the basic lipid–lipid interactions is not known in detail, although it is thought that the proteins found associated with the interior of the membrane have a high percentage of the more hydrophobic amino acids. It is also interesting to note that there is evidence indicating that the hydrocarbon interior of the membrane behaves much as a liquid hydrocarbon does and that the proteins are somewhat free to move in the plane of the membrane. If covalent or other very strong forces helped stabilize membranes instead of VDW forces, the membranes would not have these properties and would probably be static rather than dynamic in nature. Another factor adding to the fluid nature of the membrane's inner structure is the prevalence of cis configuration double bonds in naturally occurring fatty acids. Because the cis configuration fatty acids have a lower melting point than the corresponding trans fatty acids, they tend to remain liquid. If trans fatty acids populated membranes, then the membrane's dynamic nature might change. Whatever the detailed structure of the membrane is, it should be clear that VDW forces are strong enough to contribute stability to their structure and help give membranes their unique characteristics.

ENZYME–SUBSTRATE INTERACTIONS

As was seen in the previous section, VDW forces play an important role in conserving the structure of membranes; however, this is not the only biological situation where VDW forces are important. In this and the following section two more examples will be briefly discussed to illustrate the universal importance of VDW forces in molecular biology.

In our earlier discussion of the complexing of the enzyme acetylcholinesterase with its substrate, it was indicated that a charge–charge interaction was responsible for most of the binding strength between the two molecules. However, it has also been recognized that VDW forces are important in the binding of acetylcholine to the enzyme. This has been determined by interacting acetylcholinesterase with several different chemical compounds that are similar in structure to acetylcholine. In many cases these compounds are inhibitors of normal enzymatic activity because of competition with acetylcholine for space on the active site of the enzyme. If the structure of the inhibitor is only slightly different than the normal substrate acetylcholine, then it may be inferred that this structural difference is important in producing a different interaction with the enzyme itself. By reacting compounds of the form $N(CH_3)_4^+$ with acetylcholinesterase, it was found experimentally that by replacing the various methyl groups by hydrogen atoms, the binding energy between substrate and enzyme could be affected (Wilson, 1952). If one such methyl group were replaced by an H, little difference was noted; but if two or more were replaced, then the binding energy would decrease. Also, it was found that $N(C_2H_5)_4^+$ showed increased binding strength over $N(CH_3)_4^+$, and that cations such as K^+, NH_4^+, and Na^+ did not bind as well to the enzyme. What these results indicate is that the nonpolar methyl groups can interact through VDW forces with a similar nonpolar group in the active site, thus increasing the binding energy. As the methyl groups are replaced by H or ethyl groups, the VDW interactions decrease or increase, respectively. For the acetylcholinesterase system, then, it appears that the major factor responsible for aligning the enzyme and substrate is a charge–charge interaction, although VDW forces are important also. One may speculate that the charge–charge interaction serves only to roughly position the two molecules, whereas the VDW forces are important in the fine adjustment of the spatial relationship necessary for a perfect fit between enzyme and substrate.

Another enzyme–substrate example where VDW forces have been shown to play a role in binding is the case of carboxypeptidase (Smith *et al.*, 1951). Carboxypeptidase catalyzes the hydrolysis of the peptide bond adjacent to the free carboxyl group in proteins and peptides:

$$R-CONH-CHR'COOH + H_2O \xrightarrow{\text{carboxypeptidase}} R-COOH + H_2NCHR'COOH$$

The binding strength and rate of reaction depends on the nature of the R' group, and VDW interactions play a role in the total binding. This was demonstrated by measuring the difference in binding energies of several inhibitors. These studies showed that a hydrocarbon chain of sufficient length has to be attached to the terminal COO^- group for proper binding to take place. Acetate by itself (CH_3COO^-) is not tightly bound. Compounds in the series $CH_3-CH_2-COO^-$, $CH_3CH_2CH_2COO^-$, and $CH_3CH_2CH_2CH_2COO^-$ are bound to the enzyme in increasing strength; since only the hydrocarbon portion of the series is being lengthened, VDW forces must be involved with a hydrocarbon portion of the active site of the enzyme. This effect is also seen in the compounds indoleacetic, indolepropionic, and indolebutyric acid where the inhibition decreases as the ring structure gets further away from the carboxyl group. These results do indicate that VDW forces are involved in enzyme–substrate binding, but it should also be stated that the carboxyl group of the substrate is essential, and compounds without it do not bind. This implies a charge–charge interaction.

DNA

Another example where VDW forces are of importance in maintaining structural integrity of a biological molecule is in the case of DNA. As previously discussed, double-stranded DNA has two separate chains oriented in opposite directions with each forming the shape of a helix. The individual bases of each chain are on the inside of the structure and they interact with one another in a specific manner through hydrogen bonds. When this structure was first proposed by Watson and Crick, it was thought that the stability of the double helix relied mainly on the strength of the hydrogen bonds between adjacent bases on opposite chains. That is to say, the main reason that the helix did not unwind and form two separate random coils in the solvent was due to the hydrogen bond interactions between bases. Subsequently, experimental evidence indicated that this interpretation was not entirely correct; also, theoretical calculations showed that electrostatic and VDW forces between bases contributed significantly to helix stability. One set of calculations of this nature was made in 1962 by H. DeVoe and I. Tinoco.

What DeVoe and Tinoco did was to evaluate, based on theoretical grounds, the contributions from base–base electrostatic and VDW interactions to the potential energy of the helix. By calculating the dipole moments of each base, the strength of induced dipoles in each base, and the strength of VDW interactions between bases, they were able to calculate the potential energy of the helical system due to base–base electrical interactions. As always, the potential energy was calculated with respect to the reference state where all charges are at infinity, or in this case where all bases are at infinity. Table 6-5 shows the magnitude of the different interactions for bases in the same base pair. A significant feature of these numbers is that the dipole–dipole interaction is the strongest as one would expect, but it is repulsive for the AT base pair. Also, the VDW forces are, in general, the second strongest force present, and they are the strongest attractive force for the AT base pair. An even more enlightening set of calculations is shown in Table 6-6. Here, DeVoe and Tinoco have calculated the interactional energy between

TABLE 6-5

Magnitude of Interactions between the Base Pairs in DNA[a]

Base pair	Dipole–dipole	Dipole-induced dipole	VDW	Total
GC	−3.1	−0.3	−0.5	−3.9
AT	0.8	−0.1	−0.5	+0.2

[a] Units are in kcal/2 moles of base. From DeVoe and Tinoco (1962).

TABLE 6-6

Energy of One-Half the Sum of the Four Interactions between Nonpaired Bases for Two Adjacent Base Pairs in DNA[a]

Adjacent base pair	Dipole–dipole	Dipole-induced dipole	VDW	Base pair energy	Total energy
↑GC |CG↓	−5.8	−4.1	−6.0	−3.9	−19.8
GC GC	2.1	−3.3	−6.8	−3.9	−11.9
TA CG	−0.5	−2.0	−6.8	−1.9	−11.2
AT CG	−0.9	−1.4	−3.6	−1.9	−7.8
AT GC	3.3	−2.0	−6.8	−1.9	−7.4
TA GC	3.1	−2.4	−6.0	−1.9	−7.2
GC CG	4.2	−2.4	−3.6	−3.9	−5.7
TA AT	1.3	−0.7	−6.0	+0.2	−5.2
AT AT	2.2	−0.6	−6.8	+0.2	−5.0
AT TA	2.2	−0.4	−3.6	+0.2	−1.6

[a] The arrows indicate the direction of the DNA chain from the 3′ carbon to the 5′ carbon. Units are in kcal/2 moles of base (DeVoe and Tinoco, 1962). The total column gives the sum of energies shown plus the average contribution per two moles of base from the appropriate base pair energies for AT and GC given in Table 6-5.

nonpaired bases for 10 possible combinations of base sequences. What these numbers show is that the total potential energy due to nonhydrogen bonding interactions can be as high as 20 kcal/2 moles of base. Another interesting fact to come from these calculations is that the contributions to stability are dependent on base sequence. Also, the relative contributions due to VDW forces is, in general, the largest of those shown. Although DeVoe and Tinoco's calculations have been criticized, they probably do illustrate the general situation here. The point to be made is that VDW forces are responsible for a significant portion of the stability of double helical DNA, although at first inspection this does not appear to be the case. This illustrates yet another aspect of VDW forces; they are subtle.

EXTENSION OF THEORY

Although the original London theory of VDW forces laid the foundation for subsequent work in this area for many years, it did not always agree exactly with experimental data, and it also failed to adequately mention the important aspect of retarded forces. In general, London's treatment yielded values lower than experimental results, and adaptations of his work introduced several new additions to the theory in an effort to obtain better agreement. One other disadvantage was that London's theory was very clumsy to apply in determining the total VDW force operating between large bodies. The theoretical development following London's work was mainly directed toward these problem areas. In the 1950s there was introduced a completely new theory which overcame many of the limitations of London's concepts; although very complicated, this new theory may prove to be a much more inclusive description of VDW interactions.

The following two sections will deal with both an extension and an application of London's theory.

Retarded Forces

Until now the explanation for the mechanism of VDW forces has been in terms of a transient dipole of one molecule inducing a dipole in another molecule, thus creating an attractive force between the two particles. Using a model similar to this, London derived a formula to describe the distance dependence of VDW forces. In this calculation, however, it was implicitly assumed that both dipoles were oscillating in phase, i.e., both were oscillating in exactly the same fashion. By oscillating in phase the maximum force of interaction would be experienced because of the nature of the way in which dipoles are induced. The only way that this situation can develop in reality is if the electric field of one dipole can be transmitted instantaneously to the other. For molecules that are relatively close to one another, transmitted fields are detected almost instantaneously since the electric field is propagated at the velocity of light. On the other hand, if two molecules are relatively far apart, the short but finite time it takes for the field from the first dipole to traverse the required distance to reach the second molecule can be enough to cause the induced dipole to lag the first in phase, thus creating a less than maximum possible interaction. In this case the two dipoles are no longer aligned relative to one another to give a maximum interaction.

To get a clearer idea of the above principle, consider the following situation. Dipole I and II are close together, and dipole II is induced by I. They are both oscillating in phase with one another. If dipole I suddenly speeds up in its rate of oscillation, dipole II can respond quickly because the new electric field from dipole I reaches II almost instantaneously. The two dipoles are still in phase. Now consider what would happen if the two dipoles are separated by a large distance. When dipole I changes to a new frequency, dipole II will lag in its response by an interval equal to the time it takes for the electric field to traverse the distance separating the two dipoles; hence the second dipole will be out of phase with the first from the very start (Fig. 6-11). While the original dipole (I) may rotate $90°$ in a certain time, if the induced dipole (II) does not respond instantaneously to I's movement, dipole II may only rotate through an angle θ. Check Eq. (5-15) to see how the energy of interaction varies depending on whether or not the two dipoles are exactly aligned or are at an angle with one another. In the case illustrated here, dipole II lags I by an angle $90° - \theta°$. If the dipoles are out of phase with one another, the maximum force of interaction is not attained. What this implies then is that as the distance of separation between dipoles increases, the

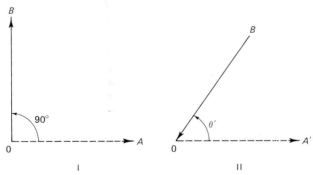

Fig. 6-11 While the original dipole (I) may rotate $90°$ in a certain time, if the induced dipole (II) does not respond instantaneously to dipole I's movement, dipole II may only rotate through an angle θ. Check Eq. (5-15) to see how the energy of interaction varies depending on whether or not the two dipoles are exactly aligned or are at an angle with respect to one another. In this case, dipole II lays dipole I by an angle $90° - \theta$.

VDW attractive force decreases. To put it another way, as the distance between dipoles increases, the distance dependence of VDW forces changes from its normal $1/r^7$ dependence. The exponent will become larger as separation distance increases. Under these conditions VDW forces are said to be retarded.

In 1948 H. B. G. Casimir and D. Polder advanced the principle of retarded forces and showed that retarded VDW forces exhibited a $1/r^8$ distance dependence as opposed to the $1/r^7$ dependence shown by London. Thus, with increasing separation distance, VDW forces fall off more rapidly with distance. To estimate the distance at which one would expect to see a transition from nonretarded to retarded forces, consider again our model of the H atom with one electron orbiting a proton. For this situation, it takes about 3×10^{-16} sec for the electron to complete one orbit. In this time an electric field can travel

$$(3 \times 10^{-16} \quad \text{sec})(3 \times 10^8 \quad \text{m/sec}) = 10^{-7} \quad \text{m}$$

or about 100 nm. So, if two dipoles, or molecules, are separated by more than about 50 nm, the VDW forces should be completely retarded. At this distance the first transient dipole can make a complete oscillation by the time its electric field travels to the induced dipole and returns. The transition to a retarded situation should start somewhere below this value of 50-nm distance. Experimental evidence illustrating that this effect really exists will be shown a little later in the chapter.

VDW Forces between Large Objects

Although London first considered interactions between two neutral atoms where the force varied as $1/r^7$, subsequent work showed that the VDW interaction between larger objects could have a less dramatic dependence on distance of separation. This implies that large objects can generate forces that are quite substantial. The explanation for this effect lies in the assumption that VDW forces are additive. That is to say, all atoms in one body experience a VDW attraction by all atoms in another body, and vice versa. Consider the situation depicted in Fig. 6-12, where it is required to calculate the total VDW interaction between two spheres of unequal size. Assuming additivity, one way to obtain the desired result is to find the VDW force that exists between two arbitrary infinitesimal volumes, one in each sphere, and then to sum the contributions between all such pairwise infinitesimal volumes in the two spheres. Mathematically, the energy of interaction can be represented by

$$U = - \int_{v_1} dv_1 \int dv_2 \frac{\rho_1 \rho_2 K}{R^6} \tag{6-18}$$

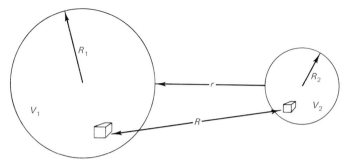

Fig. 6-12   To find the total VDW forces between two continuous spheres, it is necessary to evaluate the VDW force between two arbitrary infinitesimal volumes, and then to sum all possible pairwise interactions between the two spheres.

where ρ_1 and ρ_2 are the number of atoms per cubic centimeter in each sphere, respectively, and

$$K = 3I_1I_2\alpha_1\alpha_2/2(I_1 + I_2) \qquad\qquad (6\text{-}19)$$

is called the London constant. Evaluating Eq. (6-18) is by no means easy, and it would be even more complicated if the geometrics of the objects were more irregular than spheres. In 1937 H. C. Hamaker evaluated Eq. (6-18) for the case of two spheres and found that when the distance of separation is small, the interactional energy between the spheres varies as $1/r$, as opposed to $1/r^6$ for individual atoms. This is a dramatic difference. So, the VDW force operating between two closely spaced large spheres varies as $1/r^2$, just as Coulomb's force does. Using similar techniques, the distance dependence for bodies of other geometrics have also been calculated for the VDW forces of interaction. Figure 6-13 shows the distance dependence for several other geometric arrangements of large bodies.

The importance of VDW forces between large bodies is most pronounced in colloidal solutions where large molecules are dispersed in a fluid. An aqueous solution of a high molecular weight biological component is a good example, although many studies have been performed on inorganic particles in aqueous and nonaqueous solvents. The solid particles of a colloid are usually kept from aggregating with one another by the presence of similar charges present on the surface of the molecules; however, if these repulsive forces are diluted enough by one means or another, then the VDW forces present are frequently strong enough to cause flocculation and eventually precipitation out of solution. A common example illustrating this phenomenon will be presented in the next chapter after we have discussed a little about double charge layers. Usually, colloidal particles attain some equilibrium distance from one another based on the balance between the repulsive and the attractive forces.

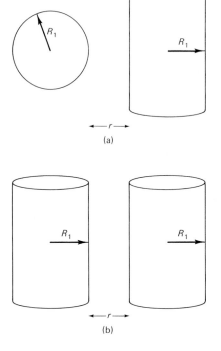

(a)

(b)

Fig. 6-13 The distance dependence for the nonretarded VDW forces between large bodies of different geometrical shapes. (a) Two crossed cylinders; Force $_{VDW} \sim 1/r^2$. (b) Two paired cylinders; Force $_{VDW} \sim 1/r^{3/2}$. Cases (a) and (b) are for the situation where $R_1 \gg r$. If $R_1 \ll r$, other relationships hold (Israelachvili, 1974).

LIFSHITZ THEORY

The London treatment and modifications thereof can be collectively classified as microscopic theories of VDW forces. This is because the interactions between individual atoms or molecules are being considered, and the total effect between large bodies is assumed to be the sum of all pairwise interactions. In order to use this type of theory, it is necessary to know detailed information about the individual atoms or structural units comprising the body of interest; i.e., one must somehow obtain information about polarizabilities. Disadvantages to the microscopic approach include the fact that this information is not necessarily easy to get and that the assumption of additivity must be made. In the 1950s the Russian physicist E. M. Lifshitz developed a description of VDW forces that is quite different from London's original approach. Lifshitz's theory is a macroscopic one in that a body or a particle is considered as a whole without emphasis being placed on its individual constitutive parts. This theory is very complicated in its mathematics, and hence its application is somewhat difficult, although its agreement with experiment is good.

The concept behind Lifshitz's approach is as follows. When one considers matter, it is realized that electrons are in constant motion and that this menagerie of electronic movement will set up a fluctuating electric field that is characteristic of these electronic orbits and of the particular material itself. This fluctuating field, called the zero point field, extends a short distance outside the surface of a solid object; Lifshitz showed that it can cause an attractive force between two objects if they are separated by a small gap. The advantage to this approach is that it is unnecessary to make the assumption of additivity in calculating the VDW force between two large objects, as was necessary with the London model. It also avoids the previous assumption that the VDW force between two atoms is unaffected by the presence of a third atom. With Lifshitz's theory, the retarded and non-retarded forces appear as special cases, and the previously discussed distance dependences are also verified. Application of Lifshitz's theory is dependent on macroscopic properties of the material under consideration; in fact, a generalized form of the dielectric constant is the main parameter needed. Lifshitz's approach is also capable of treating the case where a third medium separates the principal interacting particles, which is a case of real practical interest since all biomolecules are immersed in aqueous solutions. In general, this macroscopic theory seems to hold many advantages over previous work, and present-day descriptions of VDW interactions are relying on it more often.

EXPERIMENTAL VERIFICATIONS

The simplest and most direct way of measuring the VDW forces between two materials is to place a smooth surface of each material in close proximity to the other and then measure the strength of the interaction as a function of distance. Smooth surfaces are necessary so that the interspatial distances can be measured with accuracy and irregular projections from either surface do not interfere with a close approach. A number of experiments of this general nature were performed in the 1950s and 1960s. In a typical experiment an optically polished piece of glass would be held on the end of a very sensitive spring balance, while another piece of glass, fixed to a solid support, would be brought close to the first. The magnitude of the VDW forces could then be detected by measuring the deflection of the spring balance from its equilibrium position. The power law dependence on distance of separation could be obtained by performing the experiment as a function of the gap separation which, in turn, could be determined by using one of several optical interference techniques. In general, the experimental values obtained from this type of measurement agreed with theoretical predictions, and the correct power dependence on distance was observed for gaps between about 25 and 1200 nm.

The problems with the above type of experiment was that only the retarded VDW forces could be measured because the gap separations were so large. Even though the glass surfaces used were optically polished, they were not smooth enough to permit gaps small enough that normal VDW forces predominated. To measure these normal forces, gaps on the order of 10 nm were needed, but on the glass surfaces projection spikes of 20 nm were not uncommon. To achieve smaller gaps, smoother surfaces were necessary. It was subsequently found that if muscovite mica were cleaved in the proper manner, atomically smooth surfaces could be made. These surfaces, then, were ideal for use in producing the small gap distances necessary for studying the normal forces. In one set of experiments thin layers of mica were prepared and glued to the convex surfaces of glass cylinders for support. The glass cylinders were then mounted perpendicularly to one another in a manner to allow gaps of 1 nm or less to be obtained. The gap distances were controlled by having one glass cylinder attached to a piezoelectric device. This device has the property of expanding or contracting a small distance when stimulated by an electrical signal. In one set of experiments performed by J. N. Israelachvili and D. Tabor (1972), it was found that for gap distances of 2–12 nm, the VDW forces are normal; whereas above a separation of 50 nm, they are retarded. The region between 12 and 50 nm thus represents a transition region between the two types of forces. By referring to Fig. 6-13, it is seen that the normal VDW force between two crossed cylinders varies as $1/r^2$ where r is the gap distance. For the retarded case, the force varies as $1/r^3$. In the experiment just described one would expect to see a $1/r^2$ dependence for gaps of 2–12 nm and a $1/r^3$ dependence for gaps greater than 50 nm. Figure 6-14 shows the combined results of several experiments illustrating the transition from normal to retarded forces as a function of gap distance where this transition actually takes place.

Israelachivili and Tabor also performed an experiment where a monolayer of calcium stearate was absorbed to the mica surfaces:

$$CH_3—(CH_2)_{16}—COOC_a$$

calcium stearate

The molecules of calcium stearate were oriented with their long axes perpendicular to the mica surface with the hydrocarbon tail facing outward some 2.5 nm. With this arrangement it can be reasoned that the VDW forces of the calcium stearate should be due to the hydrocarbon tail of the molecule. The results indicated that for separations greater than 2.5 nm, the observed VDW forces were characteristic of the bulk mica; however for smaller gaps, the force tended to be characteristic of the stearate monolayer.

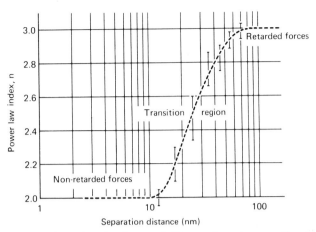

Fig. 6-14 Results illustrating the change of distance dependence as the gap separation between two surfaces increases. The experiment is for two surfaces of mica mounted on crossed cylinders (Israelachvili, 1974).

Even though these experiments are not directly related to biological applications, the reader should get a better feeling for the type of experimental work going on in the field. It is certainly easy to appreciate how difficult it would be to perform a comparable experiment on a biological system. It should also be noted that even though the concept of VDW forces is relatively old, the experiments described here are fairly recent.

SUMMARY

Van der Waals forces are yet another way that matter holds itself together. These forces are universally present and are most important between molecules that are nonionic or nonpolar. Their origin lies in the transient dipoles that are periodically formed as electrons race around their orbits in molecules. These transient dipole induce dipoles in nearby molecules in such a way as to cause an attractive force. In 1930 London presented a theoretical explanation of VDW forces that became the basis for much work in this area. In London's treatment the total VDW force between large bodies is assumed to be the sum of all pairwise interactions; i.e., all atoms of one molecule attract all atoms of another molecule, and the total effect is cumulative. The distance dependence of VDW forces depends on the gap separation of the particles of interest. For particles in close proximity, the distance dependence of the force varies as $1/r^7$ for the interaction between two atoms, whereas it varies as $1/r^8$ for larger separations. VDW forces are important in a variety of biological situations on the molecular level, such as in stabilizing the structures of micelles, membranes, and DNA. They are also known to help in enzyme–substrate recognition and in determining the physical properties of colloidal solutions.

REFERENCES

Casimir, H. B. G., and Polder, D. (1948). *Phys. Rev.* **73**, 360.
DeVoe, H., and Tinoco, I. (1962). *J. Mol. Biol.* **4**, 500.
Hamaker, H. C. (1937). The London-van der Vaals attraction between spherical particles, *Physica.* **4**, 1058.
Israelachvili, J. N. (1974). The nature of van der Waals forces, *Contemporary Phys.* **15**, 159.
Israelachvili, J. N., and Tabor, D. (1972). The measurement of van der Waals dispersion forces in the range of 1.5 to 130 nm, *Proc. Roy. Soc. Ser. A* **331**, 17.
Israelachvili, J. N., and Tabor, D. (1973). Van der Waals forces: theory and experiment, *Progr. Surface Membrane Sci.* **7**, 1.
London, F. (1937). The general theory of molecular forces, *Trans. Faraday Soc.* **33**, 8.
Salem, L. (1962). The role of long-range forces in the cohesion of lipoproteins, *Can. J. Biochem. Physiol.* **40**, 1287.
Smith, E. L., Lumry, R., and Polglase, W. J. (1951). The van der Waals factor in carboxypeptidase interaction with inhibitors and substrates, *J. Phys. Chem.* **55**, 125.
Wilson, I. B. (1952). Acetylcholinesterase: further studies of binding forces, *J. Biol. Chem.* **197**, 215.
Winterton, R. H. S. (1970). Van der Waals forces, *Contemporary Phys.* **11**, 559.

7

DEBYE–HÜCKEL THEORY

INTRODUCTION

In considering the electrical interactions between biological macromolecules, we first discussed the strength of these interactions in a vacuum. A few simple calculations showed that the electrical forces were potentially of a very large magnitude, so large in fact that without some method of diluting these forces, molecules of opposite charge would surely stick to one another and be very hard to separate. One possible mechanism that nature has of reducing the magnitude of these forces was subsequently discussed, namely the placing of the whole biological system into an aqueous environment. With its high dielectric constant, water could then act to dilute the magnitude of electrical forces. It was also mentioned that it was probably no accident that life first started in the sea, for here was a huge aqueous reservoir. In this chapter we shall discuss yet another way in which nature reduces or dilutes the strength of electrical forces between molecules. This second method has to do with the population of small inorganic ions that are present in many natural solutions. Ions such as Mg^{+2}, Ca^{+2}, Na^{+1}, K^{+1}, PO_3^{-2}, SO_4^{-2}, Cl^{-1}, etc. are commonly found in the sea and other natural waters. One may wonder why they are there, whether their presence is really necessary, and further how the presence of these charged atoms affects the electrical interactions among macromolecules. Other natural solutions containing biological molecules, such as blood, urine, or spinal fluid, all have definite concentrations of inorganic charged ions. Furthermore, it should be noted that when a researcher works with a solution of biomolecules, the molecules are not dispersed in pure water, but that the solution inevitably contains some amount of small ions since otherwise, one or a variety of artifacts may appear in experiments and cloud the interpretation of results. The reader should therefore be impressed that these charged atoms or small ions are a natural constituent of biological systems and that their total absence would be a rare situation indeed.

With the introduction and consideration of these ions another complication has been introduced in the quest to better understand the electrical interactions among biological molecules, the first having been the presence of a dielectric medium. However, on the other hand, the situation is becoming more realistic and descriptive of the real world. One must now consider a solution having water, macromolecules, and ions (Fig. 7-1).

The question, or rather questions, to be asked now are, If you have a charged biomolecule in water along with a hoard of small charged ions, how do these ions interact with the macromolecule and subsequently, how does this affect the interaction of one molecule with another? One would intuitively guess that if a macromolecule has, say, a net positive charge Q, then negative ions in the solution will tend to be closer to the macromolecule than the positive ions present, just because of simple coulombic interactions. But since we are dealing with a solution, subject to thermally disrupting and randomizing influences such as brownian motion, this picture will be true only on the average. That is to say, at any one instant in time a positive ion may be closer to the positively charged macromolecule than any of the negative ions, but that on the average this situation will not prevail. These counterions, as they are called, form a sort of "cloud" around the macromolecule, and their attraction is usually so strong that they tend to move with the macromolecule as part of a unit. The counterionic cloud will then

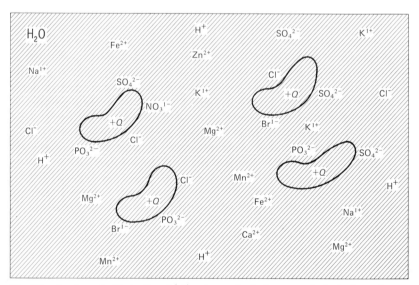

Fig. 7-1   A schematic illustrating the fact that real solutions contain both charged macromolecules $(+Q)$ and also small inorganic ions of positive and negative charge in addition to water.

alter the net charge of the macromolecule as seen by the rest of the solution. This "cloud" affects not only the electrical, but also the hydrodynamic properties of the biomolecules because the cloud in effect increases the size of the parent macromolecule. Referring again to Fig. 7-1, it is seen that the negative counterions have been drawn somewhat closer to the macromolecule than the positive ions to illustrate the points just discussed.

What will be attempted in this chapter is to describe in quantitative detail the exact distribution of this counterionic cloud and to show how it affects the electrical potential V of the charged macromolecule at a point r. This problem was originally solved by P. Debye and E. Hückel in 1923. In the discussion presented here it is necessary to introduce some second-order differential equations which on first sight, seem rather formidable. The reader unfamiliar with this type of mathematics should take care not to lose sight of what the equations are saying. Once the problem is clear enough to be put into equation form, the science is complete and all that remains is to solve the mathematics. The important aspects are the physics of the situation, while the mathematics is a vehicle.

Before trying to solve the problem outlined above, let us define the situation a little more specifically, or in other words, construct a model. Consider then a charged macromolecule in an aqueous solution surrounded by counterions. It will be assumed that the solution is dilute in the macromolecule, so that each macromolecule is essentially independent of the others and the effect produced by many macromolecules is simply the sum of their individual effects. For simplicity of the mathematics, the macromolecules will be assumed to have a spherical shape with the charge evenly distributed over the entire outer surface. The counterions will be initially assumed to be singly charged, either positively or negatively, and their distance of closest approach to the macromolecule will be limited by the finite sizes of the two particles. These assumptions are displayed schematically in Fig. 7-2 for a single macromolecule.

In considering the situation depicted in Fig. 7-2 it will be assumed that the origin of the coordinate system is fixed at the center of the macromolecule. For the time being, normal cartesian coordinates will be used, although later spherical coordinates will prove more useful. With this situation we are in a position to determine the density distribution of the counterion charges surrounding the macromolecule, i.e., we can determine the spatial concentration of counterions surrounding the macromolecule. The counterion charge density is needed because from it we can determine the potential at a point due to both the macromolecular charge and the charges of the counterions. Once this potential is obtained, it is then possible to see how the counterions have changed the potential of the isolated charged macromolecule.

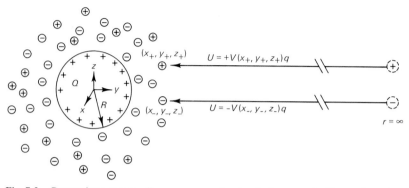

Fig. 7-2 Counterions surrounding a macromolecule of charge $+Q$. The surface charge density of the macromolecule is $\sigma = Q/4\pi R^2$. The negative counterions tend to cluster closer to the macromolecule than do the positive ones. The energy needed to bring a counterion from infinity to a position (x, y, z) is equal to $V(x, y, z)q$ where $V(x, y, z)$ is the potential at the point in question and q is the charge of the counterion. The macromolecule is assumed to be spherical and the origin of the coordinate system is taken at its center.

DENSITY OF COUNTERIONS

In trying to determine the concentration of the counterions as a function of position, several things should be realized. First, the concentration of counterions will not be constant at any one spot due to thermal fluctuations. Secondly, the concentration of counterions for locations far from the macromolecule should equal that of the bulk counterion concentration, and also the concentration of positive counterions should equal the concentration of negative counterions at distances far from the macromolecule. At locations close to the macromolecule the concentration of oppositely charged counterions should be greater than that for the counterions with the same type of charge as the macromolecule. When we finally arrive at an equation for the counterion density distribution, the mathematics has to include all of the above conditions in order to be correct.

The basic assumption made in order to calculate the charge distribution is that it can be determined by using the Boltzmann distribution. It is assumed that the concentration of negative counterions near the macromolecule relative to those far from the macromolecule can be given by an expression e^{-U/k_BT} where U is the electrical potential energy of the counterion at a particular location. This energy is equal to the work necessary to bring the counterion in from infinity to its final location. The quantity k_B is Boltzmann's constant and T is the absolute temperature. A similar expression is also assumed to be true for the positive counterions. The counterions then take up positions with respect to the macromolecule that are determined by the Boltzmann distribution of their potential energies.

Remembering Eq. (2-35), the work necessary to bring a charge q from infinity to a particular location is

$$W_{ab} = \Delta V_{ab} q \qquad (2\text{-}35)$$

but the potential at infinity is zero by definition, so the potential energy of any one counterion is

$$U = W = Vq \qquad (7\text{-}1)$$

where V is the potential at the final location of the counterion, and q is the counterion's charge (see Fig. 7-2). For positive ions, $U = Vq$; and for negative counterions, $U = -Vq$, where V is a function of location (x, y, z). The concentration of positive ions is then given as

$$C_+ = C_0 e^{-qV/k_B T} \qquad (7\text{-}2)$$

and for negative ions

$$C_- = C_0 e^{qV/k_B T} \qquad (7\text{-}3)$$

where C_0 is the bulk concentration of ions measured in ions per cubic centimeter. Equations (7-2) and (7-3) give the counterion concentrations as a function of location, although the location dependence is not explicit, but is implicit since V is a function of (x, y, z).

The net charge density ρ at any location can now be written as the difference between the concentration of positive and negative counterions, or

$$\rho = C_+ q - C_- q \qquad (7\text{-}4)$$

Each term on the right-hand side of Eq. (7-4) is the product of the ion's charge times its concentration to give the total charge. The term $C_+ q$ gives the total positive charge concentration, whereas $C_- q$ is the total negative charge concentration. If we now wish our model to be general and include provisions for more than just one electrolytic salt (two ions) and for ions with multiple charges, Eq. (7-4) is rewritten as

$$\rho = \sum_i \varepsilon N_i C_i \qquad (7\text{-}5)$$

where the summation is over all charged ions present, N_i is the number of electronic charges ε of ionic species i, and C_i is the concentration of ion i. N can be positive or negative depending on the charge and $q = N\varepsilon$. The application of Eq. (7-5) to the special case of two equal but oppositely charged univalent ions in solutions yields Eq. (7-4) where $q = \varepsilon$ or $-\varepsilon$.

If C_i in Eq. (7-5) is now represented by the Boltzmann distribution, we have

$$\rho = \varepsilon \sum_i N_i C_{0_i} e^{-\varepsilon N_i V/k_B T} \qquad (7\text{-}6)$$

where C_{0i} is the bulk concentration of ionic specie i. Equation (7-6) states that the net charge density ρ is zero at infinity because here $V = 0$; and $\sum_i N_i \varepsilon C_{0_i}$ also equals zero because the sum of all ions in solution must add up to zero. Equation (7-6) also states that ρ is negative close to the central molecule if the central molecule has a positive charge. Since we know that both these conditions must be true, we can have confidence that Eq. (7-6) correctly describes the charge density.

Equation (7-6) can be simplified by expanding the exponential in an infinite series. An exponential can be expanded according to

$$e^x = 1 + x + \frac{x^2}{2!} + \frac{x^3}{3!} + \cdots + \frac{x^n}{n!} + \cdots \tag{7-7}$$

so Eq. (7-6) becomes

$$\rho = \varepsilon \sum_i N_i C_{0_i} \left[1 - \frac{\varepsilon N_i V}{k_B T} + \frac{1}{2} \left(\frac{\varepsilon N_i V}{k_B T} \right)^2 + \cdots \right] \tag{7-8}$$

where now the assumption that $k_B T \gg \varepsilon N_i V$ is made. This is the same assumption that was made in Chapter 4, namely that the Boltzmann energy is much greater than the electrical potential energy between the field and the ion. Using this assumption, all terms in the expansion beyond the second term can be ignored as too small to count. The net result is

$$\rho = \sum_i \varepsilon N_i C_{0_i} - \sum_i \varepsilon N_i C_{0_i} \left(\frac{\varepsilon N_i V}{k_B T} \right) \tag{7-9}$$

but the first term of Eq. (7-9) equals zero because again all charged counterions have to add up to give a zero net charge for the solution. The net charge density is then

$$\rho = -\sum_i \frac{\varepsilon^2 V C_{0_i} N_i^2}{k_B T} = -\frac{\varepsilon^2 V}{k_B T} \sum_i C_{0_i} N_i^2 \tag{7-10}$$

This equation is a generalized one, describing the net charge density of counterions at a particular location. The spatial dependence of ρ is not explicitly seen in Eq. (7-10), but is instead implicit since V is a function of (x, y, z); hence ρ is also a function of (x, y, z). For the case of a single salt that dissociates into two monovalent ions ($N = 1$), Eq. (7-10) reduces to

$$\rho = -\frac{2 C_0 \varepsilon^2 V}{k_B T} \tag{7-10a}$$

IONIC STRENGTH

In deriving Eq. (7-10) the units of C_{0_i} were the number of ions per cubic centimeter. A much more convenient unit is that of molar concentration

C_{m0_i}. The conversion between the two is

$$C_{m0_i} = \frac{1000C_{0_i}}{\eta_A} \tag{7-11}$$

where η_A is Avogadro's number. The factor of 1000 changes 1 cm^3 to 1000 cm^3 or 1 liter. Substituting Eq. (7-11) into (7-10) gives

$$\rho = -\frac{\varepsilon^2 V \eta_A}{k_B T 1000} \sum_i C_{m0_i} N_i^2 \tag{7-12}$$

If we now go back to our original conditions assuming only monovalent counterions, Eq. (7-12) reduces

$$\rho = -\frac{\varepsilon^2 V \eta_A}{1000 k_B T} 2C_{m0_i} \tag{7-13}$$

Comparing Eqs. (7-12) and (7-13), it is seen that the summation in (7-12) is replaced by a single term representing concentration in Eq. (7-13). Because of this resemblance, the summation term in Eq. (7-12) is looked upon as a general measure of the ionic concentration of a solution.

The ionic strength I of a solution is defined as

$$I = \frac{1}{2} \sum_i C_{m0_i} N_i^2 \tag{7-14}$$

This new quantity I is another way of quantifying the concentration of counterions in solution, rather than just using the bulk concentration C_{m0}. What the ionic strength equation does is to consider not only the concentration of a particular ion, but also its charge. With this new definition, the charge density is finally given by

$$\rho = -\frac{2\varepsilon^2 V \eta_A I}{1000 k_B T} \tag{7-15}$$

As an example of how to calculate the ionic strength of a solution, consider the case of 0.1 M NaCl solution. In solution NaCl will ionize according to

$$\text{NaCl} \rightarrow \text{Na}^+ + \text{Cl}^- \tag{7-16}$$

hence there are two charged ions of equal concentration. The ionic strength is given as

$$I_{\text{NaCl}} = \tfrac{1}{2}[(0.1)(1)^2 + (0.1)(1)^2] = 0.1 \quad M \tag{7-17}$$

In this case the ionic strength is equal to the molarity concentration. Now consider a solution that is 0.3 M in K_2SO_4. Here the ionization looks like

$$K_2SO_4 \rightarrow 2K^+ + SO_4^{-2} \tag{7-18}$$

and the ionic strength is calculated to be

$$I_{K_2SO_4} = \tfrac{1}{2}[(0.6)(1)^2 + (0.3)(-2)^2] = 0.9 \quad M \qquad (7\text{-}19)$$

which is distinctly different from the regular molar concentration. A solution that is 0.1 M NaCl and also 0.3 M in K_2SO_4 will have an ionic strength of 1.0 M. For univalent ions, the ionic strength concentration equals the regular molar concentration; however, this is not true for multivalent ions in solution.

The reason for introducing the ionic strength is due to the fact that it is a better measurement of electrolytic concentration than is just plain molar concentration. Two solutions that have identical ionic strengths are more similar from an electrical standpoint than are two solutions that have equal molar concentrations of different salts. Ionic strength gives a better indication of the total amount of counterions present. It should also be pointed out that the ionic strength is a property of the solution as a whole, and not of any one particular ion or ionic species; i.e., a solution has an ionic strength, but a salt does not.

An equation describing the net counterion charge concentration as a function of location has now been derived and the ionic strength made a part of it. This completes the first step in solving our problem. It should be noticed that ρ is a function of the potential V. What we really want is an expression for the potential in terms of the parameters of the system, i.e., we want V as a function of r, q, T, etc. Once this potential is found, it can then be compared to the simple potential function $V = Q/r$ of a charge in the absence of an ionic cloud to see the effects of the ionic cloud. The desired potential of the central molecule carrying charge Q surrounded by counterions may be obtained by making use of an equation derived in classical physics known as the Poisson equation.

POISSON EQUATION

The Poisson equation is a relationship between the net charge density of a system and that system's potential as a function of position. It is expressed as a second-order differential equation and takes the form

$$\frac{\partial^2 V}{\partial x^2} + \frac{\partial^2 V}{\partial y^2} + \frac{\partial^2 V}{\partial z^2} = -\frac{4\pi\rho}{D} \qquad (7\text{-}20)$$

where D is the dielectric constant of the medium. This equation looks rather formidable, and in some respects it is; but it should be realized that it is merely a way of stating a relationship between several physical variables. The second derivative of the electric potential at a particular spot with respect to the x coordinate plus the second derivitive with respect to the y coordinate plus the second derivitive with respect to the z coordinate equals a constant times the charge density at that point where the potential is equal to V. Remember that the potential V is a function of the three coordinate variables x, y, z, i.e., the potential can change as a function of location. What Poisson's equation is saying in words, then, is rather straight forward, although knowing that this relationship exists and is true is far from being obvious.

At this point a few words of elaboration may make Poisson's equation seem reasonable. The key to the explanation is in Gauss's law, so the reader may want to quickly review that section now. Consider an imaginary cube with one corner at the origin and its sides extending out along the three axes of the coordinate system {Fig. 7-3}. Imagine that a nonuniform electric field is passing through this box, and it is desired to find the total net flux emerging from the six sides of the cube. One way this can be done is to total the net flux emerging from opposite sets of faces, e.g., faces A and B. Let the z component of the field at face A be noted as E_z and that at face B be noted as $E_z + (\partial E_z/\partial z)\,\Delta z$. The net field out is then equal to the field in at

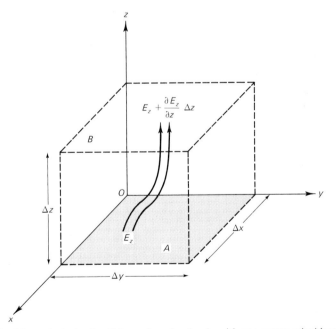

Fig. 7-3 A hypothetical cube of dimensions Δx, Δy, Δz with one corner coincident with the origin. A nonconstant electric field is passing into face A at the bottom and out from face B at the top. The value of this field at face A is E_z and its value at face B is $E_z + (\partial E_z/\partial z)\,\Delta z$.

face A subtracted from the field out at face B, or $(\partial E_z/\partial z)\Delta z$. The net flux is then $(\partial E_z/\partial z)\Delta x\,\Delta y\,\Delta z$ since $\Delta x\,\Delta y$ represents the area of faces A and B. By symmetry, the net fluxes out of the other set of faces are equal to $(\partial E_y/\partial y)\Delta x\,\Delta y\,\Delta z$ and $(\partial E_x/\partial x)\Delta x\,\Delta y\,\Delta z$, respectively. The total net flux out of the box is then equal to

$$\phi = \frac{\partial E_x}{\partial x}\Delta x\,\Delta y\,\Delta z + \frac{\partial E_y}{\partial y}\Delta x\,\Delta y\,\Delta z + \frac{\partial E_z}{\partial z}\Delta x\,\Delta y\,\Delta z \qquad (7\text{-}21)$$

If, perchance, there is a net charge density within the box of magnitude ρ, then the total charge density involved inside the box is

$$\text{total charge} = \rho\,\Delta x\,\Delta y\,\Delta z \qquad (7\text{-}22)$$

By Gauss's law the net flux must equal 4π time the enclosed charge, so

$$\frac{\partial E_x}{\partial x} + \frac{\partial E_y}{\partial y} + \frac{\partial E_z}{\partial z} = 4\pi\rho \qquad (7\text{-}23)$$

but by Eqs. (2-48) the electric field is related to the electric potential

$$E_x = -\partial V/\partial x, \qquad E_y = -\partial V/\partial y, \qquad E_z = -\partial V/\partial y \qquad (2\text{-}48)$$

Substituting these equations into Eq. (7-23) then yields

$$\frac{\partial^2 V}{\partial x^2} + \frac{\partial^2 V}{\partial y^2} + \frac{\partial^2 V}{\partial z^2} = -4\pi\rho \qquad (7\text{-}24)$$

which is Poisson's equation.

The above treatment is not an extremely rigorous derivation of Poisson's equation, but it does give insight into its origin and existence, and it should convince the reader of its validity. It should also be mentioned that for systems not in a vacuum, a factor $1/D$ should appear on the right-hand side of Eq. (7-24) to account for the dielectric medium.

Since the charge density for our particular system involving the ionic atmosphere has already been calculated, we can write

$$\frac{\partial^2 V}{\partial x^2} + \frac{\partial^2 V}{\partial y^2} + \frac{\partial^2 V}{\partial z^2} = 4\pi\left(\frac{2\varepsilon^2 V\eta_A I}{1000 k_B\,DT}\right) \qquad (7\text{-}25)$$

MATHEMATICAL SOLUTION

Once Eq. (7-25) is solved for V, we know what the potential is at any spot in terms of system parameters, and that is what we want. At this point the physics of the situation is complete, and the rest of the problem's solution is just a matter of carrying out the mathematics. Although the solution will be carried out in a fair amount of detail, the reader should try to keep things in perspective and not get completely lost in a sea of manipulations. The solution to a second-order differential equation is not easy, but it is quite understandable if a little time is spent studying it.

The first step in solving Eq. (7-25) for V is to have a very clear idea of what the physical situation is. Figure 7-4 shows an exaggerated view of the relationship between the central ion and one of the counterions. The situation depicted in Fig. 7-4 can be conveniently divided up into three regions I, II, and III, where each region represents a unique type of space. Within region I, there is no charge at all; region II has solvent, but no mobile counterions; and region III has mobile ions. This is essentially the model Debye and Hückel used in originally approaching this problem. By dividing the problem up in this way Eq. (7-25) can be simplified for several regions, and it can be made easier to solve. The approach usually taken is to solve Eq. (7-25) separately for the potential V in each of the three regions, and then to make the three separate solutions compatible with one another at each region's boundary.

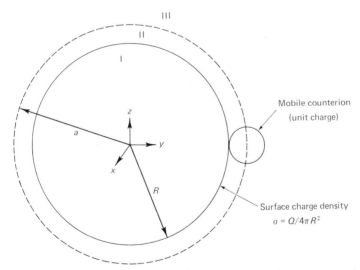

Fig. 7-4 Debye–Hückel model of a counterion's closest approach to a macromolecule of radius R. The macromolecule is assumed to be spherical and to have a uniform surface charge density $\sigma = Q/4\pi R^2$. The system is divided up into three distinct regions of space I, II, III with the cartesian coordinate system having its origin at the center of the macromolecule. Region I is totally within the macromolecule. Region II is between the outer surface of the macromolecule and the distance of closest approach of the counterion ($r = a$). Region III is beyond the distance of closest approach.

One other trick that is used to make Eq. (7-25) easier to solve is to change from rectangular coordinates to spherical coordinates. The reader will remember that both coordinates systems can be used to locate a point P in space (Fig. 7-5). The conversion between cartesian and spherical coordinates is given as

$$x = r\sin\theta\cos\varphi, \qquad y = r\sin\theta\sin\varphi, \qquad z = r\cos\theta \qquad (7\text{-}26)$$

Using the conversions expressed in Eq. (7-26), Eq. (7-25) can now be written as

$$\frac{1}{r^2}\frac{\partial}{\partial r}\left(r^2\frac{\partial V}{\partial r}\right) + \frac{1}{r^2\sin\theta}\frac{\partial}{\partial\theta}\left(\sin\theta\frac{\partial V}{\partial\theta}\right) + \frac{1}{r^2\sin^2\theta}\frac{\partial^2 V}{\partial\varphi^2} = \frac{8\pi\varepsilon^2\eta_A I V}{1000k_B\,DT} \qquad (7\text{-}27)$$

which at first seems to make things more complicated; however, due to our initial assumptions, Eq. (7-27) can be greatly simplified. Remember, it was assumed that the central macromolecule was spherical, and its charge was symmetrically spread out on its outer surface. Because of this, the electrical potential is going to be spherically symmetric also, and hence it will have no dependence on θ or φ. Therefore, the last two terms on the left-hand side of Eq. (7-27) are zero, i.e., because $\partial V/\partial\theta = \partial V/\partial\varphi = 0$. We then have

$$Region\ III: \qquad \frac{1}{r^2}\frac{d}{dr}\left(r^2\frac{dV}{dr}\right) = \frac{1}{r}\frac{d^2(rV)}{dr^2} = K^2 V \qquad (7\text{-}28)$$

where

$$K = \left(\frac{8\pi\varepsilon^2\eta_A I}{1000k_B\,DT}\right) \qquad (7\text{-}29)$$

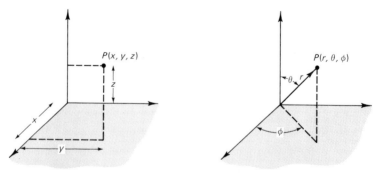

Fig. 7-5   Two coordinate systems can be used to describe the same point P in space. Different coordinate systems sometimes have an advantage in simplifying the mathematics because of the symmetry of a problem. The three parameters of a cartesian coordinate system are the familiar x, y, z, while for a spherical system, they are r, θ, φ.

is known as the Debye–Hückel parameter. The reader should verify that K has the units of inverse distance. Equation (7-28) is valid for region III since it is only here that $\rho \neq 0$. For regions I and II, there are no free mobile counterions ($\rho = 0$), so Eq. (7-28) in these cases reduces to

$$Region\ I\ and\ II: \quad \frac{1}{r}\frac{d^2(rV)}{dr^2} = 0 \qquad (7\text{-}30)$$

Equations (7-28) and (7-30) are the differential equations relating the electric potential to the parameters of the system. They have taken into account the specific aspects not only of our generalized model, but also the requirements of each region. The solution to each of these equations will describe the potential in that particular region; however, the boundaries that have been set up (Fig. 7-4) are artifical, and the potential function should be continuous throughout the system from $r = 0$ to $r = \infty$. Therefore, it is necessary to make the separate potential functions generated for each region agree with one another at the boundaries set up in the model (Fig. 7-4); e.g., the potential function for region II must agree with the potential function for region III at the boundary $r = a$. The potential functions cannot abruptly change at the boundaries since there is no physical reason for this to happen—the boundaries are artificially constructed. The process of making the potential functions in different regions agree with one another is carried out by using "boundary conditions." These conditions are a set of physical requirements that are arrived at by the use of common sense and physical theory. Boundary conditions place restrictive conditions on the form the electric potential can take. For instance, the potential function must be equal to zero at $r = \infty$, while it cannot equal ∞ at $r = 0$. At the boundaries between regions I, II, and III, the potential function must be continuous. This last requirement is necessary so that the mathematical form of V makes sense. If these concepts seem unclear at this stage, continue on for a few paragraphs and then reread this section.

For regions I and II, Eq. (7-30) is appropriate. The general solution to it can be shown to be of the form

$$V_{I,II}(r) = A + B/r \qquad (7\text{-}31)$$

where A and B are two constants of integration. The reader should verify that Eq. (7-31) is indeed a solution to Eq. (7-30) by substituting Eq. (7-31) into (7-30) and carrying out the indicated operations. Now, how do we find the values of A and B? In region I we have the point $r = 0$, if V is to remain finite there, the constant B must equal zero, otherwise the potential will blow up, which cannot be allowed. So for region I, the electrical potential has the form

$$V_I(r) = A \tag{7-32}$$

which says the potential is constant.* It remains yet to solve for A explicitly. For region II, the form

$$V_{II}(r) = A + B/r \tag{7-33}$$

is still correct, and the value of B is not necessarily equal to zero. Also, the value of A in region II is not necessarily the same as in region I. Because of this, let us write

$$V_{II}(r) = C + F/r \tag{7-34}$$

as the equation for the potential in region II where C and F are still constants. At this stage, we cannot explicitly evaluate the constants A, C, or F.

The next step is to solve for $V_{III}(r)$ from Eq. (7-28). It can be shown that a general solution to this type of differential equation takes the form

$$V_{III}(r) = \frac{Ge^{-Kr}}{r} + \frac{He^{Kr}}{r} \tag{7-35}$$

where G and H are constants of integration that have to be determined by using boundary conditions. Again, the reader should verify that Eq. (7-35) is really a solution of Eq. (7-28) by substituting back into Eq. (7-28). Because $V_{III}(r)$ must equal zero at $r = \infty$, it is necessary to impose the condition that $H = 0$, otherwise $V_{III}(r)$ would equal infinity at $r = \infty$. The form of the potential in region III is now

$$V_{III}(r) = Ge^{-Kr}/r \tag{7-36}$$

At this point we must now explicitly evaluate the constants A, C, F, and G. To do this, additional boundary conditions are needed (we have already used two). Keeping in mind that the potential function must be continuous across a boundary we impose the conditions that

$$V_I = V_{II} \qquad \text{at} \quad r = R \tag{7-37}$$

$$V_{II} = V_{III} \qquad \text{at} \quad r = a \tag{7-38}$$

This then puts restraints on the possible values of the unknown constants, and forces the potentials to be continuous across the boundaries. The other boundary conditions needed are not as obvious as the ones above. Classical

---

* This conclusion has been arrived at by applying the boundary condition that the potential V is finite at $r = 0$.

theory states that the following must be true:

$$D_{II}\frac{dV_{II}}{dr} - D_I\frac{dV_I}{dr} = 4\pi\sigma \qquad \text{at} \quad r = R \qquad (7\text{-}39)$$

$$D_{III}\frac{dV_{III}}{dr} - D_{II}\frac{dV_{II}}{dr} = 0 \qquad \text{at} \quad r = a \qquad (7\text{-}40)$$

where the Ds are the dielectric constants in the respective regions. Equation (7-40) states that across the boundary separating regions II and III the product of the derivitive of the potential and the respective dielectric constant must be continuous. This same requirement holds true for the boundary separating regions I and II, except a slight modification is needed due to the fact that this boundary contains a fixed charge density (charge/unit area). Equation (7-39) then takes this into consideration. The interested reader is referred to a textbook on electricity and magnetism for a detailed explanation of the boundary conditions expressed by Eqs. (7-39) and (7-40).

To get a feel for what these equations are saying, consider that $-dV/dr$ is equal to the electric field in the radial direction, which is also perpendicular to our artificial boundaries. The charge density on the boundary between regions I and II is $\sigma = Q/S$ where S is the area of the spherical boundary. Keeping these facts in mind we may rewrite Eq. (7-39) as

$$D_I E_{In} - D_{II} E_{IIn} = 4\pi Q/S \qquad (7\text{-}41)$$

or

$$(D_I E_{In} - D_{II} E_{IIn})S = 4\pi Q \qquad (7\text{-}42)$$

where E_n represents the component of E normal to the spherical boundary. The normal component of E, times the area, however, is just the flux of the field. We then have that the flux from one side of the boundary minus the flux at the other side (equaling the net flux) equals 4π times the charge in the boundary area. This is nothing more than another way of stating Gauss's law. If there is no fixed charge at the boundary, then $\sigma = 0$. The boundary conditions represented by Eqs. (7-39) and (7-40) are then a form of Gauss's law.

Equations (7-37)–(7-40) represent four conditions that must be imposed on the potential functions. Carrying out the indicated operations on Eqs. (7-32)–(7-34) yields

$$A = C + F/R \qquad (7\text{-}43\text{a})$$

$$Ge^{-Ka}/a = C + F/a \qquad (7\text{-}43\text{b})$$

$$DF = Q \qquad (7\text{-}43\text{c})$$

$$Ge^{-Ka}(1 + Ka) = F \qquad (7\text{-}43\text{d})$$

We now have four equations and four unknowns; hence it is possible to solve them simultaneously and determine unique values for the unknown constants. Doing this yields

$$A = \frac{Q}{DR}\left(1 - \frac{KR}{1 + Ka}\right)$$ (7-44a)

$$C = -QK/D(1 + Ka)$$ (7-44b)

$$G = Qe^{Ka}/D(1 + Ka)$$ (7-44c)

$$F = Q/D$$ (7-44d)

and finally we arrive at the explicit equations for the potential:

$$V_I = \frac{Q}{DR}\left(1 - \frac{KR}{1 + Ka}\right)$$ (7-45)

$$V_{II}(r) = \frac{Q}{Dr}\left(1 - \frac{Kr}{1 + Ka}\right)$$ (7-46)

$$V_{III}(r) = \frac{Q}{Dr}\frac{e^{Ka}}{1 + Ka}e^{-Kr}$$ (7-47)

Equations (7-45)–(7-47) are the final desired results for the potential of the system diagrammed in Fig. 7-4. Each equation is for a specific region of space, with no single function describing the overall picture. This is in contrast to the simpler situation where no counterions are present. With the derivation of these questions we are now in a position to see how the counterions affect the original potential of the charged molecule. If no counterions were included in the system, the potential of the macromolecule at some arbitrary distance r would be $V(r) = Q/Dr$; however, with counterions the analogous expression for potential without counterions is now multiplied by a constant factor $e^{Ka}/(1 + Ka)$ and also by the variable factor e^{-Kr}. This second factor has the effect of reducing the strength of the potential faster as a function of distance than if it only had a $1/r$ dependence. The conclusion is then that the electric potential function of the central molecule drops off much faster with distance when counterions are present than when they are not. The relative behavior of $1/r$ and e^{-r}/r dependence on distance is shown in Fig. 7-6 to illustrate this effect. In this context one can think of the counterions as screening the central molecule and reducing the strength of its net charge as seen by the rest of the solution. If an observer were to look at some arbitrary point $P(r > a)$ and if no counterions were present, a charge Q on the macromolecule would be seen; if counterions were present, a lesser charge would be seen because counterions of the opposite polarity tend to cluster closer to the macromolecule and "dilute" its effective charge. How much would the charge be reduced? By a ratio equal to $e^{Ka}e^{-Kr}/(1 + Ka)$. So, by having small, charged counterions in a biological system, nature has

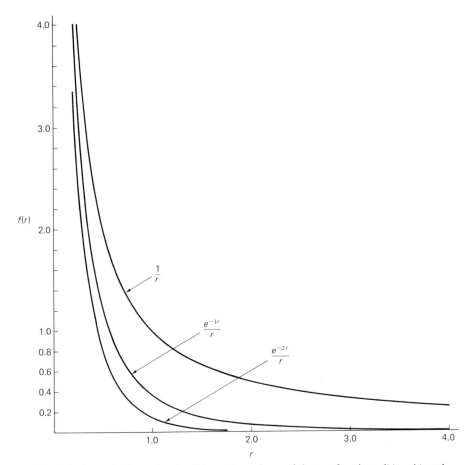

Fig. 7-6 A graph illustrating the distance dependence of the two functions $f(r) = 1/r$ and $f(r) = e^{-Kr}/r$ for two values of K. The function e^{-Kr}/r decreases faster than $1/r$.

another way of reducing the potentially immense electrical forces. By having counterions crowding around large charged macromolecules, their effective charge is reduced, as is their relative influence as a function of distance. It should also be pointed out that Eq. (7-47) is actually the sum of two electric potentials, one due to the central macromolecule itself and the other due to the counterionic atmosphere.

DEBYE LENGTH

When the charge density of the counterions was previously calculated [Eq. (7-14)], it was found that it depended on the value of the potential V;

since this potential has now been explicitly derived, it is possible to substitute back and relate charge density to the parameters of the system. Substituting Eq. (7-47) into (7-14) yields

$$\rho = -\frac{2\varepsilon^2 \eta_A I}{1000 k_B T} \frac{Q}{Dr} \frac{e^{Ka}}{1 + Ka} e^{-Kr} \qquad (7\text{-}48)$$

or

$$= \text{const.} \, (e^{-Kr}/r) \qquad (7\text{-}49)$$

where it can be seen that K has the units of inverse length. The factor K is called the Debye length, and it is used to measure the radius of the counterion cloud.

In order to get a physical understanding of what the Debye length measures, let us consider the relative density of counterions surrounding a central molecule. The counterionic charge in a spherical shell concentric with the central molecule and bounded by radius r and $r + dr$ is the charge density ρ multiplied by $4\pi r^2 \, dr$ (Fig. 7-7) or

$$\text{frac.}(r) = \text{fraction of charge} \sim (e^{-Kr}/r) 4\pi r^2 \, dr$$

$$= \text{const} \, re^{-Kr} \qquad (7\text{-}50)$$

Plotting Eq. (7-50) vs. r gives the graph shown in Fig. 7-8. It is seen that the fraction of charge reaches a maximum when $r = 1/K$. Since this graph is a

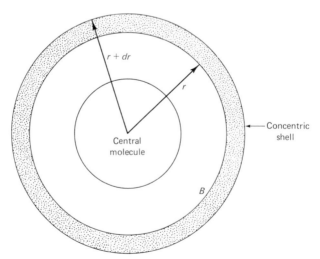

Fig. 7-7   Diagram illustrating how to find the relative amount of counterionic charge in a spherical shell concentric with the central macromolecule. The area of surface B is $4\pi r^2$; hence the volume of the shell is $4\pi r^2 \, dr$. This figure is a two-dimensional representation of a three-dimensional situation.

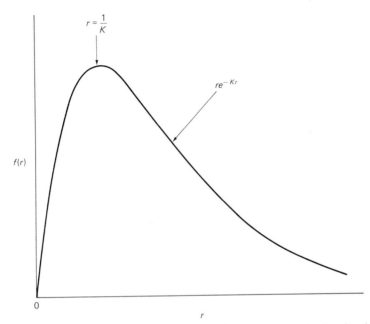

$r = \dfrac{1}{K}$

re^{-Kr}

$f(r)$

0

r

Fig. 7-8   A plot of the function $f(r) = re^{-Kr}$ vs. r. The maximum relative fractional charge is at a distance $r = 1/K$ away from the origin.

general one, the term $1/K$ is used as a distinguishing descriptive term to characterize the size of the counterionic cloud. $1/K$ is used as a measure of the radius of the counterion cloud. From Eq. (7-29) it is seen that K is related to the concentration of counterions. As this concentration increases, K increases, hence $1/K$ decreases. This means that the cloud of counterions tends to crowd proportionately closer to the central macromolecule as the concentration of counterions increases relative to the case where the counterion concentration is low. If the counterion concentration is low, the cloud is more loosely associated with the macromolecule. Another way of stating the above is as follows: For a particular central molecule, the relative percentage of counterions closeby increases as the bulk concentration of counterions increases. The importance of this effect is seen in the phenomenon known as salting out.

SALTING OUT

Biological macromolecules are isolated from their hosts organisms by a wide variety of physical and chemical techniques. The particular method(s) that works for a new system is found only by experiment. One method that is commonly used is called salting out; it relies on the existence of a counterion cloud around each macromolecule and also on VDW forces. The method works as follows. If a salt like $(NH_4)_2SO_4$ is added to a solution containing a protein, a virus, etc., then after a certain amount of the salt has been added the particular biomolecule will flocculate and precipitate out of solution. If several different proteins are present, then each particular variety will precipitate out of solution at a specific salt concentration. If these salt concentrations are far enough apart, then the different proteins can be separated from one another by removing the precipitates as they occur and redissolving them in separate solvents. Using this technique, biological molecules can be separated, purified, and concentrated without a lot of complex equipment. The following is an explanation to give some insight into this phenomenon in terms of principles we have already discussed.

As salt is added to the solution, its concentration increases, with a subsequent decrease in the Debye length of the counterionic cloud around each biomolecule. The cloud shrinks about the macromolecule making the entire unit more compact. Also, with the increasing counterion concentration comes a more effective screening of the macromolecular charge, so that the identical macromolecules in solution can now approach each other more closely, and not be repelled by the like charge that each carries. When the macromolecules approach closely enough, VDW forces become large enough to hold them together creating a larger mass, which will in turn attract even more macromolecules. Remember that the VDW forces between large bodies can be substantial because of additivity effects. When the mass becomes large enough, the entire complex will then fall out of solution forming a precipitate.

It might be asked, Why does the counterionic layer not respond to the movement of the dipoles causing the VDW forces and neutralize or shield them, much as the counterions are doing for the net charge? The answer is that the counterions would respond if they could; but remember that the dipole oscillations responsible for the VDW forces are taking place on the order of 10^{15} Hz, and the counterion distribution simply cannot keep pace. The counterion distribution cannot reorient fast enough as the dipoles oscillate from one orientation to another. The counterion cloud dilutes the net charge on the macromolecule more so than it does the VDW forces. The key to the whole explanation is the fact that as the salt concentration increases, the Debye length, which measures the radius of the shell of counterions about the macromolecule, decreases. This provides better screening of the macromolecule's net charge and allows closer approaches of one macromolecule to another. In the next chapter, on water, an additional effect will be discussed that also contributes to the salting out phenomenon.

ENERGY OF THE SYSTEM

One question that could be asked now is, How much energy does it take to charge the central macromolecule in Fig. 7-2? Or to put it another way, How does the ionic atmosphere affect the potential energy of the system? These questions can both be answered by realizing that the energy of a charge q in a potential V has energy qV [Eq. (2-59)]. In our case the central macromolecule has charge Q, and the total energy can be derived by calculating the energy necessary to bring a charge Q from infinity to the surface of the macromolecule. Alternatively, this quantity can be derived more directly by summing the incremental work necessary to increase the charge on the macromolecule by an incremental amount in a potential V. Hence, the amount of work needed to charge the macromolecule from 0 to Q is

$$U = \int_0^Q V \, dq \qquad (7\text{-}51)$$

But what potential is to be used? The answer is the potential at the charge Q, which is described by Eq. (7-45). The energy then is

$$U = \int_0^Q \frac{q}{DR}\left(1 - \frac{KR}{1 + Ka}\right) dq \qquad (7\text{-}52)$$

$$= \frac{Q^2}{2DR}\left(1 - \frac{KR}{1 + Ka}\right) \qquad (7\text{-}53)$$

which is less than for the case having no ionic atmosphere at all. For the case of no ionic atmosphere at all, the potential energy would equal $Q^2/2DR$. Equation (7-53) does reduce to this, as it should, if $K = 0$.

For some numerical examples of Eq. (7-53), consider a protein of about 40,000 molecular weight having a radius of 25 Å. At 25°C in an aqueous solvent, the energies of the system are given in Table 7-1 for various Q and I. As can be seen from the table, for a particular net charge on the macromolecule, the net energy needed to charge the macromolecule is reduced as the ionic strength increases. This makes sense based on our previous model because with the counterionic cloud it is easier (requires less energy) to bring a charge in from infinity and deposit it on the macromolecular surface. The ionic cloud shields the already existing like charge on the macromolecule; and hence this charge does not repulse the new incoming charge as much as it would were the ionic cloud not there. If our original model had been a solvent penetrable sphere, then the values in Table 7-1 would be much smaller than those shown. In fact, in this latter case U could be reduced to almost zero with a high ionic strength solution. The reason for this is due to the ionic cloud's increasing ability to shield macromolecular charges; i.e., the ionic cloud can get closer to a macromolecule

TABLE 7-1

Electrical Potential Energy for a Charged Protein in Aqueous Solution of Various Ionic Strengths[a]

Q \ I	0.001 M	0.01 M	0.05 M	0.15 M
10	6.7	4.8	3.3	2.5
20	27	19.2	13.2	9.9
30	60.5	43.1	29.7	22.2
40	108	76.8	52.8	39.6

[a] The energy values are in kilo calories per mole. Q is the charge on the macromolecule and I is the solution's ionic strength (Tanford, 1961).

if it is solvent penetrable. So, the energies one calculates depend heavily on the details of the model used.

Another way that Eq. (7-53) could be utilized is in calculating how much energy it takes to ionize one more group on a protein, as could easily occur if, say, the pH were changed. The energy involved would just be the amount of work needed to bring one more charge in from infinity and place it on the protein. To put it another way, it is the difference in energy given by Eq. (7-49) when the charge is equal to Q as opposed to the situation where the molecule has $Q + 1$ electronic charges. This change in energy can be calculated to be

$$\Delta U = \frac{(N\varepsilon + \varepsilon)^2 - (N\varepsilon)^2}{2DR}\left[1 - \frac{KR}{1 + Ka}\right] \qquad (7\text{-}54)$$

where N is the original number of electronic charges on the protein, and ε is the electronic charge. By looking at the form of Eq. (7-54) and also Table 7-1 it is concluded that the higher the ionic strength of the solution, the lower will be the energy needed to ionize one more group of the protein. At high ionic strengths very little energy will be needed for this compared to the case where a protein is in a low ionic strength solution.

ANOTHER BIOLOGICAL IMPLICATION OF THE DEBYE–HÜCKEL THEORY

It is well documented that certain biomolecular processes are very dependent on the presence of a small concentration of univalent or divalent ions. For instance, many enzymes require the presence of a metal ion before enzymatic activity is conferred. Likewise, the structural integrity of a biomolecule is often dependent on a minimum concentration of a metal ion; e.g., the structure and function of ribosomes are very dependent on the concentration of Mg^{+2}. Ribosomes are subcellular components that are the site of active protein synthesis. As a last example, consider that the conduction of a nerve impulse is dependent on the concentration of Ca^{+2} in the immediate vicinity of the nerve cell. In these cases just mentioned the small counterions play an active role in the molecular mechanisms and do not just act as a counterion cloud. In many reactions the biomolecule has specific requirements for a particular ion, not just a univalent or a divalent ion. In fact, another ion of similar charge is often inhibitory to the function of the system, as is illustrated by the fact that Mg^{+2} ions can disrupt the normal workings of nerve conduction by suppressing the release of neurotransmitters.

The usual explanation for alike ion inhibition is in terms of active site binding. This implies that there is an oppositely charged site on the macromolecule to which the ion binds. In this situation a similarly charged ion could also bond there and prevent coupling with the preferred ion. If the substituting ion does not possess all the required properties, the biomolecule may be nonfunctional. A somewhat different explanation for this phenomena can also be given in terms of the Debye-Hückel theory.

Let us assume that the most important situation in getting the correct ion to bind with a macromolecule is that the concentration of this ion in the immediate vicinity of the macromolecule be above a certain limit; i.e., the ion must have at least a minimum effective concentration. This effective concentration is not necessarily equal to the bulk concentration of that ion. If now a second similar ion (same charge) is introduced into the solution, then by Eq. (7-15) the concentration of the preferred ion must decrease near the macromolecule as the other ion's concentration increases. The effective concentration of the preferred ion is decreased, hence it has less of an opportunity to interact with the macromolecule. Equation (7-15) has no provision for distinguishing ions except by their net charge. In this situation the second ion has in effect prevented the formation of the correct macromolecule–ion complex without the need to form a complex itself.

ELECTRICAL INTERACTIONS BETWEEN TWO IONS

How does an ionic atmosphere affect the strength of electrical interactions between two charged molecules? It is known qualitatively that the strength of the electrical interactions are reduced. But how is this taken into account when making a mathematical calculation? The following example is given to illustrate this situation.

Suppose that it is desired to calculate the energy of interaction between two charged ionic groups, say an NH_4^+ and a COO^- group in a solution that is $0.16\,M$ in ionic strength. The total energy of interaction is given by a sum of several terms; charge–charge, charge-induced dipole, and VDW forces will all contribute to the interactional energy. To calculate the individual contributions due to charge–charge and charge-induced dipole we need some of the equations derived in Chapter 5; however, these equations must be corrected to take into account the ionic nature of the solvent. This is done by multiplying by the factor $e^{K(a-r)}/(1 + Ka)$, so we have

$$U = -\frac{331.9n_1n_2}{r_1\,D}\frac{e^{K(a-r)}}{1 + Ka} - \frac{165.2n_1n_2(\alpha_1 + \alpha_2)}{r^4\,D^2}\frac{e^{K(a-r)}}{1 + Ka} - \frac{143v_1v_2\alpha_1\alpha_2}{(v_1 + v_2)r^6}$$

$$(7\text{-}55)$$

as the equation describing the interactional energy between the NH_4^+ and COO^- groups in an aqueous solvent containing electrolyte salt. The two interactions involving charge in Eq. (7-55) have a multiplicative factor accounting for the ionic atmosphere, while the VDW term has no correction. This is a reflection of the fact that the atmosphere does not directly affect the strength of VDW forces. The counterions do not shield charges because none are involved in VDW interactions, and also the counterions cannot reorient fast enough to keep up with the rapid fluctuations of the dipole generating the VDW interaction.* In performing the evaluation of Eq. (7-55) it will be assumed that the dielective constant varies as $D = 6r - 7$ where r is the distance of separation in angstroms. The other values needed for this calculation are shown in Table 7-2. Table 7-3 summarizes the results for the calculation for various distances of separation and includes results assuming no ionic atmosphere. It should be noted that as r increases, the percentage difference in the energy for the ionic atmospheric and the nonionic

TABLE 7-2

Summary of Values to Be Used in Calculating the Energy of Interaction between an NH_4^+ and COO^- Group in a 0.16 Ionic Strength Solution

$D = 6r - 7$	$T = 37°C$
$\alpha\,(COO^-) = 1.87 \times 10^{-24}\ cm^3$	$I = 0.16\ M$
$\alpha\,(NH_4^+) = 2.14 \times 10^{-24}\ cm^3$	$a = 3\ Å$
$v\,(COO^-) = 1.6 \times 10^{15}\ sec$	D (bulk water) $= 75$
$v\,(NH_4^+) = 1.7 \times 10^{15}\ sec$	$K = 0.13\ Å^{-1}$

* In the Lifshitz treatment, the added salt could affect the VDW forces if the dielectric constant of the medium were altered.

TABLE 7-3

Values for the Energy of Interaction between COO^- and an NH_4^+ Group in an Aqueous Solvent of 0.16 M Ionic Strength[a]

	Ionic atmosphere				No ionic atmosphere			
r (Å)	U_{cc}	U_{ci}	U_{VDW}	U_{tot}	U_{cc}	U_{ci}	U_{VDW}	U_{tot}
3	7.2	0.04	0.65	7.9	10.0	0.07	0.65	10.72
4	3.0	0.006	0.12	3.13	4.9	0.009	0.12	5.03
5	1.6	$\sim 10^{-4}$	0.03	1.63	2.9	$\sim 10^{-4}$	0.03	2.93
10	0.18	$\sim 10^{-6}$	$\sim 10^{-4}$	0.18	0.63	$\sim 10^{-5}$	$\sim 10^{-4}$	0.63

[a] Energies are in units of kilo calories per mole.

atmospheric case increases also. This is due to the fact that as the distance of separation increases, more counterions can fill the gap between the two interacting species, and hence more effective shielding takes place. Also, the respective energies for the atmospheric case are always smaller than the corresponding one for the no ionic atmosphere case, and this is as expected.

WEAKNESSES OF THE DEBYE–HÜCKEL THEORY

The Debye-Hückel theory has shown good agreement with experiment for ionic strengths up to about 0.1 M if the counterions are monovalent. For multivalent ions, the value is an order of magnitude less. The treatment is also limited to the temperature range where $k_B T \gg qV$; otherwise, the counterionic charge density cannot be described by Eq. (7-15). The initial model of the charge on the macromolecule assumed it to be symmetrically spread out on the spherical surface of the molecule. It turns out that any spherically symmetrical distribution here would yield identical results. However, many proteins do not have a spherical shape but are better represented by an ellipsoid of revolution. Modifications of the theory for this effect have been made, but they will not be discussed here. In general, they make the treatment more complicated. The implicit assumption that the central molecule is an impenetrable sphere is not always true since many times a protein or other biological molecule is impregnated with solvent. Calculations for a penetrable sphere have also been carried out; in general, the energies of the system are less than for the impenetrable case. One final problem has to do with the waters of hydration that surround each charged molecule. According to the Debye length equation, no account of the extra radius of these waters is taken.

SUMMARY

The Debye-Hückel theory is a treatment that can lead to an expression describing the potential of a charged particle when it is surrounded by an ionic atmosphere. Since molecular biological systems are immersed in a sea of ions, it is reasonable to know how these counterions affect electrical interactions. One effect of these ions is to reduce the strength of electrical interactions by shielding or screening the charge on the macromolecules from the rest of the system, thus causing their effective net charge to be less than their real net charge. The number of ions in a solvent is quantitatively described by a quantity called the ionic strength. A high concentration of counterions can cause proteins to precipitate out of solution. Also, the concentration of counterions can influence the potential energy of the system and the amount of energy needed to ionize one more group on the central macromolecule. In general, an ionic atmosphere reduces the overall potential energy of a system compared to the identical situation without an ionic atmosphere.

REFERENCES

Setlow, R. B., and Pollard, E. C. (1962). "Molecular Biophysics." Addison-Wesley, Reading, Massachusetts.
Tanford, C. (1961). "Physical Chemistry of Macromolecules." Wiley, New York.

8

WATER

INTRODUCTION

Water is not only one of the most ubiquitous substances to be found on earth, it is also one of the most important to life systems. Different forms of life are known to live under a variety of extreme conditions, yet it is rare that an organism can live without water. The necessity for water seems to be almost a universal requirement for any form of life as we know it, and the important role of water has been demonstrated on both the macroscopic and the molecular level. It has also been suggested that primordial life first began in the water and only later moved onto the land. Water, then, is a very special liquid, and in studying the interactions among biomolecules it is quite useful to understand the properties and behavior of this remarkable substance.

The possibility that life started in an ocean and still today relies heavily upon the presence of water implies that the water molecule must have some unique properties essential to the preservation of life. In this chapter we shall be looking at the properties of water and trying to describe its structure, both as a molecule and as a liquid. Although water has been around for a very long time, it has only been relatively recently that scientists have tried to determine the physical structure associated with the various forms of water and to ascertain how liquid water can affect biomolecules. The first investigations into the microstructure of water took place in the 1930s, but it has been since the 1950s that most of our information concerning water has been gathered. Although much work has been performed in an effort to better understand the structure of water and how this structure relates to water activity, the picture is still incomplete and research into this area continues. It should be remarked at the outset that liquid water is not made up of a homogeneous mixture of individual molecules randomly blending into one another to form haphazard associations, but is instead a fluid where the molecules are associated with one another in microstructural arrangements that change with time. At the molecular level water molecules are involved in a three-dimensional structural arrangement, or several three-dimensional arrangements; and these structures are constantly changing with time.

Before looking at some of the molecular aspects of water, let us first examine a few of its macroscopic properties and see how they compare with other substances of similar organization. Figures 8-1 and 8-2 show plots of the melting points and boiling points, respectively, for various hydrides, H_2O being just one example. It will be noticed that the hydrides other than water show a consistent pattern for the parameters shown as a function of molecular weight, and that in both cases water is anomalous. The elements shown in combination with hydrogen in Figs. 8-1 and 8-2 are all from the same group in the periodic table as that of oxygen. For both the melting and boiling points, water has a much higher value than would be predicted based on the trend shown by the other hydrides. If one were to make similar plots for some common substances whose total number of electrons is identical to that of water (Ne, H_2F, NH_3, CH_4), it would be found that water still has thermal properties that are significantly different

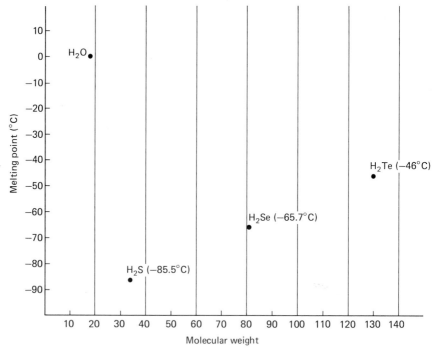

Fig. 8-1   A graph of melting point plotted vs. molecular weight for some hydrides. (Data from "Lange's Handbook of Chemistry," 11th Ed. McGraw-Hill, New York, 1973.)

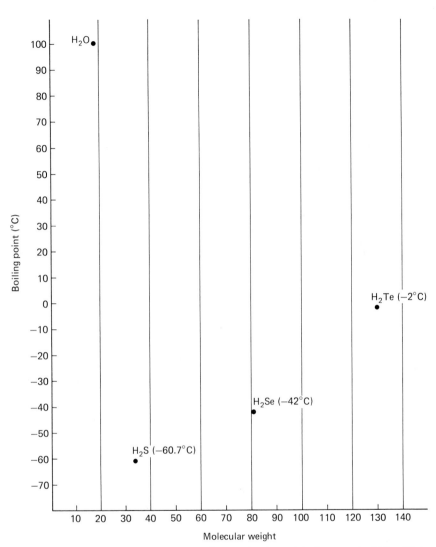

Fig. 8-2 A graph of boiling point plotted vs. molecular weight for some hydrides. (Data from "Lange's Handbook of Chemistry," 11th Ed.

than those of the rest of the group. In all these cases it should be noted that at room temperature water is a liquid, whereas the rest of the molecules are present as gases. This indicates immediately that water is more an associated substance; i.e., the individual molecules of water tend to interact with one another more so than in nonassociated liquids, and these interactions are not so easily overcome by thermal agitation. So, in comparing water with

other molecules that are seemingly similar in structure, it is seen that water
has thermal properties that make it stand out from the rest.

Another property that is interesting to examine is the specific heat, which
is defined as the amount of heat required to raise the temperature of a
substance 1°C compared to that of water. Table 8-1 shows some common
liquids and their respective specific heats. Again, it is seen that water has a
value that is anomalous compared to the other substances in the group.
The information in this table is telling us that it takes relatively more energy
to raise the temperature of water than it does for the other liquids. Although
it is not shown here, a similar trend would be seen for the heat of vaporization,
where this parameter is a measure of how much energy is needed to transform
a liquid into gas. From all these data a case can be built to advocate that
water has a rather unique set of thermal properties and that it takes relatively
large amounts of energy to raise its temperature.

One may now wonder how water's thermal characteristics are important
to life systems. Obviously, they must be or else the whole subject would not
be discussed. But how? It has been mentioned that life probably first developed in the sea, and that this was no accident. One advantage aqueous
systems had over other systems was their high dielectric constant, which was
ideal for reducing the strong electrical interactions between molecules. But
water's thermal properties were also important. Because of its high specific
heat, water has the advantage of being an ideal reservoir against temperature
fluctuations. Compared to other liquids, it takes a relatively large change in
ambient temperature to change the temperature of an aqueous reservoir.

TABLE 8-1

Specific Heats and Molecular Weights of Some Common Liquids[a]

Liquid	Specific heat (cal/g deg)	Temperature (°C)	Molecular weight
Acetic acid	0.47	0	60
Acetone	0.51	0	58
Benzene	0.39	5	78
Carbon tetrachloride	0.20	0	154
Chloroform	0.23	0	120
Ethanol	0.54	0	46
Ethyl ether	0.53	0	74
Formic acid	0.53	20	46
Methanol	0.57	0	32
Octane	0.49	0	114
Toluene	0.39	0	92
Water	1.0	0	18

[a] Data from "Lange's Handbook of Chemistry," 11th Ed.

Therefore, anything within this reservoir would enjoy a relatively constant temperature, and this would seem to be almost essential for a developing life system. If temperature extremes were frequently encountered, proteins and other biomolecules could be denatured or could experience conformational changes that would alter their biological activity, and this could easily spell doom to any living system. The reader can appreciate the importance of a relatively constant temperature by considering warm-blooded mammals. Adult humans can tolerate only about a six or seven degree rise in body temperature before serious damage occurs. Table 8-1 tells us that for the same amount of heat, a formic acid solution would rise in temperature twice as much as water, acetone three times as much, and ethanol almost twice as much. Water, then, seems to be the best substance for maintaining a relatively constant temperature for a biological system.

Water can not only protect an organism from external temperature extremes, it can also dissipate internal heat rather efficiently. In the process of normal metabolism cells generate heat; and if this heat is not conducted away somehow, it would eventually kill the cells. Since water constitutes a major percentage of cellular material, it can absorb this heat without a substantial rise in temperature. On a somewhat larger scale, animals manage to keep their temperature under an upper limit by evaporating water from their surfaces. Dogs, for example, cannot perspire over their entire body; hence they must rely on panting to allow water to evaporate from their tongue to help maintain a reasonable temperature. Water, then, plays an important role at all levels, from the molecular to the macroscopic.

By now the reader should have an appreciation of the facts that water is a very special liquid, that it has a variety of functions in biological systems, that life would be quite different without it, and that it would be difficult to study separately the interactions among biomolecules without considering some aspects of water.

The approach taken here in our examination of water will be to look first at the water molecule itself and then to examine how the molecular structure can be used to explain the various phenomena associated with water. To understand water's structure, it is useful to examine the electronic structure of oxygen. Oxygen is not only the biggest atom in water, it is also the most electronegative; its arrangement of electrons plays a major role in determining the geometric shape of the water molecule. In order to understand the electronic structure of oxygen, we shall first consider the carbon atom.

ELECTRON ORBITALS OF CARBON

Atomic carbon is in group IV and the second period of the periodic chart of the elements. It has an atomic number of six, and thus six electrons orbit its nucleus. These electrons are not aimlessly wandering about the nucleus, but instead they are confined to restricted regions of space called orbitals. There are a number of different possible kinds of orbitals surrounding each nucleus with each having a characteristic size and shape. Those orbits that are closest to the nucleus tend to hold electrons that are more stable, or less reactive, compared to those electrons in outlying orbitals. The number and kinds of orbitals that have electrons in them depends on the number of electrons an atom has to accommodate. The more electrons an atom has, the more orbitals that are filled. These orbitals can be defined by using the principles of quantum mechanics, and their geometric shapes represent those volumes of space where there is a high probability of actually finding an electron. In general, each different type of orbital has a unique shape. For instance, the 1S orbital is the closest to the nucleus and has a spherical shape. This means that the probability of finding an electron in the 1S orbital has spherical symmetry; i.e., for a particular distance, there is equal probability of finding a 1S electron above or below, to the right or left of the nucleus. For an orbital without spherical symmetry, this is not true. A 2S orbital also has spherical symmetry, but it is bigger in radius than is the 1S orbital. Electrons in a 2S orbital are a little further away from the nucleus, hence their interaction with the nucleus is less.

In elementary chemistry it is learned that the orbital containing electrons of the lowest energy is called a 1S orbital, the next is called a 2S orbital, and that each of these orbitals can hold a maximum of two electrons. The orbitals of next lowest energy are called 2P orbitals, and there are three of these designated $2P_x$, $2P_y$, and $2P_z$. Electrons in each 2P orbital have identical energies and again each can hold two electrons. The three 2P orbitals are asymmetric in shape; they have their long dimension directed along one of the three mutually perpendicular axis of a cartesian coordinate system. The other types of orbitals continue on, but for our purposes the S and P orbitals are sufficient. In terms of this organization, then, the six electrons of carbon would be placed as shown:

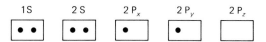

The general principles followed in placing electrons in orbitals is to fill the orbitals of lowest energy first and also to keep the electrons as far away from one another as possible. This latter principle is used to place one electron each in the $2P_x$ and $2P_y$ orbitals, instead of placing both in, say, the $2P_x$ orbital. Electrons like to be far from other electrons because of coulombic repulsion. With the configuration shown above it is seen then there are two unpaired electrons in the outer P orbitals. Since covalent bonds are formed by having orbitals with unpaired electrons overlapping similar orbitals from other atoms, it would be expected based on the above model that carbon could form a total of two covalent bonds with hydrogen. Hydrogen has one electron in a 1S orbital. The two bonds would be formed by the two 2P orbitals of carbon overlapping two separate 1S orbitals of two different hydrogen atoms. This would form the compound CH_2.

The above explanation for the bonding between carbon and hydrogen is very nice except for the fact that it does not agree with reality. In fact, carbon forms covalent bonds with four hydrogen atoms, not two. The compound formed in this case is methane CH_4. So, because the previous model of the orbitals surrounding carbon led to a conflict with what is known to be true, it is necessary to change the model. Instead of describing the carbon atom in terms of the common 2S and 2P orbitals, four new orbitals are constructed in their place. These new orbitals are called SP^3 orbitals, and they are formed by mathematically combining the characteristics of the 2S and 2P orbitals. These SP^3 orbitals are four in number, and they have the advantage of being able to put one electron into each of them as shown

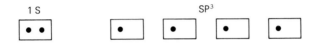

This arrangement is then capable of explaining the bonding between C and H to form CH_4. Each SP^3 orbital overlaps a 1S orbital of four separate hydrogen atoms forming four covalent bonds. Each of the four covalent bonds are also exactly equivalent to one another, as is known to be true for methane.

The four SP^3 orbitals have an important property other than the fact that they can each form a covalent bond. This has to do with their geometric arrangement in space. Each SP^3 orbital is asymmetric in shape, and the long axis of each orbital is directed toward one of the four corners of a regular tetrahedron. This is illustrated in Fig. 8-3. The angle between any two orbitals in this structure is equal to the tetrahedral angle of 109.5°. This configuration permits the orbitals to be as far away from one another as possible. If this model is correct, the angle between any two C—H bonds in methane should also be equal to 109.5°; experimental evidence does confirm this. In conclusion, when carbon forms four covalent bonds with hydrogen, or any other atom, the geometric arrangement is based on a tetrahedral structure; i.e., each bond is directed toward the corner of a tetrahedron. The reader should now be able to understand why the hydrocarbon chains in Fig. 6-4 are drawn with zigzag bond lines. This representation illustrates the fact that the C—C bonds are not all collinear, but instead are part of a tetrahedral character, and that the successive carbon bonds are not all exactly aligned in a nice straight row. As each C—H bond in methane is directed toward a corner of a tetrahedron, so too is a C—C bond.

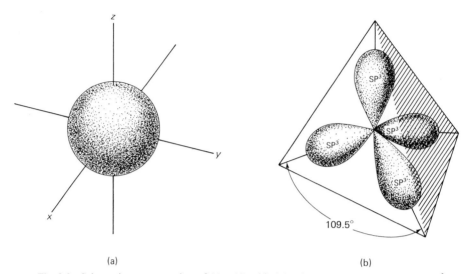

(a)   (b)

Fig. 8-3   Schematic representation of (a) a 1S orbital (nucleus at center) and (b) four SP³ orbitals. Each of the four SP³ orbitals is illustrated as a lobe, and each is directed toward one corner of a regular tetrahedron which is shown superimposed over the actual orbitals. Each SP³ orbital can overlap a 1S orbital of a hydrogen atom to form a covalent bond. Each orbital donates one electron; then both orbitals share both electrons, essentially filling the valency requirement of each orbital. The respective SP³ orbitals are directed toward the corners of a tetrahedron because this configuration allows the electrons to stay away from one another as as much as possible.

ELECTRON ORBITALS OF OXYGEN

Well, what do the above concepts have to do with water? The answer lies in the electronic arrangement of oxygen. The electron arrangement of oxygen can be explained as an extension of the theory used for carbon. Atomic oxygen has two more electrons than carbon, and the electron distribution is as shown in Fig. 8-4. Now only two SP³ orbitals have one electrons, and one would expect that atomic oxygen could form covalent bonds with two hydrogen atoms (H_2O), and this is quite correct. The other two orbitals that would normally interact with H atoms as in the case of methane are now filled by a pair of electrons that belong completely to oxygen. These two particular SP³ orbitals are not overlapping any other

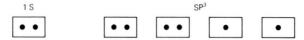

Fig. 8-4   The distribution of electrons in the SP³ orbitals for oxygen.

orbitals from other atoms in an effort to share electrons. Each of these SP^3 orbitals already has its full complement of two electrons. So, if this description is to be correct, the water molecule should have a tetrahedral structure just as methane does. In fact, the bond angles have been measured to be about 105°. This lower value compared to the perfect value of 109.5° is thought to be due to the bulky pairs of electrons compressing the bond angle. The water molecule itself can then be considered as essentially having a tetrahedral structure with the four corners of the tetrahedron being filled with either an electron pair or a hydrogen atom. This is illustrated in Fig. 8-5. If one considers only the oxygen and two hydrogen atoms, then water is a planar molecule since three points define a plane. But in considering water in three dimensions it must be remembered that it is really tetrahedral in shape. The two sets of electron pairs cannot be ignored when describing water's behavior. Having thus introduced all the necessary concepts, we are now in a position to understand some of water's remarkable properties.

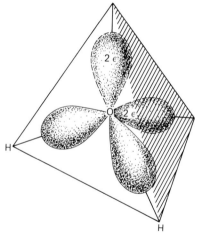

Fig. 8-5   Diagram illustrating the tetrahedral structure of a single water molecule. Each SP^3 orbital holds an electron pair alone (dotted lines) or shows a bond with a hydrogen atom.

WATER AS A MOLECULE

It has already been stated that water has a large dipole moment, and now we can see how it is formed. By referring to Fig. 8-6 it can be seen that each O—H and O—: bond contributes a small dipole moment to form the molecular dipole moment, the molecular dipole moment being the vector sum of the individual bond dipoles. If all the carbon bonds were with

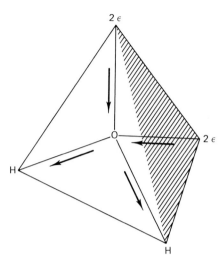

$$\mu_{H_2O} = 1.84 \text{ D}$$

Fig. 8-6   A tetrahedral water molecule with each arrow indicating the strength and direction of an individual bond dipole moment. The vector sum of all these dipoles forms the molecule dipole moment, which is shown below the molecule. A similar diagram for methane would show a molecular dipole moment of zero. Why is this?

hydrogen, as is the case with methane, the molecular dipole moment would be zero. However, since two bonds are with electron pairs, each of these contributes a large bond dipole, so when the individual bond dipoles are vectorially added, the total molecular dipole moment turns out to be quite large (1.84 D). Therefore, both sets of electron pairs are very important in a calculation of the molecular dipole moment, and they cannot be ignored.

Since water has two hydrogen atoms, it can act as a hydrogen donor in a hydrogen bond with two other separate molecules. Since water also has two electron pairs, it can also act as a hydrogen acceptor in a hydrogen bond with two other separate molecules. This means that any single water molecule can be hydrogen bonded with a maximum of four other molecules, be they other waters or completely different molecules. This property turns out to make water a very associated liquid, and accounts for its unusual thermal properties. Because a water molecule can hydrogen bond with four other waters, any single molecule seldom remains isolated, but is instead in constant contact with other water molecules. Although relatively weak, these hydrogen bonds tend to stabilize a three-dimensional, or several three-dimensional, structures among water molecules. Any heat that is supplied to a volume of water breaks a myriad of hydrogen bonds and increases the temperature of the water simultaneously. If a fixed amount of heat is applied to a volume of water, some of that heat will break hydrogen bonds and some will be used to increase the kinetic energy of water molecules, or raise the temperature. All of the heat will not be applied toward raising the temperature of the water. Because the applied heat is divided in its action, it takes a relatively larger quantity of heat to increase water's temperature compared to that of another liquid that is not hydrogen bonded. In this latter case the applied heat works only to raise the temperature. The other molecules shown in Table 8-1 do not form hydrogen bonds, or do not form them to the extent that water does; hence their thermal properties are not quite as unique. The amount of heat needed to raise a substance's temperature is a direct measure of how closely associated the individual constituent molecules are related to one another. In Chapter 5 a calculation was made indicating that the hydrogen bond between two water molecules has a strength of approximately 5 kcal/mole, which alone is not considered to be an extremely strong bond. However, in the case of water we have an excellent example of a case where a large number of hydrogen bonds can have a very great macroscopic influence, even when the individual bonds are weak. So far, the tetrahedral structures and the two sets of electron pairs have afforded a nice model to explain water's thermal properties. Next, this same model will be used in an explanation of the crystalline structure of ice.

ICE

One of the problems in studying the structure of liquid water is the fact that it is a liquid; just because of their basic nature, liquids tend to change their molecular structure relatively quickly; i.e., the spatial relationship of one molecule to another changes rapidly with time. For a solid, the situation is much more well defined because the spatial relationships between molecules exist for periods long compared to the time it takes to make measurements. This reduces the ambiguity associated with averaging molecular and atomic positions, which is necessary in liquid systems. It is therefore sometimes useful to start a study of a liquid structure by first considering the structure of the solid phase. In the case of water this means ice. It will turn out that there are similarities between the arrangements of molecules in both ice and liquid water, and that hydrogen bonding plays a large role in both structures.

One of the most unusual properties of water is that its solid form is less dense than its liquid form ($0.916\,g/cm^3$ at $0°$ vs. $1.0\,g/cm^3$ at $4°C$). This is in marked contrast to most other substances. Water has its most dense form at $4°C$, and from here, as the temperature is lowered, the density decreases.

This fact alone is of profound biological significance in that, once formed, ice will float on top of liquid water rather than sink to the bottom. If ice were more dense than water, the oceans would freeze from the bottom up, and the ice would have a tendency to remain throughout the seasons because of the insulating properties of the surrounding water, and the ice itself. The spring and summer sun would not be nearly as effective in thawing the ice as it presently is. This situation would not only reduce the total amount of free water available to living organisms, it would also create an unfavorable environment for aquatic life. Lakes and oceans would remain mostly frozen even in the summer, and less water would be available to be evaporated and redistributed as rain. Consider also, How could life systems have evolved from the sea if it froze solid every winter? Fortunately, ice does float, leaving most bodies of water in the liquid state, even in the coldest climates.

In fact, there are seven or eight different known types of crystalline ice structures which are formed depending on the pressure of the system. One form, ice VII, is stable up to a temperature of about 80°C (under high pressure); however, for our purposes, this discussion will be restricted to the common form known as ice I, which is formed at atmospheric pressure at a temperature of 0°C. It is this type of ice that will be compared to liquid water.

Much of our knowledge concerning the structure of ice and water itself comes from diffraction studies. In this type of technique the sample is bombarded with either electrons, neutrons, or X rays, and the resulting interactions are analyzed to deduce structural relationships about the sample. These methods supply structural information that would be difficult or impossible to obtain in other ways. For the common form of ice, X-ray diffraction studies have shown that the structure is essentially tetrahedral, with one water molecule hydrogen bonded to four others. An illustration of this structure is shown in Fig. 8-7. It should be noticed that the two water molecules at the left and bottom of the figure are hydrogen bonded to the electron pairs of the central water molecule. The outlying waters on the top and right are hydrogen bonded to the hydrogen atoms of the central water molecule. The structure shown in Fig. 8-7 extends essentially throughout all of the ice mass. Because of the hydrogen bonding, this tetrahedral relationship in ice encloses a relatively large volume of space compared to the situation where the individual molecules could approach one another as closely as possible without the restriction of the hydrogen bond. The hydrogen bonds here are acting as barriers, not allowing the individual water molecules to approach one another as closely as is possible. The significance of this is that the resulting structure is very open, and hence the density is low. If the hydrogen bonds were not keeping the water molecules at arm's length, so to speak, the density would be much higher. It has

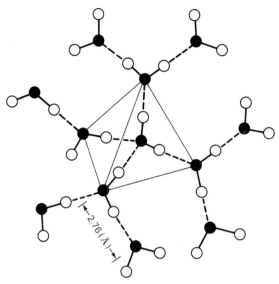

Fig. 8-7   The spatial relationship between a central water molecule and its four nearest neighbors in ice. The symbol ● represents oxygen atoms, while ○ represents hydrogen atoms. The encompassing solid lines show the tetrahedral shape of the central water molecule. This crystalline structure essentially extends throughout the entire ice mass. The dotted lines represent hydrogen bonds. The oxygen–oxygen nearest neighbor distance is 2.76 Å.

been calculated that if water molecules were free of the restrictive hydrogen bonds, the density of ice would be over 1.5 times as high as it is. The structure illustrated in Fig. 8-7 then explains why ice has a low density and can float on water, and again it is related to the tetrahedral structure and the ability of water to form hydrogen bonds.

LIQUID WATER

When ice melts as the temperature is raised, hydrogen bonds are broken, and the neat tetrahedral arrangement that exists throughout the ice is disrupted. If this were the only effect present, then the density of water should increase, and indeed it does up to a temperature of about 4°C. This density increase is due to the fact that the individual water molecules can approach one another more closely in the liquid state than in the ice form where they are held apart by the extensive system of hydrogen bonds. Above 4°C, however, random thermal energy keeps the water molecules separated to the extent that the density decreases with increasing temperature. Also, as the temperature continues to increase, more hydrogen bonds are broken, and the structure present in ice is further reduced. But what of the spatial relationship of one water molecule to another, how is that affected?

In the tetrahedral structure shown for ice, the oxygen–oxygen nearest neighbor distance is 2.76 Å, and the next oxygen–oxygen nearest neighbor's distance is 4.5 Å. In an experiment performed in 1938 two investigators utilized X-ray diffraction techniques to investigate the oxygen–oxygen nearest neighbor distance in liquid water. The results are shown in Fig. 8-8. What this graph depicts is the relative density ρ of the water molecules surrounding a central water molecule. It gives an indication of how close two water molecules in the liquid state are associated with one another. As is seen, the first peak is at 2.9 Å, which is surprisingly close to the 2.76 Å distance between oxygen atom centers in ice where there is extensive hydrogen bonding and a tetrahedral arrangement among molecules throughout the volume. The data shown in Fig. 8-8 have been improved upon in more recent years, but the essence still remains; they indicate that there is extensive hydrogen bonding in liquid water and that the tetrahedral arrangement present in ice is conserved to a large extent even when the ice melts. The reason for the 2.9 Å nearest neighbor distance in liquid has been interpreted to be that the hydrogen bond distance in the liquid is greater than in the solid. In this respect it has also been calculated that only about 15% of the hydrogen bonds present in ice are broken on melting and that it takes extremely high temperatures to break all of the hydrogen bonds present. This tetrahedral relationship among water molecules is seen also to exist at 83°C, at least for the first peak, with the second peak flattening out somewhat at this higher temperature. At the lower temperature the second nearest neighbor distance is shown to be 4.5 Å; this compares favorably with the value of 4.51 Å in ice. The conclusions one arrives at, then, are that liquid water is not structureless on the molecular level, that there is extensive hydrogen bonding, and that a tetrahedral arrangement can exist in the liquid. One main difference between ice and liquid water however is that the tetrahedral arrangement in the liquid extends outward only maybe 10 Å, whereas in

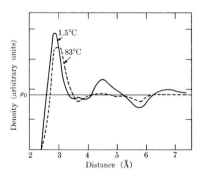

Fig. 8-8   The radial density distribution curve for liquid water. Here the relative density of molecules surrounding one water molecule is plotted as a function of distance away from that molecule (Snell *et al.*, 1965).

ice it is much more extensive and much more uniform and regular. Also, the hydrogen bonds form and break much more rapidly in the liquid than they do in ice. If they did not, then water would have a much higher viscosity than it does. Because the hydrogen bonds break and re-form so rapidly (lifetime typically 10^{-11} sec), the overall structure is flexible and flows easily; hence water has a relatively low viscosity.

Although it is generally agreed that liquid water possesses a large amount of hydrogen bonding and retains some of its tetrahedral structure, there is still controversy as to whether or not there are other structural arrangements among water molecules in the liquid state. In describing what is meant by other structural arrangements, it is questioned whether or not water molecules in the liquid state participate in any higher order structures, other than just random tetrahedral structures via hydrogen bonding. Do water molecules in the liquid state hydrogen bond with one another to form a more regular structure, or do they just experience bonding with whatever other molecule happens to be close at the time? Several different schools of thought have emerged for describing the spatial relationships of water molecules in the liquid; only one such example will be briefly discussed here to illustrate the point.

It is known that when certain inorganic compounds are placed in water, hydrates are formed in which the solvent water molecules form sort of a caged structure surrounding the foreign compound. These caged structures are well organized and in some cases are remarkably stable. Figure 8-9 illustrates a possible arrangement of one such structure. These structures are sometimes called clathrates, clusters, or icebergs. It has been proposed that liquid water may be composed of clathrates of very short lifetimes (10^{-11} sec) where the internal space within the clathrate could be either empty or filled with other water molecules. The clathrate structures themselves are relatively bulky in that they have a large volume compared to other structural arrangements. Also, each water molecule participating in a clathrate is part of a local tetrahedral arrangement. The clathrate structure then can be thought of as a higher order of structure than a simple tetrahedral relationship,

Fig. 8-9 A proposed cagelike structure of water where each corner represents a water molecule and each line represents a hydrogen bond. The center of each cage can house another water molecule or be empty. (From Pauling (1961). Copyright 1961 by the American Association for the Advancement of Science.)

much as the tertiary structure of a protein is of a higher order than the primary structure. Other versions of water structure indicate that a closed cage is not necessary and that this is closer to the actual truth. However, the clathrate theory has a basis in that inert compounds to form clathrates, but the theory also has several difficulties and is not universally accepted. Other models for the organization of water molecules in the liquid form have also been presented; the situation in general is not clear. Most theories do however retain the basic tetrahedral nature of the ice structure to some degree.

Even though the detailed structure of water is unknown, one thing is almost certain. Water structure is dynamic and not static. Water molecules are not rigidly clamped together, never to break once joined. Rather, the spatial relationship of water molecules to other water molecules is a statistical one; and on the average an organized structure exists. Bonds and micro-crystals are constantly being formed and broken, and at any one instant there is only a certain fraction of molecules involved in one type of structure or another depending on the temperature, pressure, etc. Whereas the structure of ice may be regarded as relatively static, the organization of liquid water is dynamic in nature.

HYDRATION

Hydration is the process whereby a substance interacts or combines with water. In many cases water molecules become associated and move along with the invading compound as part of a larger complex. The hydrodynamic properties of the compound are thus altered from the unhydrated state. Also, the water structure adjacent to the foreign compound is altered relative to that of bulk water. In this section we shall consider how ionic and nonionic substances affect the properties of water and discuss the various ways in which solute molecules interact with water.

Ionic Substances

Let us first consider the behavior of some macroscopic properties of water when a simple salt is added as compared to the pure liquid. If we were to plot the heat capacity of an aqueous salt solution (say $CaCl_2$) as a function of the concentration of $CaCl_2$ added, a graph similar to that shown in Fig. 8-10 would result. The horizontal line represents pure water and the sloping line represents the heat capacity of the solution. This plot states that an aqueous salt solution has a heat capacity that is less than for pure water; i.e., it takes less energy to raise the temperature of an aqueous salt solution

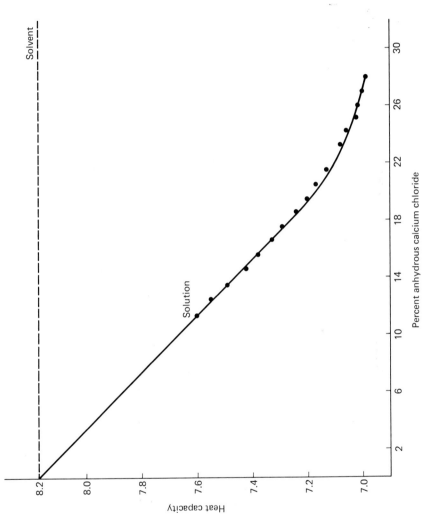

Fig. 8-10 Plot of the heat capacity of aqueous salt solutions of $CaCl_2$ vs. the concentration of $CaCl_2$ in solution. Units of heat capacity are British thermal units per gallon degree Farenheit. (Data from "Lange's Handbook of Chemistry," 11th Ed.

than it does for pure water. This fact immediately tells us that the addition of the salt alters the molecular structure of the water. Another interesting fact is that if NaCl is added to water, the effective volume decreases. The combined volume of the NaCl plus the water is larger than the volume of the aqueous NaCl solution. Along a similar line one mole of NaOH may be placed in one liter of water, and once dissolved, the total volume of the solution will be somewhat less than one liter. And finally, it is possible to experimentally measure the hydration energy when a cation such as Na^+, K^+, etc. is hydrated, and these energies are typically found to be on the order of 100 kcal/mole which is a considerable amount of energy.

Looking at the above three considerations, it can be seen that there are some very definite physical changes taking place when salts are dissolved in water. The individual properties of each substance are not simply added together to give properties for the solution. There is a definite interaction between the components of the system. In order to account for these and other phenomena, a model of hydration has been developed, and it will now be discussed.

In considering this model it will be assumed that a cation is being hydrated. A schematic diagram of the model is shown in Fig. 8-11. It should be noticed that there are three distinct regions of interest. The water molecules directly adjacent to the cation are known as the primary hydration layer; the number of water molecules in this volume will vary depending on the cation. For Na^+ or K^+, this number will typically be on the order of four to six. These primary waters are greatly influenced by the cation and tend to become associated with it and to move as a unit with it. Because of the close proximity, charge–dipole and VDW forces between the cation and the waters keep the waters nearby and oriented for maximum interaction. In region I the primary water molecules do not really have the same relationship to one another as they would in bulk water. The strong electric field of the cation essentially influences them to align with the field. However, even though they are held fairly rigidly by the electrical interaction between them and the cation, the lifetime of any one particular water molecule in the primary hydration layer is short compared to macroscopic times. For instance, if the cation is Mn^{+2}, the mean lifetime of any primary water is 10^{-7} sec; for Cu^{+2} it is 10^{-6} sec. So, the primary waters are constantly exchanging with other waters in regions II and III making the entire model a dynamic structure. It should also be mentioned at this stage that the orientation of the primary waters will be different depending on whether the ion is a cation or anion. This is illustrated in Fig. 8-12.

Region III is relatively far from the cation, and this space is occupied by bulk water with its normal structure. Region III is relatively uninfluenced by the presence of the ion. The distance away from the ion needed to reach

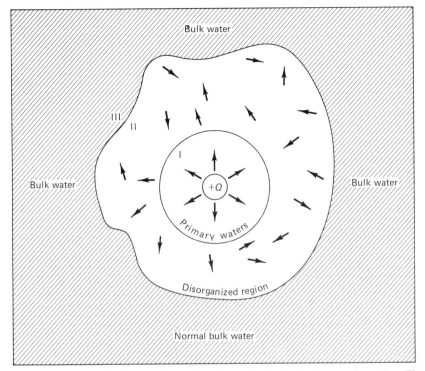

Fig. 8-11 A schematic diagram illustrating the model for the hydration of a cation. The various water molecules are represented by their respective dipoles. In this model there are three regions with the waters in each region having different relationships to each other and also to the cation. Region III is comprised of bulk water and is least influenced by the cation. Region I is composed of the primary waters of hydration and is most influenced by the ion. Region II has an organization that is intermediate between those of regions I and III. The water molecules here are generally considered to be in a disorganized state. This diagram is patterned after the model presented by Frank and Wen (1957).

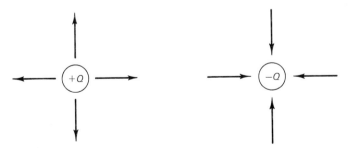

Fig. 8-12 Water molecules will align differently in the primary layer depending on whether the ion has a positive or negative charge. This is due to the way a dipole aligns in an electric field. In one case the hydrogens are presented to the bulk solution, whereas in the other case the oxygen atoms are. The water molecules are represented by their respective dipoles.

region III naturally depends on the ion itself since a larger electric charge will spread its influence over a larger distance. The water molecules in region III represent the opposite extreme from those in region I where the molecules are maximally influenced by the ion. Between regions I and III, region II represents a space where the waters do not have the structure of bulk water, nor are they aligned like the primary waters of hydration. This region is thought to have a less organized structure than either of the other two regions. It is considered to be composed of relatively disorganized waters because the two regions on either side of it are constantly trying to draw water molecules out and incorporate them into their own structures. The cation tends to align the water dipoles toward it in order to gain favorable electrical interactions, and the bulk water region would like to see the water in region II fit into the regular structural arrangements of bulk water. The waters in region II then act as a transition from the primary waters of one type of structure to the bulk water having an entirely different type of structure. Both extreme regions are effectively competing to influence the waters in region II thus causing a "disorganized" region.

By creating an area of disorganized structure in the water the cation has made it easier for the solution's temperature to be raised because part of the input heat goes into breaking up the regular structure of the bulk water. In this case the cation has already performed part of this function by its very existence in solution. Part of the bulk water structure has already been broken and any added heat can go into raising the temperature of the water. When the concentration of the cation increases, the percentage amount of disorganized water increases also, thus explaining the negative slope of the graph in Fig. 8-12.

Keeping the above model in mind, and with the equations developed in previous chapters, it is now possible to make a rough calculation of the energy of hydration for a cation. For a concrete case, K^+ will be used for the cation, and only the primary hydration layer will be considered in detail. Six waters of hydration will be assumed. This calculation can be made by considering the sum of all the separate interactions occurring between the K^+ ion and the water molecules. The interactions to be considered are as follows: charge–dipole attraction between cation and water molecules, VDW attraction between cation and water, the VDW attractions between the separate water molecules in the primary layer, and finally the dipole–dipole repulsive interaction between the similarly aligned water dipoles in the primary hydration layer. Each of these interactions will be considered separately. In making this calculation it will be assumed that the primary waters are aligned radially with the K^+ ion and that they are equally spaced around the cation in a plane (Fig. 8-13). This last assumption is not completely correct; however, it makes the calculations easier.

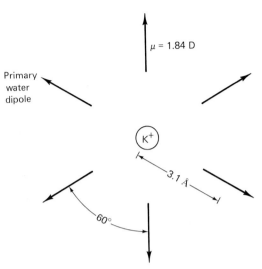

Fig. 8-13 A K$^+$ ion surrounded by six primary waters of hydration. Each water is represented by its dipole moment. For purpose of the calculation, it is assumed that the water molecules are equally spaced around the K$^+$ ion and that all lie in a plane.

Charge–Dipole

This is the major interaction between the cation and the primary waters.
Its strength is calculated using the corrected form of Eq. (5-8). Using the
parameters

$$\mu_w = 1.84\,\text{D} \qquad r = 3.1\,\text{Å}$$
$$\cos \theta = 1.0 \qquad D = 1$$
$$n = 1.0 \qquad \text{six primary waters}$$

we find for a single cation–dipole interaction

$$U = (-57.6 n \mu_w \cos \theta)/r^2 D = -11.0 \quad \text{kcal/mole}$$

Assuming additivity, the total interactional energy for six primary waters
is $U = -66.0$ kcal/mole. The value of $D = 1$ is a pretty safe assumption
here because there is essentially nothing in the space between the cation
and the primary waters.

Charge-Induced Dipole

This interaction is due to the cation's ability to induce a dipole in the
primary water molecules in addition to the permanent one already present.
Using the above values and a value of $\alpha_w = 1.44 \times 10^{-24}$ cm^3, it can be

calculated that

$$U = -110.1\alpha_w n^2/r^4 = -1.72 \quad \text{kcal/mole}$$

and the total charge-induced dipole energy of interaction for all six primary waters is $U = -10.3$ kcal/mole.

VDW Interaction

The first VDW interaction considered is that between the cation and the primary waters of hydration. Using the values $\alpha_{K^+} = 0.87 \times 10^{-24}$ cm^3, $v_w = 2.1 \times 10^{15}$ sec^{-1}, $v_{K^+} = 10^{15}$ sec^{-1}, it is found that

$$U = -143 v_w v_K \alpha_w \alpha_K /(v_w + v_K) r^6 = -0.13 \quad \text{kcal/mole}$$

and for all six waters, the final energy is $U = -0.78$ kcal/mole.

The other VDW interaction is between adjacent primary water molecules. Since there are six such molecules, and it has been assumed that they are uniformly distributed around the cation in a plane, the distance of separation between dipole centers is equal to 3.1 Å.* For this calculation and the next, it will be assumed that there are three separate pairs of water molecules where each water interacts only with its partner. This is not strictly correct, but it does make the calculation easier. The VDW energy between two water molecules is then

$$U = -143 v_w v_w \alpha_w \alpha_w /2 v_w r^6 = -0.33 \quad \text{kcal/mole}$$

and for all three pairs, the total energy is $U = -0.99$ kcal/mole. Since both types of VDW interactions add to the stability of the system, the total VDW contribution is -1.77 kcal/mole.

Dipole–Dipole Repulsion between Waters

Since all of the primary water molecules in our model are perfectly aligned with respect to the cation, all their dipoles are aligned in a like manner. They are all essentially parallel to one another. This will give rise to a repulsive term in the energy considerations due to the dipole–dipole interaction. The magnitude of this interaction can be calculated by using the corrected form of Eq. (5-16). For a single pair, we have

$$U = 7.17 \mu_w \mu_w /r^3 = +0.8 \quad \text{kcal/mole}$$

and for all six primary waters, $U = +2.4$ kcal/mole.

Now, the final energy of hydration can be calculated by adding all of the energies in the separate interactions. Doing this yields $U = -75.7$ kcal/mole,

---

* This value can be arrived at by using the law of cosines where the known angle between dipoles is 60° and two sides of the triangle are the cation–dipole distance. The distance from one dipole center to another is then the final side of the triangle and is equal to 3.1 Å.

which is actually very close to the experimental value of $U = -76$ kcal/mole. The fact that the potential energy is negative indicates that the hydration arrangement is a favorable one from the energy standpoint. The system thus liberates energy for every cation hydrated, and the arrangement of cation and waters of hydration has a lower electrical potential energy than do the separate systems of cation and water.

At this point the reader should be impressed with two points. The first is that this calculation demonstrates the relative contribution of each type of interaction between the cation and the primary waters and that the initial assumption that $D = 1$ must be reasonably correct. In the model used the most important interaction by far is that between the cation's charge and the water's dipole. This implies that a multivalent cation should have a larger hydration energy, and this is the actual case. However, the size of the cation is also important since it must be able to accommodate the waters around it. The second point is that this type of calculation is not by any means easy. In fact, there are several interactions that have been completely ignored. For instance, the interaction between the cation and the waters in region II should contribute negative U to the total energy, whereas the hydrogen bonds the primary waters had to break to become primary waters would contribute a positive U to the total energy. Such a seemingly simple thing as calculating the hydration energy of a monovalent salt is actually quite tricky; it depends heavily on the details of the model used, not to mention the accuracy of the physical parameters used in the calculations.

Hydrophobic Substances

Another group of compounds whose hydration characteristics should also be described fall under the general category of hydrophobic groups. In many respects the hydration of hydrophobic groups is quite different from the hydration of ionic or dipolar species. For hydrophobic compounds, the basic interactions with water are different than the interactions between water and ionic compounds, and in some situations nonelectrical interactions or motivating forces are responsible for the characteristic behavior of hydrophobic compounds.

Just as an inorganic salt tends to partially disorganize the structure of bulk water upon hydration, the hydration of a nonpolar compound has just the opposite effect. It is generally believed that upon placing a nonpolar group in an aqueous solution, the water structure in the immediate vicinity of the nonpolar group becomes more organized than that of regular bulk water. In fact, the structural hydrates that are formed in the case of tetra-butylammonium molecules have a melting temperature that can be substantially above that of ice. For tetrabutylammonium hydroxide $(C_4H_9)_4NOH$ the melting temperature is about $30°C$. Another interesting fact is that the heats of hydration for a number of simple nonpolar compounds have been reported to be surprisingly similar, independent of their chemical composition. The implication of this fact is that the hydration energies of these nonpolar molecules, e.g., CH_4, C_2H_2, CH_3, SH, Cl_2, are not as dependent on the molecule itself as with the case of the cation discussed previously. It can then be further deduced that the hydration energy can have a significant contribution from the water structure itself that is formed in response to the introduction of the apolar molecule. It is definitely known that in some cases apolar molecules can have a clathrate or cagelike structure formed around them, and that these structures can exhibit great symmetry and structure, not to mention stability. The geometric arrangements of water molecules is not always the same in these clathrates, but the overall cage effect is real, with the apolar molecule(s) being enclosed. So, as an ion can cause the disruption of water structure, apolar molecules can cause its organization.

One may well wonder at this point why clathrate structures would want to form around an apolar molecule. Remember that the system (water + apolar molecule) would like to get into as energically favorable a position as possible; i.e., U is to be a minimum. For the hydration of an ion, this means a maximum number of strong interactions between the ion and the dipolar waters; but an apolar molecule cannot interact with water in the same fashion as an ion. One way of looking at this situation is to consider that since the apolar molecule cannot interact strongly with the waters, it is best if the waters interact more with themselves via hydrogen bonding to form a clathrate or cage structure around the apolar molecule. In so doing these water molecules will have to break their original hydrogen bonds in order to form new ones in the cage structure, but the overall result will be a net decrease in electrical potential energy, thus making the process energetically favorable. Usually the decrease in U is not relatively large, but it does decrease. Another way of looking at the situation is to think of the apolar compound as forming a hole in the bulk water structure because the water molecules would rather interact with themselves than interact with the apolar compound. This is manifested in the general chemical principle that like dissolves like; i.e., polar compounds are best dissoved or solvated by polar liquids, and apolar compounds are best solvated by apolar solvents.

One consequence of this process of organizing the structure of water is that the entropy of the system decreases; i.e., with respect to the separate apolar and water systems, respectively, the hydrated apolar molecule system has a lower entropy. Another way of saying this is that the hydrated apolar molecule system is more organized as a result of the apolar molecule becoming hydrated. This process goes against a basic principle of thermodynamics, which says that all reactions tend to become more disorganized with time, with entropy being a measure of this disorganization. A detailed analysis of the apolar hydration process, then, would say that apolar molecules are not readily solvable in aqueous solutions because the entropy decreases too much, instead of increasing as it does in most other reactions. In this particular case both the change in electrical potential energy and the entropy must be taken into consideration for a complete description. The hydration of an apolar compound cannot be completely understood in terms of only electrical interactions. The reader is referred to one of the references at the end of the chapter for a more in-depth analysis of these points.

WATER AND BIOMOLECULES

At this time it is reasonable to ask the question, How does the structure of water affect biomolecules? Or conversely, how do biomolecules affect the structure of water? The two questions are hard to separate since each can influence the other. The answers to these questions are still far from complete, but it is certain that different biomolecules will generate varied responses, and that specific changes in the neighboring water molecules will depend heavily on the outlying chemical groups of the biomolecule. Biomolecules contain both polar and apolar chemical groups, both charged and uncharged groups. Proteins in particular have hydrophobic and hydrophillic amino acid side chains; and if, for instance, a charged group were exposed on the surface of one protein, it is expected that that spot would become hydrated in a fashion similar to that of the K^+ ion. Likewise, a hydrophobic chain sticking out from the surface of a protein would be expected to create a local hole in the bulk water structure. So, in the same biomolecule different chemical groups will be influencing water structure in completely opposite ways.

When an ionic and hydrophobic group are in close proximity to one another, the resulting water structure will depend on several factors. Consider, for example, a short chained fatty acid of two or three carbons. Here, the charged COO^- group will be strongly hydrated by primary waters, and the hydrophobic portion will have relatively little influence or interaction with the water. Because of the steric hindrance of the hydrophobic part, the COO^- group will probably only have two or three primary waters; i.e., the water molecules do not have complete freedom to hydrate on all sides of the charge. For a long-chained fatty acid, the above situation will be changed somewhat because the hydrophobic portion now makes up a relatively large percentage of the total molecules. This apolar part will extend a considerable distance away from the charged head group. In this case the water structure along the hydrocarbon backbone would be expected to be more typical of a clathrate or cagelike structure.

Another point to be made is that by themselves the hydrophobic portions of proteins could not be solvated to the extent that they are in the proteins. One of the main reasons that apolar groups can go into solution to the extent they do in proteins is due to the fact that they are attached to polar or dipolar groups. These polar groups can interact so well with water and decrease the electrical potential energy so much that hydrophobic groups can be attached and the system can still be hydrated. In effect the polar constitutents of a molecule are "pulling" the hydrophobic groups into solution regardless of the fact that these hydrophobic groups do not dissolve well by themselves.

Although it is not absolute, there is a general principle that proteins tend to have their hydrophobic side chains located in the interior of the protein structure, while the polar and dipolar amino acids occupy the exterior parts. This is due to the fact that the polar side chains can interact much more strongly with the surrounding water molecules than can the hydrophobic groups. The hydrophobic groups are also not readily soluble in aqueous solutions, as was just discussed. In general, then, the interior parts of a protein will have a slightly different character than the exterior. For one thing, there will be fewer water molecules in the interior; this has the consequence that some of the titratable groups there will ionize at pH values that are different than if they were on the surface and had free access to water. Another example already discussed where these general principles apply is the case of a lipid bilayer. Here again the hydrophobic hydrocarbons are located in the interior, whereas the charged head group resides on the outside to interact with the neighboring waters.

Because of the structural complexity of most biomolecules, the exact state of water structure in the vicinity of the biomolecule is difficult to detail; yet, it will surely be different from that of bulk water. The interactions between the charged and polar groups on the surface of most biomolecules and the water molecules are of sufficient magnitude that an aqueous layer constantly surrounds the biomolecules and moves with them as part of a large hydrodynamic unit. So much water typically surrounds proteins that it is not unusual to have 1 g of water per gram of protein for a hydration value. Also, this layer of water has to be considered when biomolecules interact with one another. If two molecules have an equilibrium distance that is greater than their extended hydration layers, then the dielectric constant cannot safely be assumed to be 1. Conversely, if the two biomolecules have an equilibrium distance that is closer than the hydration layers allow, then these layers have to be displaced at the expense of some energy source, and also the dielectric constant will be reduced from that of bulk water.

As an example of how the structure of water can influence biological behavior on a macroscopic level, the case of general anesthetics will be discussed. As was pointed out by Linus Pauling in 1961, a number of general anesthetics are compounds that have very little reactive chemistry; i.e., they form relatively few complexes with other chemicals. Typical examples of these anesthetics are $CHCl_3$, N_2O, CO_2, C_2H_2, C_3H_6, N_2, Ar, and Xe. Looking at this group of compounds, there is hardly any property that all of them have in common with one another. What Pauling suggested was that their mode of action in acting as anesthetics is their ability to change the structure of water in the vicinity of nerve cells, thus disrupting the normal flow of ions necessary for nerve conduction. It was proposed that these compounds may form a clathrate type of structure, or another type for that matter, that is different than the normal water structure near nerve cells. This change of water structure would then inhibit the normal flow of ions into a nerve cell as an impulse is transmitted from one cell to another.

In 1973 a paper was published indicating that local anesthetics may well work by a similar principle. Using the luminescense produced by the firefly luciferin–luciferase reaction, the interaction of several local anesthetics with proteins was investigated. In the presence of an energy source this luciferin–luciferase enzyme system is responsible for the glow emitted in the tail of fireflies. This model system was chosen because the inhibitory effect of the anesthetics could be measured by the amount of luminescence emitted. The anesthetics used had a chemical structure with a hydrophobic benzene ring at one end of the molecule and a hydrophillic tertiary amine at the other end. For this particular system, it was suggested that the hydrophobic portion of the anesthetic interacts with the hydrophobic interior part of a protein thus causing a conformational change in which surface positive and negative charges neutralize one another by being moved close together and in which the interior hydrophobic groups become more exposed. Since the surface charges on the protein originally held the primary waters in one configuration, their neutralization, along with the increased exposure of the hydrophobic interior, will change this water structure. Possibly clathrate structures will then surround the newly exposed hydrophobic regions. The net effect is then that the water structure in the immediate vicinity of the protein is changed. If this effect were to also occur with the proteins associated with the surface of nerve cell membranes, the transport properties of specific ions, i.e., Ca^{+2}, would be disrupted and nerve conduction impaired.

The attractive part of this explanation is that it is supported by several other pieces of evidence. When the primary waters are being held by the surface charged groups of the protein, they are probably arranged in a structure that has a reduced volume compared to a clathrate configuration or that of bulk water. This is due to the strong electrostatic attractive forces exerted by the surface charges. The volume of this water structure (compared to bulk water) is decreased under the influence of strong ionic forces, and this phenomenon is known as electrostriction. When this reduced volume structure is replaced by a clathrate type structure, the volume of the water structure increases, which implies that any mechanism that resists the increase in volume of the water structure should be able to block the effect of the anesthetic. One such mechanism capable of doing this is pressure; it has been shown that for several systems, an increase in the external pressure can negate or reverse the effect of the anesthetic. This is quite in line with the above explanation. The increase in pressure makes it more difficult to form clathrate type water structures because more work must be done in order to counteract the increased pressure to expand the volume to the clathrate water structure. The water system thus finds it harder to push outward against an increased external pressure in order to form clathrates.

One final example of how water can influence biological systems will be illustrated by again discussing precipitation of proteins by ammonium sulfate. In Chapter 7 an explanation for this phenomenon was discussed in which the Debye length and VDW forces were integral parts. What was stated there is true, but it may not be the complete story. By solvating high concentrations of ammonium sulfate, which is essentially an ionic species and which will be hydrated as such, large numbers of water molecules will be sequested for this process. If this water is pulled away from hydrating the dissolved proteins, then there will be less water for this purpose, and the proteins will have a decreased ability to be hydrated. This will increase the tendency of the protein to fall out of solution. The ammonium sulfate precipitation then can be partially due to the decreased availability of water to hydrate the proteins present.

By considering the possibilities of the above situations, it can be easily seen that the structure of water can have pronounced biological effects. However, the full extent of water's influence on biological systems is not yet fully understood. In earlier chapters it was stated that water's high dielectric constant was quite significant; here we see that its basic structure, even if it is only statistical in nature, can also have a pronounced influence on biomolecules. In considering a complete description for the action of biomolecules from both a physical and a biological standpoint, water structure and its influence cannot be ignored.

SUMMARY

Water is a very ubiquitous substance and is also very important to life systems as we know them. It not only has a high dielectric constant, it also has important thermal properties that tend to minimize the impact of temperature fluctuations. These properties are beneficial to life systems, ensuring that they will have a relatively stable temperature. The structure of a water molecule is tetrahedral in shape with each corner of the tetrahedron holding either an electron pair or a hydrogen atom. Water molecules are capable of hydrogen bonding with a maximum of four other water molecules; these hydrogen bonds are responsible for many of water's macroscopic properties. In normal ice the water molecules are hydrogen bonded to other water molecules in a tetrahedral structure that essentially extends throughout the volume of the ice. When melted, liquid water retains much of this tetrahedral arrangement. When an ion is placed in water, the water structure in the immediate vicinity of the ion is altered, as is the case when a hydrophobic group is placed in water. These effects can influence the structure of proteins and can also affect macroscopic behavior.

REFERENCES

Berendsen, H. J. (1967). Water structure, *in* "Theoretical and Experimental Biophysics," Vol 1 (A. Cole, ed.). Dekker, New York.

Buswell, A. M., and Rodebush, W. H. (1956). Water, *Scientific American*, April, p. 77.

Frank, H. S., and Wen, Wen-Yang (1957). Structural aspects of ion-solvent interaction in aqueous solutions: a suggested picture of water structure, *Discuss. Faraday Soc.* **24**, 133.

Klotz, I. M. (1958). Protein hydration and behavior, *Science* **128**, 815.

Klotz, I. M. (1962). Water, *in* "Horizons in Biochemistry" (M. Kasha and B. Pullman, eds.). Academic Press, New York.

Pauling, L. (1961). A molecular theory of general anesthesia, *Science* **134**, 15.

Rich. A., and N. Davidson, eds. (1968). "Structural Chemistry and Molecular Biology." Freeman, San Francisco, California.

Snell, F. M., Shulman, S., Spencer, R. P., and Moos, C. (1965). "Biophysical Principles of Structure and Function." Addison-Wesley, Reading, Massachusetts.

Ueda, I., H. Kamaya, and H. Eyring (1976). Molecular mechanism of inhibition of firefly luminescence by local anesthetics, *Proc. Nat. Acad. Sci. USA* **73**, 481.

9

EXPERIMENTAL ELECTRICAL TECHNIQUES

INTRODUCTION

Through the previous chapters we have mainly focused our attention on the theoretical aspects of electrical interactions among biological macro-molecules. This chapter will deal with a variety of experimental techniques that rely on the charge characteristics of these molecules. Each technique will be described in summary since detailed descriptions would (and do) require whole volumes. It has been seen that biological molecules can have a net charge, either positive or negative; and even if the net charge is zero, they can have a dipole moment. The fact that each type of individual mole-cule has a unique charge structure can be used as a means of characterizing, purifying, or separating it from other biomolecules; this is of utmost im-portance in experimental molecular biology. If individual types of molecules cannot be physically separated into essentially homogeneous solutions, then it is quite difficult to unravel the mysteries of their workings. The techniques to be discussed here may all differ in their basic principles of operation, but they all rely on the simple physical fact that biomolecules carry electrical charge, either in the form of net charge or in the form of a dipole moment.

ELECTROPHORESIS

If you place a particle in a uniform electrical field, it will do one of several things, depending on that particle's charge. If there is a net charge, the particle will begin to move under the influence of the field. If the net charge is positive, the particle will migrate toward the cathode; if negative, the particle moves toward the anode. The transport of a charged particle under the influence of an electric field is called electrophoresis and is by far the most prominent and widely used of all the electrical techniques. Without the technique of electrophoresis, our present-day knowledge of molecular biology would not be as advanced as it is. It is one of the major tools the research scientist uses in the description of molecular systems. The technique itself was developed in this century, although the phenomenon has been known for over 150 years. The technique of electrophoresis can take a number of forms with each variety having a set of advantages and disadvantages. It should also be remarked at the outset that electrophoresis is not only an experimental technique used in the laboratory, but that it is routinely used in hospitals in the diagnosis of many disease states. It can therefore be used as a technique of practical, real-world interest.

The fact that colloidal particles can move through a liquid under the influence of an electric field was first observed by the Russian physicist Alexander Reuss in 1807. This finding was further investigated by Michael Faraday and others, and it was demonstrated that the rate of movement of charged particles was dependent on their net charge, with positively charged particles moving toward the cathode and negatively charged particles moving toward the anode. This migration of particles based on net charge thus initiated a new technique for separating a group of particles from a common solution. If each separate species exhibited a different charge, then in principle they could be separated from one another by electrophoresis. The resulting homogeneous groups could then not only be studied independently, but the relative abundance of any group could be determined with respect to the total. For proteins in blood serum, this is quite significant since the relative amount of various proteins present is indicative of a person's health.

Over the years a number of different electrophoretic techniques have been developed. There are basically three distinct types: microscopic technique, moving boundary, and zone electrophoresis. Whichever technique is being discussed, there are certain elements common to all techniques. First, it is necessary to have a charged particle that is free to move. Second, a fluid is needed that is compatible with the particles and that will support enough of an electric field to move the particles, yet not conduct too much to cause excessive heating. Thirdly, a source of electrical power is required. Fourth, a means of observing the movement or detecting the final resting place of each component is needed; and finally, a means of stabilizing the separated components is required to prevent random mixing due to convection currents or other disrupting effects while the electrophoresis is taking place. The various forms of electrophoresis commonly employed today all have these elements, although they differ in form and in the principles used to achieve the success of the techniques.

ELECTROPHORETIC MOBILITY

If a particle with a net charge is in a fluid and it is subjected to the influence of an electric field, the particle will tend to move due to the electrical force exerted on it. This force will depend on both the particle's charge and the strength of the field E and is given by

$$F_{el} = qE \qquad (9\text{-}1)$$

Since the particle is in a fluid medium, as it moves it will also experience a frictional retarding force. This force will depend on the fluid and the geometry of the particle in question, but in general it is usually assumed that the frictional force is proportional to the particle's velocity or

$$F_{fr} = fV \qquad (9\text{-}2)$$

where f is a constant of proportionality called the frictional coefficient. When the electric potential is first applied, the particle will accelerate, but only for a short while. It will then quickly reach an equilibrium velocity where the motivating electrical force equals the retarding frictional force, since at equilibrium Newton's second law of motion states that the sum of the forces acting on a body in equilibrium add up to zero. This is illustrated in Fig. 9-1. Since frictional forces always resist motion, we may equate the two forces. In the case depicted in Fig. 9-1 the forces point in opposite directions, so at equilibrium

$$fV = qE \qquad (9\text{-}3)$$

or

$$M = V/E = q/f \qquad (9\text{-}4)$$

where the ratio V/E is defined as the electrophoretic mobility. The electrophoretic mobility then is a measure of a particle's velocity per unit field strength. It is a physical parameter that is characteristic of a molecule under the conditions in which it was measured. It is dependent on the molecule's charge and also on its geometric shape since different shapes will result in different frictional drags. The net charge, you will remember, depends on the pH, and also possibly on the temperature and the solvent. The quantity M can likewise be affected by all of these parameters. If the particle in question can be assumed to be spherical, it has been shown that

$$f = 6\pi\eta_v r \qquad (9\text{-}5)$$

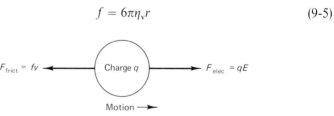

Fig. 9-1   After a brief initial acceleration a particle of charge q in a uniform electric field reaches a constant velocity. In this equilibrium condition the frictional force equals the electrical force. Frictional forces always retard motion.

where η_v is the viscosity of the solvent and r is the sphere's radius. The frictional coefficient can be similarly developed for other geometries, but this is much more difficult, and the mathematics more complicated. From looking at the definition of mobility, it can be seen that its units are square centimeters per volt second. For small univalent ions, M has values of around 6×10^{-4} cm^2/V sec; and for proteins, values of M typically range between 0.1×10^{-4} and 1.0×10^{-4} cm^2/V sec.

The above simple model allows one to gain an understanding of the definition of electrophoretic mobility and to gain insight into those factors that may influence it; however, it is not easy to develop this system in any more detail because several difficulties arise. One problem is that of the counterions surrounding each particle. This layer attenuates the applied field so the charged molecule does not experience the same field that is present at, say, the electrodes themselves. A mathematical correction for this effect has been worked out as a function of solvent ionic strength using the Debye–Hückel theory, but it will not be discussed here. Another problem arises due to the distortion of the counterion cloud as the particle experiences shearing forces when it moves through the fluid. Yet another problem has to do with the distortion of the counterion cloud due to the applied field. As the reader may easily see, a complete theoretical description relating electrophoretic mobility to common parameters of the system would be extremely difficult, and in fact no complete description has yet proven adequate. Because of this, electrophoresis has several disadvantages as a precise quantitative technique. It is even difficult to determine the net charge on a macromolecule unambiguously due to uncertainties in the theory. However, the real value of electrophoresis is not in its analytical nature but in its ability to physically separate molecules that have a different net charge. Most of the electrophoretic techniques in use today apply this fact, and they are analytical to the extent that the percentage of each component in a solution may be determined.

Three completely separate types of electrophoresis (microscopic, moving boundary, and zone) will now be discussed, with emphasis being placed on the last two.

MICROSCOPIC ELECTROPHORESIS

In this variation of electrophoresis an electric field is applied so that the particles in solution move in a horizontal direction on a glass slide. A microscope is used to focus on one or several particles; their motion is observed as a function of time. Mobilities are calculated by measuring the time a particle takes to move a prescribed distance and by knowing the value of the electric field. Naturally, this technique is limited to fairly large particles and is not as useful as other techniques in studying molecular motion. Biological cells and large colloidal particles can be studied with this technique. Microscopic electrophoresis is seldom used today, mainly because other methods are more convenient; hence it will not be discussed in any more detail.

MOVING BOUNDARY ELECTROPHORESIS

Much of the development of this technique was carried out by the Swedish chemist A. W. K. Tiselius who was awarded the Nobel prize for his efforts. This particular form of electrophoresis was introduced in the late 1930s; it was one of the first forms in which precise analytical measurements of electrophoretic mobility could be determined. It was also the technique that ushered in electrophoresis as a major scientific tool since it was with this technique that noteworthy advances were made in the detection and separation of blood serum proteins. Although moving boundary electrophoresis is seldom used today because of the ease with which zone electrophoresis can be done, it is still important to understand the role that moving boundary techniques played in the development of electrophoresis.

Tiselius Cell

As the name implies, the moving boundary technique relies on observing the movement of a boundary under the influence of an electric field. The boundary referred to is one between a macromolecular solution and the pure solvent alone. The actual technique is performed in a U-shaped tube where a sharp boundary between solution and solvent is artificially created at the beginning of the experiment. An electric field is then applied, and the boundary is observed as a function of time. Under the influence of the electric field, the macromolecules in the solution will migrate into the pure solvent thus changing the position of the boundary as a function of time. This is illustrated in Fig. 9-2. Although it is not shown in Fig. 9-2, one arm of the apparatus supports the positive electrode, and the other supports the negative electrode. When the electric field is applied, the boundary in one arm ascends while the other decends as the charged molecules in solution migrate toward the electrode of opposite polarity—this is assuming the molecular solution is homogeneous, i.e., having only one type of biomolecule. If the solution under test is heterogeneous in its composition, then the individual components will each migrate toward the appropriate electrode. By following both the distance the boundary migrates and the elapsed time, the

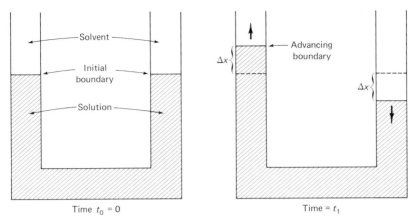

Fig. 9-2   Schematic of a Tiselius U tube used in moving boundary electrophoresis. The two situations are shown at two different times: before the start of the experiment $t = t_0$, and at a later time t_1. Under the influence of the electric field the boundary separating the solution and the solvent moves a distance Δx in time $t_1 - t_0$, so the velocity of the boundary may be calculated. By measurement and calculation, the strength of the applied electric field can also be obtained; the electrophoretic mobility may be finally calculated by a simple division. The electrodes supporting the electric field are not shown.

velocity of a boundary can be determined. By concurrently measuring the strength of the electric field, the mobility of the boundary can be found by simple division; this is assumed to be equal to the mobility of the biomolecule contained within that boundary. If a solution contained several types of molecules, then the respective boundaries would have to be monitored separately.

The actual apparatus that was originally designed by Tiselius is shown in Fig. 9-3. The large reservoirs on either side are filled with a buffer solvent which completely immerses the two electrodes and helps maintain electrical contact throughout the system. The electrodes are located relatively far from the actual moving boundaries to prevent any products formed at the electrodes from interfering with the boundary itself. The U tube part of the apparatus is located in the middle near the bottom. It has flat parallel windows to allow the boundaries to be observed with an optical system, which is not shown. Although it is not obvious in Fig. 9-3, the U tube portion of the Tiselius cell is made up of three sections that can slide to the right or left with respect to one another. This facet was incorporated to help facilitate the experimenter in creating the initial boundary between solvent and solution. The initial boundary's sharpness is of critical importance because if it is diffuse, it would be difficult to locate it accurately. The boundary will naturally tend to broaden as a function of time due to the effects of diffusion, with or without the application of the electric field.

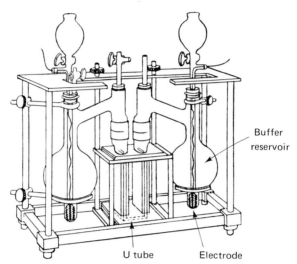

Fig. 9-3 Drawing of the Tiselius electrophoresis cell. The actual U tube where the electrophoresis takes place is in the center bottom of the apparatus. The large reservoirs on either side are for holding buffers, and the respective electrodes are seen immersed there. The whole apparatus was immersed in a water bath for temperature control (Alberty, 1948).

Boundary Stability

One of the problems associated with the moving boundary technique was that various elements tended to disrupt the stability of the boundaries. Mechanical vibrations and convection currents both worked to blur and dissipate the sharpness of the boundary. The convection currents were caused by the inhomogeneous heating of the solution and solvent as the electrical current passed through. Because of these problems, the entire Tiselius cell was placed in a water bath maintained at a temperature close to 4°C. At this temperature water has its maximum density and also its change in density with respect to temperature is a minimum. The reasoning behind this argued that if the density of the solution had a maximum stability, then convection currents based on a density differential would be less likely to occur. Remember that as the electrical current heated up various parts of the solution, its density would decrease, and these portions of solution would then have a tendency to rise. With the rising of certain columns of fluid, convection currents would be created. Tiselius made his U tube very long and narrow with a lot of surface area to help the external bath keep a constant temperature within the tube. The effect of immersing the cell in a water bath was to keep the boundaries sharper and to increase the ability of the technique to distinguish between two closely spaced boundaries, thus to better resolve the different mobilities between two components in the solution.

Buffers

The pH of the electrophoretic solvent and solution had to be maintained rather rigidly because as the pH changed, so too did the charge on the bio-molecules. If a particle's net charge changed, then its mobility would change also, and the measurement of the mobility would become hopelessly com-plicated. To prevent this the macromolecule's solvent was a buffer, as was the pure solvent itself, and in reality the solution was extensively dialyzed* against the solvent to be used in the experiment so as to provide as constant an environment as possible for the biomolecules. The importance of this is seen in that the particles under electrophoresis are initially in one buffer

* Dialysis is a process whereby the composition of a solution containing biomolecules can be changed, in particular the original solvent is exchanged for another. Typically, the process proceeds as follows. The original solution is placed in a bag made of a semipermeable membrane, and then the whole bag is immersed in the new solvent. The pores of the dialysis bag are just large enough to allow solvent molecules to pass through, but retain the larger solute molecules. In this way the macromolecular component stays in the bag. By allowing time for an equilib-rium condition to be obtained, it can be ensured that the solvent inside the dialysis bag has the same composition as the solvent outside the bag (see below). This process is usually repeated several times before the first aqueous solvent is completely replaced by another (see figure). Dialyzing a solution against its solvent ensures an exact match between solvents which is im-portant in moving boundary electrophoresis.

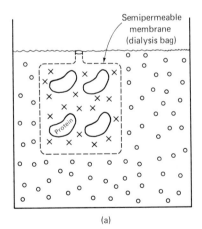
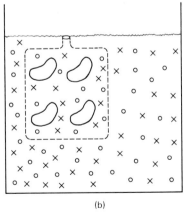

(a) The original solution in solvent X is placed in a dialysis bag and immersed in solvent ◯. (b) After the time for equilibrium has elapsed, the solvent inside the bag is the same as outside. The original solvent has been diluted by solvent ◯. After several more changes of buffer ◯, the original solvent is completely replaced by solvent ◯.

or solvent at a specific pH; if they then move into another buffer with a different composition, their charge could change. Hence their mobilities would differ from one solvent to another. Therefore, it was important to have the solvents above and below the boundary be as identical as possible. Another consideration in buffer composition was its reactivity with the macromolecular components. If the interaction caused sufficient change in the molecular characteristics, then the mobilities measured would be different from that of the unreacted molecule. In reporting mobilities it is common practice then to detail the pH and buffer systems used in the actual measurements.

Optical System

Once the initial boundary is formed and the electric field is applied, the boundaries migrate consistent with the charge of the molecules present. This boundary movement is followed by a rather sophisticated optical system known as a schlieren system which operates on the following principle. When a parallel light beam passes through a cell that has a refractive index gradient dn/dx, the light beam is deflected through an angle θ which is proportional to the magnitude of the gradient (Fig. 9-4). The schlieren system essentially detects the angle through which the light has deviated compared to that of the undeviated beam. Since the index of refraction is directly proportional to the concentration of solute present at any location, the index of refraction gradient measured by the schlieren system is actually a measure of the concentration gradient of the solute in the cell. The location where the deviation of light is maximum indicates the highest value of dn/dx, and this corresponds to the boundary between solvent and solution. By following this point as a function of time, the boundary velocity can be obtained. The schlieren system thus permits one to observe the index of refraction gradient as a function of time. This is shown in Fig. 9-5 where we consider the case of an electrophoresis experiment involving a solution containing two types of macromolecules, one having a greater mobility than the other. As electrophoresis progresses, the faster solute component migrates ahead of the bulk solution forming a new boundary, one side of which is pure buffer and the other side is pure component one. Meanwhile, the slower component also migrates, forming a new boundary; but this boundary has the faster component 1 on both sides of it. Nevertheless, there is a distinct location that characterizes the forward advanced position of the slower component. At this spot the concentration of the solute is greater than at the boundary of the faster component; hence the refractive index gradient should be higher here. The total solute concentration is depicted as a function of distance in Fig. 9-5c. This is also a plot of the total index of

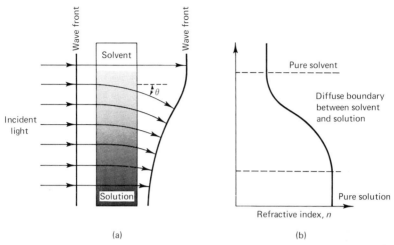

Fig. 9-4 (a) A parallel beam of light is incident on a cell containing a gradient in the index of refraction dn/dx. The highest value of n is in the bottom of the cell. The gradient is made by allowing a sharp boundary between solvent and solution to diffuse or by having the solution undergo electrophoresis into the pure solvent. The index of refraction gradient is directly proportional to solute concentration gradient; therefore n is directly proportional to the solute concentration at any point. As the light passes through the cell, it is deviated through an angle θ, thus distorting the wavefront. The angle of deviation θ is proportional to the gradient dn/dx, so pure solvent or pure solution will not bend the incident light. However, the light passing through the pure solution will not travel as fast as that going through the pure solvent. (b) A corresponding plot of the index of refraction vs. location in the cell. It is seen that in the boundary region (space between dotted lines) the value of dn/dx has a maximum. The schlerien optical system actually measures this dn/dx maximum as a function of time.

refraction vs. distance. The schlerien system is, however, sensitive to the gradient of refractive index, and a plot of dn/dx is shown in Fig. 9-5d. As can be seen, the derivative of the total solute concentration (or total n) has a peak shape; the centers of the peak are at the respective boundaries. By observing the movement of the schlieren peaks as a function of time, electrophoretic mobilities can be calculated. Every component in the solution will have its own corresponding schlieren peak; and if the mobilities are sufficiently different, the peaks will be separated from one another as in Fig. 9-5d. Two components whose peaks are not separated from one another are said to be unresolved and will appear as one large peak, possibly with a shoulder.

Since the schlerien peaks are a measure of concentration gradient, the area of each peak should be proportional to the concentration of the respective components; hence by measuring the area of each peak the percent concentration of each component in a mixed solution can be calculated by

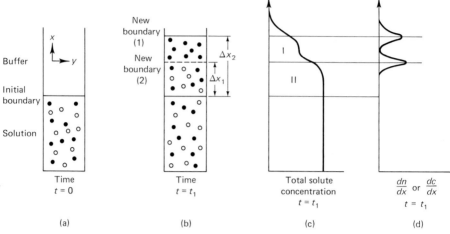

Fig. 9-5 (a) A two-component system ready for electrophoresis at time $t = 0$. A sharp boundary exists between solution and buffer solvent. (b) The same system during electrophoresis at same time later $t = t_1$. The ● solute molecules have moved ahead of the ○ molecules to form boundary 1. The ○ solute molecules have moved to form boundary 2 which is at the forward edge of that component's movement. (c) A plot of total solute concentration vs. distance at time $t = t_1$. The concentration and index of refraction in region I is due to the ● component alone, while in region II both components contribute. The shape of the curve in (c) is also the same for a plot of n vs. distance. (d) A plot of dn/dx vs. distance. This is essentially the derivative of part (c). The peaks shown in (d) are the actual shape seen in a schlieren pattern. These peaks move as the boundaries move, and mobilities are calculated from these data.

dividing by the sum of all the areas, or

$$\%C_i = \frac{A_i}{\sum_i A_i} \tag{9-6}$$

where A_i is the measured area of peak i, $\sum_i A_i$ is the sum of all areas, and $\%C_i$ is the percent of the component i in the solution.

Electric Field

In order to calculate the electrophoretic mobility M it is required to know the magnitude of the electric field moving the particles. For the Tiselius apparatus, this quantity can be ascertained as follows. If we make use of Ohm's law ($V = iR$) and consider a cross-sectional slab of the U tube, then the potential difference dV across the slab is given by

$$dV = -i\,dR \tag{9-7}$$

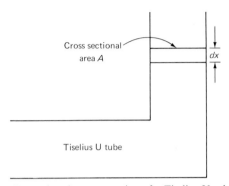

Cross sectional
area A

dx

Tiselius U tube

Fig. 9-6 A diagram illustrating the cross section of a Tiselius U tube. The cross-sectional volume shown has thickness dx and area A.

where dR is the resistance of the thin slab and i is the current across the slab (see Fig. 9-6). The minus sign is included to remind us that the current flows from the higher potential to the lower. If the resistance is now expressed in terms of the specific condutivity K,* Eq. (9-7) becomes

$$dV = -i\frac{dx}{KA} \qquad (9\text{-}8)$$

where A is the cross-sectional area of the slab. Since dV/dx is equal to the electric field, we have

$$E = i/KA \qquad (9\text{-}9)$$

Thus, by knowing the cross-sectional area of the cell, the current, and the specific conductance of the solution, E can be calculated. The current can be measured at any point in the circuit because it is the same everywhere, and K is measured in an independent experiment.

As an example of the calculations involved in the determinations of a mobility, let us consider the case of bovine serum albumin in a 0.1 ionic strength buffer of pH 8.6 where $K = 3.1 \times 10^{-3}\,\Omega^{-1}\,cm^{-1}$. If 15 mA of current are used and the schlerien peak moves a total of 4.0 cm in 180 min in a cell that is 0.76 cm^2 in cross section, it is found that $E = 6.4\,V/cm$, and

---

* The reciprocal of the resistance $1/R = L$ is called the conductance and has the units of Ω^{-1} or mhos. The conductance of a solution is a measure of its ability to pass a current and is usually measured in a conductivity cell where the conductivity $L = KA/l$. The quantity l is the length of the cell, A is its cross-sectional area, and K is the specific conductivity, which is a characteristic of the solution itself. Conductivity cells are usually calibrated so that l and A do not enter into the calibration directly. Using the above definitions, we have $1/R = KA/l$, and also $dR = dx/KA$ which is used to derive Eq. (9-9).

$M = 5.8 \times 10^{-5}\,cm^2/V\,sec$. The reader should realize that the electric field in this example is not particularly large and that in an electric field of unit strength the albumin molecules will move at a velocity of $5.8 \times 10^{-5}\,cm/sec$, which is not exceedingly fast.

Isoelectric Point

Besides allowing one to calculate the electrophoretic mobility of a molecule at a specific pH, the moving boundary method can also be used to determine isoelectric points. Remember that the isoelectric point is the pH of a solution at which there is no net movement of a biomolecule. By calculating the mobilities at different pH's, that pH at which $M = 0$ can be found. This naturally involves a series of experiments over a range of pH values with possibly different buffer systems. An example of this type of experiment is shown in Fig. 9-7 for the milk protein β-lactoglobulin. In this case the mobility equals zero at a pH of about 5.2.

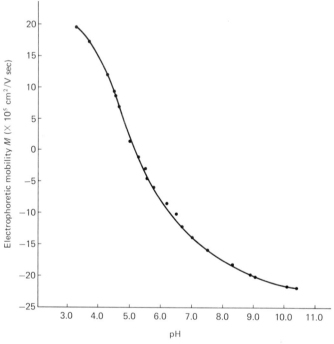

Fig. 9-7   The electrophoretic mobility of the milk protein β-lactoglobulin plotted vs. pH. Three different buffer systems were used to cover the pH range shown. The ionic strength is $I = 0.02\,M$ (Pedersen, 1936).

The isoelectric point is important because it is a physical constant used to describe a particular protein and also because it can give some insight into the chemical composition of the protein. Suppose that the electrical charges on a protein are due only to the ionization of NH_2 or $COOH$ groups in the amino acid side chains. If a particular protein, for instance, has an abundance of $COOH$ groups, the net charge would be very negative at high pH values, less so at low pH values. So, a low isoelectric point would indicate more $COOH$ than NH_2 groups in the side chain composition. This type of interpretation is however not without its pitfalls since an abnormally titrating group can cloud the issue. Table 9-1 shows the isoelectric points of a number of common proteins. From this table it is seen that isoelectric points for proteins vary throughout the entire pH range.

TABLE 9-1

Isoelectric pH for a Number of Common Proteins

Protein	pH	Protein	pH
Serum albumin	4.9	Cytochrome C	10.7
egg albumin	4.6	Hemoglobin	6.8
β-lactoglobulin	5.2	Myoglobin	7.0
Chymotrypsinogen	9.5	Pepsin	1.0

Protein Separations

One of the major macromolecular groups to be analyzed via electrophoresis is that of the proteins, and one of the first systems to be characterized was that of the proteins found in blood serum and plasma. In early work Tiselius and others used electrophoresis to separate the various protein groups in blood plasma. Figure 9-8 shows a typical schlerien pattern for normal human plasma proteins separated electrophoretically. The major protein groups are albumin, α_1, α_2, β, fibrinogen, ϕ, and the γ globulins. Blood serum would be similar, but without the ϕ component. By measuring the areas of the various peaks the relative concentrations of each component can be obtained. These are given in Table 9-2. It has been found from experiment that the resolution of the five peaks shown in Fig. 9-8 is very dependent on the buffer system used and also on the pH. Experience has shown that a pH of 8.6 is best for human plasma and that a diethylbarbiturate buffer gives maximum resolution. Horse plasma shows better results with a phosphate buffer at pH 7.7. From the electrophoretic pattern shown in Fig. 9-8 it is seen that the albumin fraction has the largest mobility and hence must have the largest negative charge. This is true only if we make the assumption

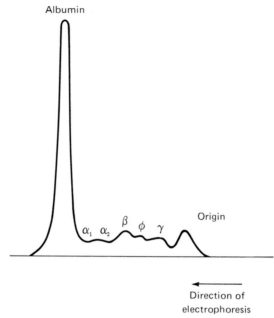

Fig. 9-8   A typical schlieren pattern for the separation of human blood plasma proteins. Each peak represents a different group of proteins within the plasma, and several protein types may be represented by each peak. The albumin group has the highest mobility, for the arrow indicates the direction of electrophoresis.

TABLE 9-2

Relative Concentrations of the Major Protein Groups in Human Blood Plasma and Their Respective Mobilities[a]

Component	Concentration (%)	Mobility (cm²/V sec)
Albumin	60	-6×10^{-5}
α_1	5	-5.1×10^{-5}
α_2	7	-4.1×10^{-5}
β	12	-2.8×10^{-5}
ϕ	5	-2.1×10^{-5}
γ	11	-1.0×10^{-5}

[a] The pH is 8.6. All the proteins have a net negative charge, hence the negative sign for the mobilities.

that all the proteins have similar hydrodynamic properties. It is possible for one protein to have less charge but a higher mobility than another if its frictional coefficient is proportionally less; however, this is not the case with serum albumin. The γ globulin fraction has the lowest negative charge, and also the lowest mobility. Included in the globulin fraction are those proteins commonly known as antibodies, which are an integral part of a person's immune system. The relative abundance of this fraction in any plasma sample can be used to diagnose a disease state. Today it is known that there are many more than six components in blood serum; the reason they do not show up in the experiment depicted in Fig. 9-8 is that the system was not sufficiently sensitive to resolve the other components.

Sickle Cell Anemia

Another example illustrating the power of electrophoresis can be seen by its ability to distinguish between macromolecules in a solution that are seemingly identical. This property can be used to check the homogeneity of a particular solution and to determine whether the solution is really composed of identical molecules or is instead a mixture of very similar structures. The most famous case illustrating this is the case of hemoglobin. The molecule hemoglobin (Hb) is a protein that is found in abundance in red blood cells (10^8/cell) and whose purpose it is to transport O_2 from the lungs to the individual cells in the organism's normal metabolism. Even a small change in its ability to transport O_2 can mean immediate death for the host. However, one should not imply that all hemoglobin molecules are identical. The molecule not only has a slightly different primary structure depending on the animal species of its origin, but the amino acid sequence can also vary within the species itself. Within the human species, there are over a 100 different types of hemoglobin. Sometimes these differences are benign and unnoticed, while others can be quite lethal.

Figure 9-9 shows a plot of the electrophoretic mobility vs. pH for two types of hemoglobin that were isolated from humans. Type Hb-A is a normal hemoglobin, while Hb-S is abnormal. From this plot it is seen that Hb-A has a slightly lower isoelectric point than Hb-S and that the mobility of Hb-S is uniformly more positive than that of Hb-A. Both these facts indicate that Hb-S has more of a positive charge than Hb-A.

A diagram of hemoglobin is shown in Fig. 9-10. It is seen that the whole protein is composed of two α and two β subunits, each of which are about 70% helix. The two α subunits have 141 amino acids each, while the two β subunits have 146 amino acids. Biochemical analysis showed that the amino acid sequence of the α chains were identical in both the Hb-A and the Hb-S hemoglobins. However, the β chains showed a small difference

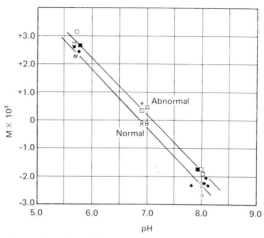

Fig. 9-9 The electrophoretic mobility plotted vs. pH for two types of hemoglobin. Hb-A is normal hemoglobin and Hb-S is abnormal (Pauling *et al.*, 1949).

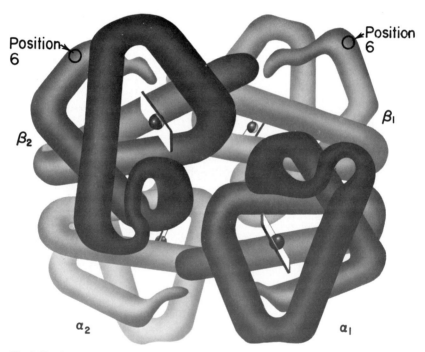

Fig. 9-10 A representation of the hemoglobin molecule. The location of the amino acid replacements causing sickle cell hemoglobin are position 6. The heme groups are shown as a flat disk. (Reprinted from "The Structure and Action of Proteins" by R. E. Dickerson and I. Geis, W. A. Benjamin Inc., Menlo Park, California, Publisher. Copyright 1969 by Dickerson and Geis.)

from the normal to the mutant. The residue at the sixth position in the normal hemoglobin was found to be glutamic acid, whereas in Hb-S it is valine; otherwise, the β chains are identical between the two hemoglobins. This small difference of two residue's difference in a total of 574 residues may at first seem insignificant, but this is far from the case. Glutamic acid, remember, has a COOH group in its side chain, and hence it can add a contribution to the net charge of the protein; however, valine cannot. Normal hemoglobin, then, will differ in charge from Hb-S by a maximum electronic charge of -2, which thus accounts for the differences in the respective mobilities.

The loss of charge in Hb-S is apparently quite critical to the life of the person having this type of hemoglobin because after the Hb-S molecules deposit their O_2 they have a tendency to aggregate with one another within the red blood cell, consequently altering the shape of the cell from its normal configuration to that of a sickle shape. This occurs because the aggregates are too large to be accommodated by the cell. This change in shape of the red blood cell in turn affects their ability to move through blood vessels. The disease state associated with this condition is called sickle cell anemia, and it is often fatal.

The above situation is one of the best cases illustrating the importance of electrical charge and how a minor change in charge on a molecular level can have significant consequences for the host organism. It also illustrates how electrophoresis can show the difference between two molecules that are identical except for a small difference in charge. Several other types of hemoglobin found in man are Hb-C and Hb-F. In Hb-C the same glutamic acid residue in the β chain that is replaced by valine to form sickle cell hemoglobin is now replaced by lysine. This type of hemoglobin can frequently cause difficulties to a woman who is pregnant, especially in the third trimester. H b-F is known as fetal hemoglobin and is present in the fetus. In Hb-F the β chains are completely replaced by two new chains, called γ. Fetal hemoglobin has the ability of being able to pick up and deliver O_2 when the O_2 tension is very low. This trait is quite advantageous to a growing fetus. Also, after birth fetal hemoglobin is slowly replaced by Hb-A. Another type of hemoglobin is termed A-2 and occurs where the normal β chains are replaced by δ chains.

The point to be made here is that all these different types of hemoglobin have characteristic electrophoretic mobilities, and they can be separated from one another in a common sample. In this situation electrophoresis can recognize a normal vs. an abnormal hemoglobin pattern and thus can aid a physician in deciding how to treat a patient. Used in this fashion, electrophoresis is not just a laboratory technique used by the researcher, but it is a routine procedure used in medicine.

ZONE ELECTROPHORESIS

In the moving boundary technique the macromolecules to undergo electrophoresis are supported by and move through a completely liquid medium. Concentration gradients are set up in response to the electric field, and their progress as a function of time is followed optically. With this procedure great care is needed to keep the boundaries sharp. In zone electrophoresis these macromolecules move through a solid or semisolid supporting medium. The reason for running on a solid medium is to maximize the stability of the various zones against the disruptive convection currents experienced in the moving boundary technique. In this fashion the different components in a solution can often be completely separated from one another in areas called zones, hence the name for the technique. The advantages of zone electrophoresis over moving boundary are: It is relatively simple to perform; only very small quantities of solution are needed; several solutions can be analyzed simultaneously; the methods of detecting the separated zones are straightforward; and in combination with molecular sieving effects, it can have remarkable resolution. Zone electrophoresis, then, is a natural outgrowth from the moving boundary technique and is an effort to overcome several of the major disadvantages of the earlier technique.

A typical zone electrophoresis apparatus is diagrammed in Fig. 9-11. In this apparatus a shallow tank is divided into two compartments, both of

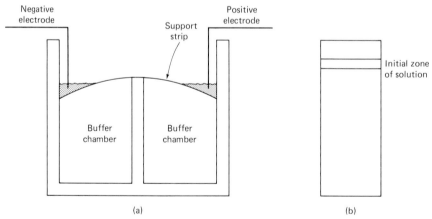

Fig. 9-11   (a) A side view for one type of configuration of a zone electrophoresis apparatus. Two separate buffer chambers are isolated from each other by a wall. Electrodes of opposite polarity are placed in each half, and the strip actually supporting the electrophoresis is placed so each end contacts the buffer in either half. (b) A top view of the support strip where the initial thin zone of solution is placed on the strip with a special applicator. The separate components in the initial solution then migrate down the length of the strip and separate from one another according to their respective mobilities.

which are filled with the buffer that is to maintain the ionic strength and the pH throughout the experiment. Electrodes of opposite polarity are inserted in each compartment, respectively, and the support strip with a thin zone of the test solution is placed so that one end makes electrical contact with each chamber. If the solution being analyzed has macromolecules of unknown mobilities, then the initial zone is placed at the center of the strip to allow for a range of motion in either direction for the components. Usually the buffer solution is at such a pH that all molecules will have the same charge, and hence they will all move in the same direction. In this case the initial zone is placed at one end of the strip, thus providing more opportunity for the components in the solution to be completely separated from one another in electrophoresis along the whole length of the strip.

The resolution of components in a solution via zone electrophoresis depends on, among other variables, the linear distance the zones travel. In general, the farther they travel, the better is the resolution. The main disadvantage with using long electrophoresis times is that diffusion will broaden the bands; if two zones are close together, the effects of diffusion will cause them to overlap. When the electrophoresis is completed, the electric field is turned off, the strips are removed and stained to show the location of the various components, which are otherwise invisible. The stains commonly used react with the proteins, but not with the strips. The amount of stain taken up by each zone is also proportional to the total amount of material in that zone; hence, the percent concentration of each component can be calculated by measuring the color intensity of each band.

The support strip itself can be made from a variety of materials. Those that are popular today are paper, cellulose acetate, agar, agarose, starch, and polyacrylamide gel. The support strips can also be constructed from a combination of the above materials; e.g., an agarose gel may be spread on top of a strip of cellulose acetate which acts as a substrate to keep the agarose gel shaped properly and to prevent its collapsing on itself. On such supporting materials the electrophoresis can be performed either vertically or horizontally. By placing both ends of the strip in buffer solution the strip will stay wet throughout its length, thus providing a constant ionic strength and pH environment for the molecules during electrophoresis.

Let us now look at an idealized version of what happens to the initial thin band in zone electrophoresis. The experimenter first mechanically applies a thin band of the macromolecular solution in question to one end of the support strip (Fig. 9-12). Ideally this band should be as narrow as possible. If no electric field were to be applied, this initial band would spread due to diffusion; the band width would get wider with time. If an electric field is now applied, each component in the solution would migrate

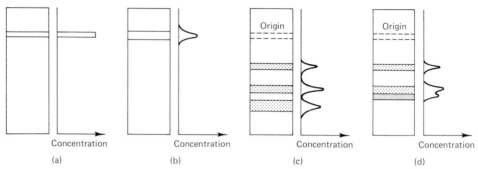

Fig. 9-12   (a) The test solution is initially placed on the support strip in a thin zone. This is done mechanically with a special applicator. A plot of solution concentration vs. distance along the strip is also shown. (b) illustrates what happens to the starting zone if it is allowed to just sit; it broadens due to diffusion. (c) Representation of three components completed separated or resolved by electrophoresis. The three components were initially all present in the thin starting band. (d) illustrates the electrophoresis of three components, two of which are not completely separated from one another. The peaks drawn on the concentration plot are a measure of the separate zone's color intensity after staining.

at a rate corresponding to its electrophoretic mobility. If each component has a different M, separate zones will be formed where each zone contains molecules with the same mobility. If the mobilities are different enough, and the time of electrophoresis long enough, the individual zones will be completely separated from one another as shown in Fig. 9-12c. However, since the molecules in each zone continue to diffuse, the zones will spread with time, creating overlap and decreasing resolution; so resolution is a function of both time and diffusion.

Figure 9-13a shows one type of zone electrophoresis apparatus that is commercially available. This system has the ability to hold four separate cellulose acetate strips, the ends of which rest in reservoirs filled with buffer. Once the strips are positioned in the holder, 0.25 μliters of blood serum is applied to one end of each strip (if replicates are desired) in a thin starting zone with a special applicator. The cover is then placed on the holder to lessen evaporation, and the unit is attached to a power supply. For the apparatus shown in Fig. 9-13a, electric fields of 20 V/cm are typical. This is in the range known as low-voltage electrophoresis. Higher voltages will separate the components faster, but they also generate more heat and cause greater disruptive forces. The shorter times used in high voltage electrophoresis have the advantage of minimizing the time for diffusion of the zones. After the electrophoresis is completed, the cellulose acetate strips are removed and stained, the intensity of stain in each zone being proportional to the amount of protein present. The last step is to make the cellulose acetate strip transparent and to measure the intensity of each stained zone using a densitometer or other instrument that can measure the intensity

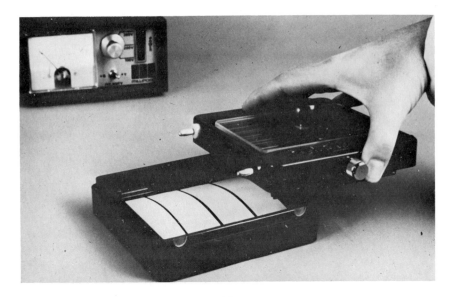

(a)

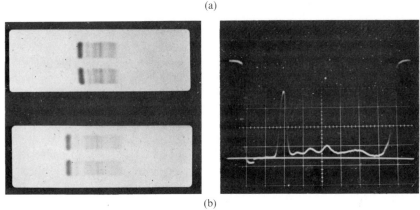

(b)

Fig. 9-13 (a) One type of commercially available zone electrophoresis apparatus. Here four support strips are placed in the chamber, and each slide can handle two separate solutions. A cover is placed over the whole apparatus to prevent evaporation, and then the two electrodes are plugged into the power supply shown in the background. (b) Typical stained slides after electrophoresing along with their densitometer readings for a serum sample. The pattern is for serum proteins. Two patterns are shown per support strip. (Courtesy of Millipore Corporation.)

of each zone. Figure 9-13b shows a typical stained slide along with its densitometer trace for a serum sample.

Another example where electrophoresis can be of medical significance is in the case of isoenzymes. Isoenzymes are enzymes that have a slightly different structure from one another, but which perform essentially the same function. The most notable example are the isoenzymes of lactic dehydrogenase (LDH). LDH is an enzyme with a molecular weight of about 134,000; it transforms pyruvate to lactate. Two different polypeptide chains called H and M, make up the enzyme which is a tetramer. There are five LDH isoenzymes commonly found in humans, and they can all be separated from one another by electrophoresis. Table 9-3 shows the com-

TABLE 9-3

Chain Composition of the Five Isoenzymes of Lactic Dehydrogenase (a Tetrameter)

Isoenzyme	Chain composition	Location
LDH-1	4H chains	Heart tissue (high O_2 concentration)
LDH-2	3H, 1M chain	
LDH-3	2H, 2M chains	
LDH-4	1M, 3M chains	
LDH-5	4M chains	Skeletal muscle (low O_2 concentration)

position of the five different isoenzymes. LDH-1 shows a perference for heart tissue, whereas LDH-5 has a relatively high concentration in skeletal muscles. LDH is not a normal component of blood, but a certain amount is usually found in serum. A normal electrophoretic pattern shows LDH-1 to be the fastest isoenzyme with a relative concentration of about 20%, and LDH-5 to be the slowest with a relative concentration of about 5%. As with the case of hemoglobin, abnormal electrophoretic patterns of LDH can give information about disease states. For instance, in a myocardial infarction the heart will release quantities of LDH-1, and the subsequent electrophoretic pattern will show an elevated level of LDH-1. Damage to skeletal muscle tissue will likewise result in an elevated level of LDH-5 in the pattern, and other disease states will similarly elevate the levels of different isoenzymes of LDH. It should be emphasized here that separation of these types of molecules must be based on their electrical properties because their hydrodynamic properties are almost identical.

DISK GEL ELECTROPHORESIS

Polyacrylamide disk gel electrophoresis is a special type of zone electrophoresis capable of achieving great resolution. It was first introduced in the early and mid 1960s and since then has become a powerful analytical tool in detecting macromolecular fractions in solution. Its uniqueness lies in an ingenious method for concentrating the various fractions into very thin bands which are stacked on top of one another prior to the actual electro-phoresis. By starting with each component in a thin layer, increased resolu-tion is possible over the situation where the solution is manually placed on top of the electrophoresis support strip, even if care is taken to make the starting zone as thin as mechanically possible. By placing the entire process in a gel, the formation of each initial ultrathin band causes minimal con-vective disruptions due to both the heating by the current and the rearrange-ment of the individual molecules. The gel is also formed in such a way that a molecular sieving effect enhances the separation via electrophoresis; i.e., larger molecules will be retarded more than smaller ones by the porous structure of the gel matrix itself. These concepts will now be discussed in detail.

Acrylamide has the formula

$$CH_2{=}CH{-}\overset{\displaystyle O}{\overset{\|}{C}}{-}NH_2$$

In the presence of a certain free radical initiator (a molecule with one un-paired electron) one acrylamide molecule will combine with another, which in turn will combine with yet another, etc., to form a long polymerized chain. By adding selected amounts of a cross-linking polymer, methyl-bisacrylamide

$$CH_2{=}CH{-}\overset{\displaystyle O}{\overset{\|}{C}}{-}NH{-}CH_2{-}NH{-}\overset{\displaystyle O}{\overset{\|}{C}}{-}CH{=}CH_2$$

methyl-bisacrylamide

a three-dimensional polymer network results, with the pore sizes of the resulting gel being determined by the relative concentrations of the acryl-amide and the cross linker. A two-dimensional representation of this struc-ture is shown in Fig. 9-14. If acrylamide alone were allowed to polymerize, only a very viscous solution would result; the cross linking is necessary for the formation of a gel and of the pores through which the macromolecules migrate under the influence of the electric field. The gel, then, has a twofold purpose: to prevent convective disturbances and to form a three-dimensional matrix that can be used as a molecular sieve.

A schematic diagram of a typical disk gel apparatus is shown in Fig. 9-15. The electrophoresis takes place from top to bottom in cylindrical glass tubes which contain the gels and which make electrical contact with the power supply through the upper and lower reservoirs which are filled with buffers. Two separate gel systems are needed for this process; the upper one has a relatively low concentration of cross linker, and hence it has large pore sizes compared to the lower gel. The stacking of the various components into ultrathin bands takes place in the upper gel, and the actual electrophoresis takes place in the lower gel. The lower reservoir contains a buffering system

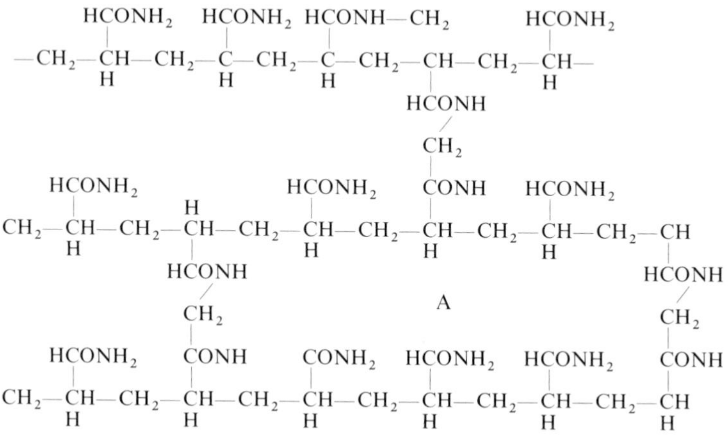

$$\begin{array}{c}
HCONH_2 \quad HCONH_2 \quad HCONH-CH_2 \qquad HCONH_2 \\
| \qquad\qquad | \qquad\qquad | \qquad\qquad\qquad | \\
-CH_2-CH-CH_2-C-CH_2-C-CH_2-CH-CH_2-CH- \\
\quad\;\; H \qquad\qquad H \qquad\qquad H \qquad\qquad\qquad\qquad H \\
HCONH \\
/ \\
CH_2 \\
| \\
\end{array}$$

$$\begin{array}{c}
HCONH_2 \qquad\qquad\qquad HCONH_2 \quad CONH \quad HCONH_2 \\
| \qquad\quad H \qquad\qquad\qquad | \qquad\qquad | \qquad\qquad | \\
CH_2-CH-CH_2-CH-CH_2-CH-CH_2-CH-CH_2-CH-CH_2-CH \\
\quad\; H \qquad\qquad\qquad\qquad\qquad H \qquad\; H \qquad\quad H \\
HCONH \qquad\qquad\qquad\qquad\qquad\qquad\qquad\qquad\qquad HCONH \\
/ \qquad\qquad\qquad\qquad\qquad A \qquad\qquad\qquad\qquad / \\
CH_2 \qquad\qquad\qquad\qquad\qquad\qquad\qquad\qquad\qquad CH_2 \\
| \qquad\qquad\qquad\qquad\qquad\qquad\qquad\qquad\qquad\qquad | \\
\end{array}$$

$$\begin{array}{c}
HCONH_2 \quad CONH \quad CONH_2 \quad HCONH_2 \quad HCONH_2 \quad CONH \\
| \qquad\qquad | \qquad\qquad | \qquad\qquad | \qquad\qquad | \qquad\qquad | \\
CH_2-CH-CH_2-CH-CH_2-CH-CH_2-CH-CH_2-CH-CH_2-CH \\
\quad\; H \qquad\; H \qquad\quad H \qquad\quad H \qquad\quad H \qquad\quad H
\end{array}$$

Fig. 9-14 A schematic of the matrix evolved when acrylamide and cross linker are polymerized together. The pores (space A) of the gel are formed by the different polymers linking to one another, and their size is determined by the relative amounts of acrylamide and cross linker. The more linker there is, the smaller are the pores.

at one pH, while the upper one contains another system at a different pH. Because the pH changes, as does the gel composition, from the top of the tube to the bottom, the technique is called discontinuous gel electrophoresis, or disk gel for short. The gel changes its composition abruptly as indicated in the figure. Polyacrylamide is used as a gel because it has good working properties, and different pore structures can be formed rather easily over a wide range. Polyacrylamide is also optically transparent and is relatively inert chemically.

To see how the stacking condition works in forming the various fractions into ultrathin zones in the upper gel prior to electrophoresis, consider the following. Suppose you have two charged species that have the same sign, say X^- and Y^-. Assume Y^- has a larger mobility than X^- (because it has a more compact structure than X^-) and that a solution of X^- is layered on top of a zone of Y^- in a cylindrical tube. If this tube is then connected to the poles of a power source such that both species X^- and Y^- move downward, then it is known that a sharp boundary will be maintained between the two species, i.e., the boundary between the two ions will not blur significantly with time. Why is this? Remember that both ions are part of the same electrical circuit, and each zone must support the same current lest there be a build up or a loss of charge somewhere in the circuit. In order to maintain a constant current the electric field in the vicinity of the X^- ions has to be greater than that in the vicinity of the Y^- ions because it is harder to get the X^- ions to move quickly (they have a lower mobility than the Y^- ions).

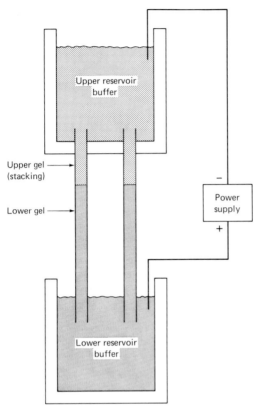

Fig. 9-15 A diagram of a disk gel electrophoretic apparatus. Only two vertical glass cylinders are shown, where in fact six or more may be run at any one time. The upper gel is used for stacking the various components into ultrathin bands that are layered on top of one another. It has pores that are relatively large compared to the lower gel. The lower gel is actually used for the electrophoresis. The various components are stacked in the upper gel and then they pass into the lower gel where they are separated by electrophoresis. Upper and lower reservoirs contain buffers and are connected to the electrodes.

The voltage drop across the X^- ion zone has to be larger than the voltage drop across the Y^- ion zone. The presence of this discontinuous electric field actually preserves the boundary between the two species (see Fig. 9-16). If a slower mobility X^- ion were to diffuse across the boundary into the Y^- ion zone, then the X^- ion would find itself in a region of relatively low electric field (compared to that found in the X^- ion zone); hence it would slow down and the boundary would overtake it. Once the boundary caught up with it, the X^- ion would travel at the same speed as the boundary. Conversely, if a Y^- ion were to diffuse into the X^- ion region, the Y^- ion would find itself in a region of relatively large electric field; hence it would

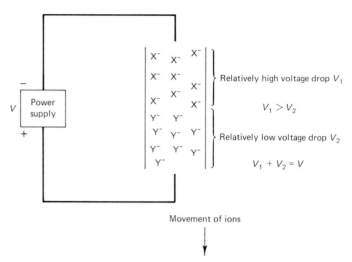

Fig. 9-16   If two ions of like sign (X^- and Y^-) are carefully layered over one another creating a sharp boundary and if the one having the larger mobility (Y^-) is on the bottom, the boundary will be preserved if the motion is downward. Because both zones are electrically in series, the current in both is identical; therefore the potential gradient must be greater across the top zone because these ions are harder to move. If an X^- ion diffuses into the lower zone, it will experience a lesser potential gradient and will subsequently move more slowly than if it had remained in its own zone. The boundary will catch up to it and then this X^- ion will move with the boundary. A similar result will happen if the Y^- ion diffuses into the upper zone.

speed up and catch the boundary, thereafter moving with the boundary. In this manner a sharp boundary is maintained between the two ionic types. If now other species with mobilities whose magnitudes are between those of X^- and Y^- are placed between the X^- and Y^- zones, they will also form sharp boundaries between adjacent zones for the same reasons as described above. The net result is a stack of thin bands or zones (much like a stack of records), the lower most, Y^-, having the largest mobility, and the succeeding ones having successively lower mobilities until the X^- ion zone is reached, where the X^- ion has the lowest mobility and trails everything. The individual bands are thin because intermediate species will have to concentrate enough so that each can carry as much current as the Y^- species zone. This is required because all zones are electrically connected in series. Figure 9-17 shows a typical stacked condition in the upper gel where three additional ions are included.

In a typical disk electrophoresis experiment on serum proteins the trailing anion (X^-) is chosen to be glycinate NH_2—CH_2—COO^-, the leading ion (Y^-) is chloride Cl^-, and the intermediate ions are the serum proteins themselves. Under identical conditions the Cl^- ion always has a larger

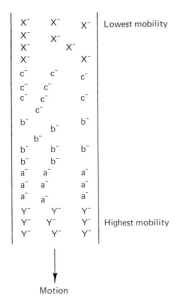

X⁻   X⁻   X⁻   Lowest mobility

Fig. 9-17 If three different anionic species (a⁻, b⁻, and c⁻) are introduced between the X⁻ and Y⁻ anions, they will form thin zones; the respective boundaries will be maintained because of the voltage discontinuities across each zone. If these anions are introduced as a mixture, they will rearrange themselves to form layers or zones with the anion of highest mobility leading and where successive zones have lesser mobilities.

Highest mobility

Motion

mobility than the glycinate because of its compact shape. Also, both of these ions are constituents of buffering systems to keep the pH constant.

A disk gel experiment requires several buffers. The upper stacking gel is bathed in a buffer that is different from the buffer used to make the lower gel. The reason for this is as follows. In the upper stacking gel the pH is kept lower than in the lower gel. Because the pH is low here, the average negative charge on the glycinate ion will be reduced (why is this?), and hence its mobility will be lowered also. By regulating this pH, it can be guaranteed that glycinate will trail all the proteins; hence they will stack as shown in Fig. 9-17. After the proteins are stacked and the whole pattern moves into the running gel, the various species find themselves in a region of higher pH where the effective negative charge on the proteins and glycinate increases. The mobility of the glycinate ions is now larger than the mobilities of the serum proteins, and hence the glycinate ions will overtake and pass the proteins but not the Cl⁻ ion zone. The proteins are now left in a single buffer solution, and electrophoresis proceeds as in regular zone electrophoresis. The advantage of this system over regular zone electrophoresis is that the initial zone widths in disk gel are extremely small (see Fig. 9-18).

The system just described will not work for separating all types of molecules. The buffers and pH must be changed to suit the particular system of interest, or to obtain a specific result. Glycinate ion was initially chosen for the upper buffer in the separation of serum proteins because it had the proper mobility so that it would trail the serum proteins in the stacking

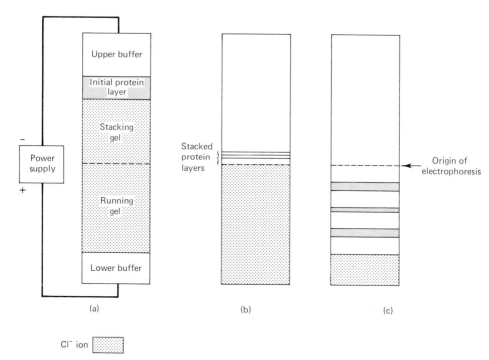

Cl⁻ ion

Fig. 9-18   (a) Initial setup for disk gel electrophoresis. The protein mixture is laid on top of the stacking gel. The buffer in the upper reservoir has glycinate ions while the gel has Cl⁻ ions. (b) Stacking is complete, and the zones are about to enter the running or electrophoresis gel which has a higher pH than the stacking gel. The proteins are then separated based on their electrophoretic mobilities and the sieving action of the smaller pore size lower gel. (c) The protein fractions are separated. In the running gel the glycinate's mobility becomes greater than that of the serum proteins and the glycinate ions overtake and pass the proteins, finally taking a position behind the Cl⁻ ion zones which cannot be passed. In the running gel the proteins are bathed only in the glycinate buffer and electrophoresis proceeds as in a regular zone experiment.

gel and then move ahead of them in the lower gel where the pH is higher. The zones themselves once stacked in the upper gel have heights of $1-10~\mu m$, which is considerably smaller than can be accomplished by a manual application of solution. The concentration of acrylamide in each gel also depends on the particular system; for small solute molecules, relatively large concentrations of acrylamide (up to 30%) are used in the lower running gel to ensure small enough pore sizes for proper sieving effects. The concentration of acrylamide in the stacking gel is not as critical since its main function is to dampen convection currents. In an actual experiment a visible dye would be included in the system to undergo electrophoresis along with everything else. The movement of the dye tells the experimenter where to stop the migration so the protein zones do not pass into the lower reservoir. A dye is chosen so that it moves ahead of the regular zones but remains behind the Cl^- zone.

Once the electrophoresis is finished, the gel is removed from the cylindrical tube and stained. The amount of stain retained by each zone is then proportional to the amount of material there. The absorbance of the gel as a function of its length is then measured in a special densitometer, where again the area of each peak is proportional to the amount of biomaterial present. Figure 9-19 shows one type of disk gel electrophoresis equipment com-

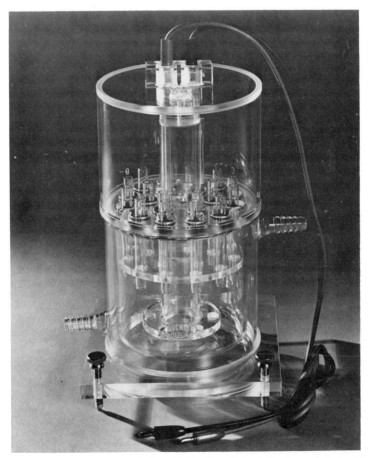

Fig. 9-19 One type of a commercially available disc gel electrophoresis apparatus. Eighteen tubes (8 mm O.D. and up to 250 mm long) can be filled and electrophoresed simultaneously. For this particular model, sample capacity is on the order of 0.005 mg to 0.2 mg per tube. Buffers are placed in the upper and lower reservoirs, and the tubes may be temperature controlled by a circulating fluid. After setup is complete, the unit is attached to a power supply. (Courtesy of Buchler Instruments.)

mercially available, and Fig. 9-20 shows a typical stained gel. The large width of each zone after electrophoresis is primarily due to diffusion. Broadening of zones will occur regardless of how thin the stacking zones are; but the thinner they are to start with, the less chance there is of zone overlap due to diffusion later on. Hence resolution is increased. The initial stacking of components in the upper gel is then one of the major advantages and characteristics of disk gel electrophoresis.

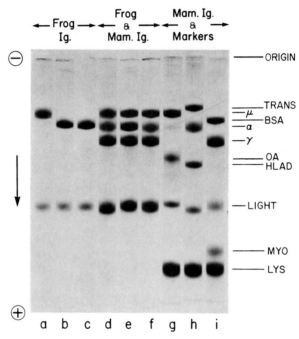

Fig. 9-20 Nine typically stained gels after a SDS gel electrophoresis experiment. The electrophoresis is from top to bottom, and each band represents the component labelled on the side. The darkness of each band indicates its relative concentration. The distance of each band from the origin is a measure of molecular weight. The labels at the top refer to the system under investigation in the experiment. The markers are proteins of known molecular weight used as a reference. (Reprinted with permission from C. Green and L. Steiner, *J. Immunol.* **117**, 368. ©1971 The Williams & Wilkins Co., Baltimore, Maryland.)

SDS GEL ELECTROPHORESIS

In the description of disk gel electrophoresis it was mentioned that the pore size of the lower gel is usually chosen to be of such a size that it can act as a molecular sieve, separating molecules based on their size as they migrate under the influence of the electric field. This principle is amplified in the technique known as SDS gel electrophoresis. SDS is sodium dodecyl sulfate, and its structure is

$$CH_3(CH_2)_{10}CH_2SO_4Na$$

It is essentially an ionic detergent, and it has the property of being able to complex with a wide variety of proteins in a rather unique fashion. But before discussing SDS gel electrophoresis, it is first necessary to understand how SDS itself interacts with proteins.

In aqueous solution by itself SDS can exist as either a monomer or as part of a micelle, the ratio being determined by the temperature and the ionic strength of the solution. When both protein and SDS are present in aqueous solution, the SDS monomer will interact with the protein, causing it to undergo a conformational change. These conformational changes have been shown to be rather similar regardless of the protein, and the resulting polypeptide chain has an extended conformation compared to that of the original form. Physical measurements indicate that the length of this extended form is about half that of the fully extended polypeptide chain, thus indicating that the chain is folded on itself. The amount of α-helix in the structure is also quite high. A diagram illustrating the interaction between SDS and a protein is shown in Fig. 9-21.

One interesting aspect of this complex is that different proteins bind identical amounts of SDS on a gram per gram basis. For instance, if the SDS monomer concentration is between 5 and 8×10^{-4} molar, then 1.4 g of SDS are bound to every gram of protein. This means that the complex is going to have a decidedly negative charge regardless of the protein because so many SDS molecules are bound. This has the added implications that in the complex the charge per unit mass of protein is going to be approximately constant and that the hydrodynamic properties of the complex will

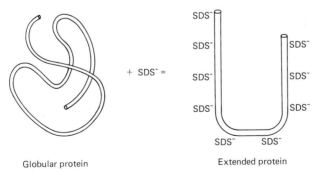

Globular protein   Extended protein

Fig. 9-21 Schematic representation of the conformational change undergone by a globular protein when reacted with an aqueous SDS solution. The initial globular conformation of the polypeptide chain is stretched out to a length roughly half that of the whole chain. The extended conformation has a high percentage of α-helix.

be mainly a function of the protein's molecular length (or molecular weight). The binding of SDS to proteins has been demonstrated to be nonspecific, and the final complex conformation seems to be independent of the protein's initial native shape. What this all means, then, is that by reacting a series of different proteins with an aqueous SDS solution, the proteins will all be forced to assume a similar shape with identical charge to mass ratios. The only difference in the various protein–SDS complexes will be in the length of the final conformation. Each distinct protein species will have a unique length and hydrodynamic properties. Now we shall relate these findings to SDS gel electrophoresis.

If in a gel electrophoresis experiment the pore sizes in the lower gel are small enough and a mixture of proteins is to undergo electrophoresis, the smaller proteins will in general encounter less resistance from the acrylamide matrix and will move farther than the larger proteins. In this case the distance covered in the electrophoresis will be related to the molecular weight of the various proteins. This is the principle behind SDS gel electrophoresis. In the late 1960s it was realized that if proteins were first reacted with SDS and then subjected to gel electrophoresis, the distance they migrated in the lower gel would be proportional to their respective molecular weights. SDS gel electrophoresis, then, not only separates molecular components, it also gives an estimate of their molecular size (see Fig. 9-22).

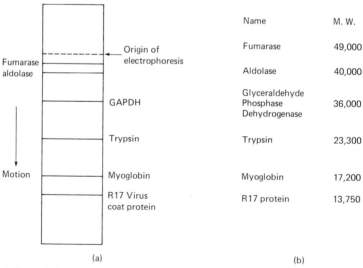

Name	M. W.
Fumarase	49,000
Aldolase	40,000
Glyceraldehyde Phosphase Dehydrogenase	36,000
Trypsin	23,300
Myoglobin	17,200
R17 protein	13,750

(a)   (b)

Fig. 9-22   (a) Schematic of a disk gel tube, illustrating the principle that small proteins electrophoresis further in a given time period than do large proteins if both had been reacted with SDS prior to the experiment. The direction of electrophoresis is indicated by the arrow. (b) Table showing the molecular weights of the proteins.

When the distance migrated by different proteins was plotted vs. their known molecular weight, it was discovered that the points fell on a smooth curve. An example of such a plot is shown in Fig. 9-23.

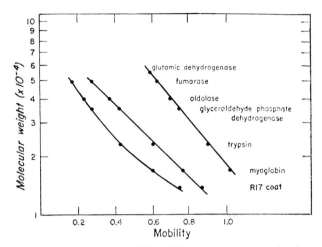

Fig. 9-23   SDS gel electrophoretic mobilities of six proteins plotted vs. their known molecular weight. The middle curve is for a normal amount of cross linker, the left curve is for twice as much, and the right one is for one-half as much. The mobility is found by measuring the relative distance a band has traveled relative to the boundary between the upper and lower gel. Notice that the ordinate is a logarithmic scale, not a linear one. Because this type of plot is linear, it indicates that the mathematical relationship between molecular weight (MW) and distance of migration (x) is $MW = d10^{-bx}$ where b is the slope and d is the intercept (Weber and Osborn, 1969).

SDS electrophoresis has since become a standard method of determining the molecular weight of proteins, and it can be routinely used for this purpose. Typically, an unknown protein undergoes electrophoresis along with several proteins of known molecular weight, which act as "markers" and are used to set up a calibration curve for the particular conditions used in the experiment. An example of this can be seen in the work of G. Nakos and L. Mortenson to determine the molecular weight of azoferrodoxin. Azoferrodoxin is a protein that can be isolated from the bacterium *Clostridium pasteurianum*, and it is involved in the ability of this bacterium to metabolize atmospheric N_2. It is composed of two identical subunits. The two scientists used SDS gel electrophoresis to determine the molecular weight of the subunits. Their results are shown in Fig. 9-24. Six proteins of known molecular weight underwent electrophoresis as markers so that the molecular weight vs. distance curve could be determined. The molecular weight of the azoferrodoxin protein is seen to be 27,500 daltons by its position on the curve.

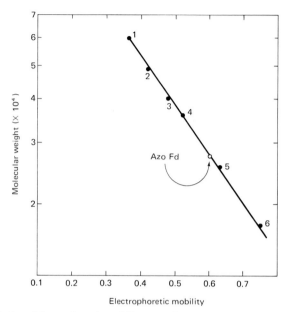

Fig. 9-24   A plot of electrophoretic mobility vs. molecular weight for azoferredoxin subunits and six other proteins used as markers. The azoferrodoxin falls on the calibration curve at a spot indicating it has a molecular weight of 27,500 D. The marker proteins are (1) catalase, (2) fumarase, (3) aldolase, (4) glyceraldehyde phosphate dehydrogenase, (5) α-chymotrypsinogen A, and (6) myoglobin. (Reprinted with permission from G. Nakos and L. Mortenson, *Biochemistry* **10**, 455 (1971). Copyright by the American Chemical Society.)

DIELECTROPHORESIS

In electrophoresis particles must have a net charge in order to move in the applied electric field. The absence of a net charge implies that the particle will not be affected by the field. Neither of these conditions is true in the relatively new technique called dielectrophoresis. In normal electrophoresis the molecules of interest are located in a homogeneous electric field of relatively low strength; in dielectrophoresis the interaction is between the external electric field and the molecule's dipole moment. Here, the electric field is not homogeneous but changes in intensity over the distance of the dipole's length.

If placed in an inhomogeneous electric field, a molecule with no net charge but with a permanent dipole moment μ will move toward the region of maximum electric field due to a force of magnitude $\mu(dE/dx)$ where dE/dx is the value of the electric field gradient [see Eq. (3-20)] over the distance of the dipole. If this inhomogeneous field is maintained, a concentration gradient will build up around the source of the field. This concentration gradient represents an equilibrium between the dielectrophoretic force $\mu(dE/dx)$ and diffusion. This is illustrated in Fig. 9-25. The dielectrophoretic force will cause the molecules to migrate toward the central electrode, whereas diffusion will tend to disperse the gradient.

The electrophoretic force acting in a homogenous field is given by $F = qE$, and for a singly charged molecule in a field of 10 V/cm, this force equals

$$(4.8 \times 10^{-10} \quad \text{esu})(10 \quad \text{V/cm})\left(\frac{1}{300} \frac{\text{dyn/esu}}{\text{V/cm}}\right) = 1.6 \times 10^{-11} \quad \text{dyn} \quad (9\text{-}10)$$

To achieve a similar force, what field gradient is needed in a dielectrophoretic experiment? Let us assume that a protein has a dipole moment of 250 D (equivalent to a proton and electron separated by 0.52×10^{-6} cm). To obtain a force of 1.6×10^{-11} dyn, the needed field gradient is

$$\frac{dE}{dx} = \frac{1.6 \times 10^{-11}}{2.5 \times 10^{-16}} \frac{\text{dyn}}{\text{esu cm}} \cong 6.4 \times 10^4 \quad \text{V/cm}^2 \quad (9\text{-}11)$$

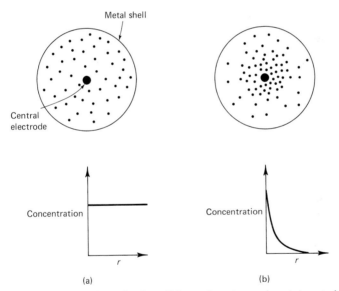

Metal shell

Central electrode

Concentration

Concentration

r

r

(a)

(b)

Fig. 9-25 (a) A typical configuration for a dielectrophoresis experiment. A central electrode wire is surrounded by a cylindrical metal shell with the solution of interest between. A voltage is applied between the shell and the central electrode creating a field that changes rapidly with distance. The biomolecules in solution are initially evenly dispersed throughout the cell before the field is applied. (b) The concentric concentration gradient existing some time after the particles have migrated toward the inner electrode. At equilibrium the dielectrophoretic force is opposed by diffusive forces which tend to disrupt the gradient. Graphs of concentration vs. distance are also shown for both cases.

which is extremely large. Normally to achieve such an electric field gradient would require very high voltages, which in turn would generate a great deal of heat. An alternative is to use the relatively intense field generated close to an electrode having a very small radius of curvature such as the central wire shown in Fig. 9-25. Here the field gradient is large and the dielectrophoretic effect can take place. It can be shown that for this situation, the electric field gradient can be given by

$$\frac{dE}{dr} = \frac{V_0}{2.3r^2 \log(r_2/r_1)} \tag{9-12}$$

where V_0 is the voltage and r_1 and r_2 are the radii of the central electrode wire and cylindrical shell, respectively. The variable r is the radial distance between the two electrodes. By examining Eq. (9-12) it is seen that dE/dr will be large only for small values of r, or to put it another way only close to the central electrode.

One type of dielectrophoretic cell is that illustrated in Fig. 9-25. A fine wire electrode (typically 10 μm in diameter) is surrounded by a concentric metal shell which acts as the electrode of opposite polarity. The solution is placed between the two electrodes, and a voltage is applied across them creating a large field gradient, especially within one or two radii of the wire electrode. This configuration is actually a cylindrical capacitor where the capacitance changes as a function of time while the concentration gradient builds up. To prevent molecules with a net charge from migrating to one electrode or the other, an alternating voltage is used. If the frequency used is low enough, the molecular dipoles will be able to reorient every time the field oscillates, and electrophoresis of the molecules will be negated; only dielectrophoresis will occur. While the voltage is applied, the concentration gradient will build to an equilibrium state as measured by the capacitance change; when the voltage is cut off, the gradient will decay via diffusion, and the capacitance will again change with time. At equilibrium the dielectrophoretic electrical force is opposed by the diffusive forces.

Now that the experimental set up has been described somewhat it is time to ask what kind of useful information can come from a dielectrophoretic experiment. The answer is that several pieces of information can be obtained. Since the concentration gradient decays by a diffusion mechanism, the diffusion coefficient of the molecules may be deduced from data on the change in capacitance as a function of time once the electric field is turned off. The diffusion coefficient is a physical parameter that is associated with a molecule and measures the molecules ability to move through a fluid via a diffusive mechanism. The diffusion coefficient is dependent upon both the size and shape of a molecule. Large particles have relatively low diffusion coefficients, while for small particles they are relatively large. Experiments have shown that the diffusion coefficients obtained in dielectrophoresis experiments compare favorably with ones obtained by classical techniques for several poly-α-amino acids.

Another piece of information that can be derived is the anisotropic frictional coefficient. In the build up of the concentration gradient, the molecules will align themselves with the electric field so that their dipoles parallel the field lines, so the molecules migrate with a preferred orientation as opposed to the random tumbling experienced when they diffuse outward after the voltage is turned off. By being able to distinguish between the anisotropic and regular frictional coefficient, it is possible to get an idea of how nonspherical a particular biomolecule is. Another piece of information that has been measured from dielectrophoretic experiments is the extent of solution homogeneity. If a dielectrophoretic experiment is performed at several different frequencies for the ac field, it is possible to deduce the molecular weight distribution of any given sample. The principle of this lies with the fact that larger molecules in a mixture will behave differently in response to the field reversals than will small molecules. Larger molecules will not be able to keep up with high frequency field reversals as well as small molecules. One of the biggest potentials that can be obtained through dielectrophoresis, however, is the ability to differentiate between similar molecules based on their dipole moments. Electrophoresis is able to distinguish between normal and sickle cell hemoglobin because an amino acid substitution changes the net charge of normal hemoglobin. Likewise, if an amino acid substitution takes place in a mutant protein, and it does not involve a charge, then it is quite possible that either the conformation or the charge distribution or both will change, resulting in a different dipole moment from the original. Regular electrophoresis would most likely not be able to detect a difference between the two cases, whereas dielectrophoresis could. In looking at the number of amino acids whose side chains can support a charge, one concludes that just on probability alone one would expect to see more amino acid substitutions where changes in charge do not occur than where a charge is involved. A mutant protein would have a greater chance of having its dipole moment changed with respect to the normal form than it would having its charge changed.

Dielectrophoresis can be performed on particles that are electrically neutral. In fact it is not even necessary that the particles have permanent dipole moments at all; induced dipoles will suffice. The technique of dielectrophoresis so far has not been limited to measurements on strictly molecular systems. One study showed that yeast cells of the strain *Saccharomyces cerevisiae* collected at the central electrode depended on such factors as cell age, thermal treatment of the cells, and chemical poisoning of the cells. Compared to electrophoresis, dielectrophoresis is pretty much restricted to the experimental laboratory; however it does possess the potential for distinguishing between very fine differences in the electrical configuration of biomolecules.

ION EXCHANGE CHROMATOGRAPHY

Chromatography is an experimental technique used for separating components from complex mixtures of gases and liquids. In a typical chromatography experiment a fluid having several components is passed over a stationary solid, and the different components interact to varying degrees with the stationary phase. If the affinity of one component for the solid phase is strong enough, it will remain bound, and the other components will pass on through, effectively separating that one component. If all the various components of the fluid mixture exhibit weak but sufficiently different affinities for the solid phase, then the order in which they leave the solid phase will be inversely proportional to their affinities; i.e., those fraction(s) interacting weakly will pass through relatively unhindered, whereas those interacting strongly will be delayed or completely retained. By collecting small samples on the downstream side of the solid phase, aliquots containing only one component or minus one component may be obtained. In this fashion the various components of a fluid mixture can be separated into pure fractions. It should also be remarked that the fluid can be either a gas or a liquid, although our interests will be exclusively with liquid chromatography.

Ion exchange is a specific type of chromatography in which the interaction between liquid and solid phase is via a coulombic charge–charge interaction. In the actual experiment the solid phase is usually supported in a closed, vertical, plastic or glass cylindrical tube, and the liquid phase is passed into and out of the cylinder by means of attached tubing. The solid phase is insoluble in aqueous solution, and it contains chemical groups that are charged either positively or negatively. If the solid phase carries a positive charge, then it is known as an anion exchanger and will then attract, or exchange, negative ions from the mobile liquid phase. If the solid matrix is charged negatively, it is termed a cation exchanger. Chemical groups that are commonly employed as sources of charge for the solid phase are phenolic, hydroxyl, carboxyl, and sulfonic for cationic exchangers, and aromatic amino and aliphatic amino groups for anionic exchangers. The type of functional group used in the solid phase determines the strength of interaction, while the concentration of these groups determines the capacity of the solid matrix to extract counterions from the mobile phase. Other factors such as solid phase conformation and porosity are also important and affect the capacity and efficiency of the column.

The separation of molecular components in a solution via ion exchange chromatography depends on the components having different affinities for the charged stationary groups. In the case depicted in Fig. 9-26 the uncharged species is not attracted by the solid matrix, whereas the negative species is firmly bound and retained. The effluent from the downstream side contains only the uncharged particles. A separation of particles has been accomplished, and a pure solution of uncharged particles can be collected. When all the positively charged groups on the solid matrix are filled with anions from the mobile phase, the anions will also flow on through. At this point the capacity of the column has been saturated. Suppose also that it is desired to have a pure solution of the charged species. To release the negatively charged species from the solid phase, it is necessary to weaken the charge–charge force holding them together. This can be accomplished in one of two ways.

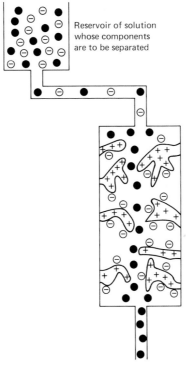

Reservoir of solution whose components are to be separated

Fig. 9-26 A diagram illustrating one application of ion exchange chromatography. A solution containing a negatively charged protein specie ○ and an uncharged particle ● is flowed through a cylindrical glass tube containing a stationary solid matrix which has a positive charge. The negative species in the mobile phase is attracted and held by the matrix, while the uncharged specie flows relatively unhindered, thus effecting a separation. The solid phase is not actually connected to the glass column itself as is indicated in the figure.

The net negative charge itself can be reduced to zero by changing the pH of the solution, i.e., by lowering the pH. If the magnitude of the negative charge is reduced the attractive force holding it to the solid matrix can be reduced to the extent that the counterion will pass through the solid phase and exit the column. This change in pH is accomplished by introducing a new solvent (or series of solvents) into the top of the column that have lower and lower pH values. As these new solvents flow by the anions, the new pH will affect their net charge, and the anions will begin to flow with the new solvent. The extent to which this is done is dependent on the initial pH and the new pH. The change in pH necessary to uncouple the biomolecule is always such that the biomolecule moves closer to its isoelectric point.

The second method of eluting components from the column is through the use of ionic strength gradients. Here the pH is kept constant, but the ionic strength of the eluting solution is gradually raised. As the ionic strength is increased, there are more charged species present that can compete for spaces on the solid matrix, thus displacing the original occupants. Also, as we learned from Debye–Hückel theory, as the ionic strength increases, the energy of interaction between two charges decreases. The net result is that at sufficiently high salt concentration, the original component in the solution is released from the solid phase, and it passes out of the bottom of the column to be collected. The salt concentration needed to rupture the bonds between the original anion and the matrix is dependent on the pH and also on the strength of the net charge.

Now suppose it is desired to use ion exchange chromatography to separate the components of a solution where each component has a different charge, but the same sign. First, a solid matrix is chosen that has a charge of opposite polarity. Next, the original solution is passed through the column and all the components become attached to the solid phase. To elute the components, either a pH or ionic strength gradient is used. Since all the components have different affinities for the solid phase, as the solvent conditions are gradually changed, the components will be eluted at characteristically different times; e.g., the component having the weakest force of interaction with the solid phase will be eluted first and that component will pass out of the column first, followed by that component whose attraction for the solid phase is the next weakest, etc. In this way all the components of the original solution can be separated from one another. The resolution of components depends however on the rate at which the solvent gradient changes and on the flow rate of the column. As the steepness of the gradient decreases, the components will be eluted further and further apart; the slower the flow rate, the better is the resolution.

Figure 9-27 shows the elution profile of a solution containing various types of hemoglobin. The elution was via a pH gradient. It is seen that all

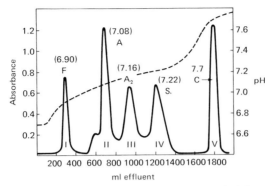

Fig. 9-27 Elution profile from an ion exchange column of an artificially constructed solution containing five different forms of hemoglobin: Hb-A, Hb-F, Hb-A2, Hb-S, and Hb-C. The dotted line shows the pH gradient used to elute the proteins. The flow rate was 20 ml/hr (Dozy and Huisman, 1969).

the various peaks are resolved from one another. The peaks shown in the figure represent the amount of material present in each component. Once the fractions are collected from the column, their optical absorbance is measured, or some other means of determining concentration is used to actually construct the plot of concentration vs. elution volume. The point to be made here is that it is possible to physically separate the various fractions of hemoglobin on an ion exchange column. Figure 9-28 shows the resolution of components for a solution containing the five isoenzymes of lactic dehydrogenase. Here the elution profile was obtained by using an ionic strength gradient.

Looking at the examples illustrated in Fig. 9-27 and 9-28, the reader should get the correct impression that ion exchange chromatography rivals electrophoresis as a means of separating the components of a solution. Both

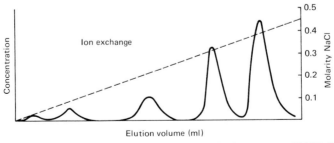

Fig. 9-28 Elution profile of a solution containing the five isoenzymes of LDH. The dotted line indicates the molarity of the NaCl solution used in the eluting ionic strength gradient (Wachsmuth and Pfleiderer, 1963).

techniques make use of the electrical properties of biomolecules, and both are routinely used in laboratory work. Whether one method or the other is superior for a particular separation depends on the system itself and exactly what is desired by the experimenter.

It should also be mentioned that another type of column chromatography called adsorption chromatography also makes use of the electrical properties of biomolecules. In this case the charge–dipole, the dipole–dipole, and the VDW forces are relied upon in creating differential interactions between the different components of the mobile phase and the solid matrix.

ISOELECTRIC PRECIPITATION

Like ion exchange chromatography, isoelectric precipitation is mainly a laboratory technique in which the worker can separate or purify one or a number of components from a complex mixture. Isoelectric precipitation in general does not have the precise quantitative appeal of other techniques but can still be quite useful in a number of situations. The technique is based on the fact that at a protein's isoelectric point, the net charge is zero and the individual particles do not have as large a tendency to repel one another via charge–charge repulsion. In fact, many globular proteins will aggregate to such an extent that they fall out of solution in the form of a precipitate. This precipitate can then be collected from the rest of the original solution as an essentially pure fraction. If a solution contains a number of protein components and if each has a sufficiently different isoelectric point, then by slowly changing the pH it is possible to precipitate them out of solution one at a time. The resolution between separations will naturally depend on how far apart the respective isoelectric points are; the closer together the isoelectric electric points, the poorer is the resolution between components.

A good example of how solubility varies as a function of pH is illustrated in Fig. 9-29 which shows the solubility of the milk protein β-lactoglobulin as a function of pH and salt concentration. From the graph it is seen that almost independently of salt concentration, the minimum solubility of β-lactoglobulin is in the region of pH 5.2–5.4. Therefore, it would be expected that this protein would have the best chance of precipitating out of solution in this pH range. The isoelectric point of β-lactoglobulin is 5.2. Figure 9-29 also tells us that β-lactoglobulin would be precipitated in the narrowest pH range when the salt concentration is high; i.e., if other components are also present in a solution containing β-lactoglobulin, the best resolution could be obtained at the higher salt concentrations shown in the figure. It should also be remarked that even at the point of minimum solubility, some β-lactoglobulin will still be in solution since the solubility is not zero here.

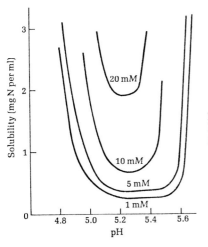

Fig. 9-29 The solubility of β-lactoglobulin vs. pH for different salt concentrations. (From A. L. Lehninger, "Biochemistry," p. 133. Worth Publishers, New York, 1970.)

As an example where the technique of isoelectric precipitation in protein isolation is routinely used, consider the preparation of tobacco mosaic virus (TMV) protein. Recall from Fig. 1-11 that TMV is composed of a helically wound strand of RNA which is in turn surrounded by a collection of protein subunits; it is these subunits that this preparation scheme seeks to isolate. In this procedure the virus is first isolated from infected tobacco plants. The procedure begins by grinding the leaves and then by performing a series of differential centrifugation steps to first centrifuge out large particulate contaminates like the leaf pulp, and then the virus itself. To separate out the RNA from the protein subunits, the virus solution is made 67% in acetic acid. The resulting precipitate turns out to be the RNA, and it can be centrifuged out and discarded. Next, the solution containing mainly solubilized protein subunits is placed in dialysis bags and dialyzed against several changes of water. This has the effect of letting the acetic acid leave the dialysis bag and letting water enter, thus raising the pH. As the pH approaches the isoelectric point, the formerly clear solution becomes turbid, then opaque with precipitated protein. From this step the precipitated protein is centrifuged to the bottom of a test tube and redissolved in a solvent of the experimenter's choice. The last step of the procedure has the double advantage of concentrating the precipitated protein, and also of being able to disperse it into a new solvent. If the protein is not in a precipitated form, it would be more difficult to concentrate it by centrifugation because it is so small and would require very large centrifugal forces to spin it down.

The whole technique of isoelectric precipitation here relies on the gradual rise of the pH within the dialysis bag. As the pH approaches the isoelectric point of TMV protein, the individual subunits begin to aggregate, thus causing the turbidity. As the pH gets closer and closer to the pK value, the aggregates become larger and finally fall out of solution yielding an opaque solution.

ISOELECTRIC FOCUSING

When performing electrophoresis, it is usual to maintain a constant pH throughout any one part of the system. Under these conditions the motivating force is the interaction between the electric field and the charge on the macromolecule that is in electrophoresis. Again, this charge is assumed to be constant. In the technique of isoelectric focusing the pH is not kept constant; the charge on the macromolecule varies depending on its location, as does the interaction between the molecule and the electric field. Like electrophoresis, isoelectric focusing is capable of resolving components from a complex mixture, only it does so based on the difference in the isoelectric pH values of the components, not their difference in net charge.

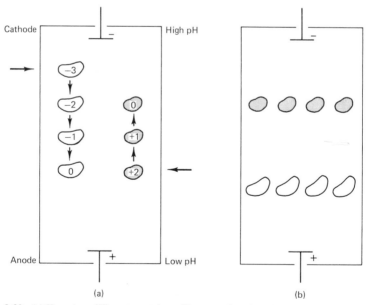

Fig. 9-30   (a) How two different proteins will react when introduced into a pH gradient. One particle has an initial net charge of -3, while the other has an initial $+2$ charge. Both will move toward the electrode of opposite polarity. As they move, however, their respective net charges will change due to the change in pH. Their respective initial positions before the field is turned on are indicated by the heavy arrows. (b) The equilibrium situation with the two particles banded at their respective isoelectric pH values. At these spots both particles have a net charge of zero.

Consider the situation depicted in Fig. 9-30a. A cylindrical container has electrodes at either end and a pH gradient is formed between them with the most acidic pH near the anode and the most basic pH near the cathode. If a protein is now introduced at an arbitrary location within the system, it would assume a charge consistent with the pH at that location. At a relatively high pH the protein will take on a negative charge, at a low pH it will have a positive charge. If a potential difference is also applied across the electrodes, the molecules will experience a motivating force moving them toward the electrode of opposite polarity. But, as the proteins move, their net charge will change due to the fact that the pH charges via the pH gradient. As each protein molecule approaches the electrode of opposite polarity, the net charge on that protein will approach zero. If a molecule is initially charged positively, it will move toward the cathode; but the pH gradient increases in that direction, so the net charge on the protein becomes less positive the closer it gets to the electrode. When the net charge becomes zero, the molecule is located at its isoelectric point, and there is essentially no net interaction between the protein and the applied electric field. A similar situation will exist for molecules with an initial negative charge. The result is that all molecules with identical isoelectric points will be banded at particular values of the pH in the pH gradient. If the test solution consists of several components with different isoelectric points, they will be separated into unique bands similar to those formed in gel electrophoresis (Fig. 9-30b).

One of the early problems inhibiting the development of this technique was the inability to form stable pH gradients. These pH gradients must be maintained by an electrolyte that meets a number of exacting requirements. These properties were first described by the pioneering work of H. Svensson, who laid the groundwork for much of the technique. To give a few examples, the electrolyte must possess multiple charges, it must have good conductivity, it must be soluble in water, and it must have a high buffering capacity at the isoelectric point. It should also have a conductivity that does not vary with pH, otherwise the potential drop across the system will not be smooth, but stepwise. Furthermore, the electrolyte must show minimal interaction with biomolecules, and it should be compatible with the method used for stabilizing the system against convection currents. During the early years of development almost no electrolyte could be found that satisfied these requirements, and it remained for such compounds to be synthesized. One of the first compounds to be made formed a group called aliphatic polyamino-polycarboxyl acids which are schematically shown in Fig. 9-31.

Another early problem associated with isoelectric focusing was the stabilizing of the system against convection currents. Just as in electrophoresis, when fine separations are desired, it is necessary to ensure that the various disturbances caused by the moving molecules themselves do not disrupt the pH gradient or the already separated molecules. In isoelectric focusing

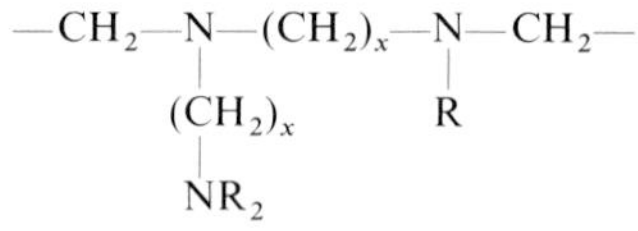

$$-CH_2-N-(CH_2)_x-N-CH_2-$$

$$(CH_2)_x \qquad R$$

$$NR_2$$

$x = 2$ or 3

$R = H$ or $-(CH_2)_x COOH$

Fig. 9-31   Structure of a polyaminopolycarboxyl acid.

there are essentially two ways this stabilization can be accomplished. The whole process can be carried out in a polyacrylamide gel similar to that used in disk gel electrophoresis, or a sucrose density gradient can be used. In the latter method a sucrose solution of continuously decreasing density is carefully layered into the glass focusing tube with the dense part on the bottom. This technique prevents convection based on the principle that a less dense solution will tend to float on top of a more dense solution and not mix, as would two solutions of identical density. In this case the focusing tube supports both a pH and a sucrose density gradient.

Isoelectric focusing can be used to determine isoelectric points of bio-molecules or to separate components from a complex mixture. If it is desired to physically isolate components, it is also necessary to separate the bio-molecules from the sucrose or the polyacrylamide gel, which complicates the technique. If a sucrose density gradient is used, the entire focusing tube can be drained drop by drop through a valve in the bottom. Each drop, or collection of drops, can then be collected in separate test tubes; each test tube then represents a fraction of the whole system. The desired fractions can then be dialyzed against any desired solvent to eliminate the sucrose. The resolution of components from a solution will depend on the steepness of the pH gradient and also on the separation of the isoelectric pH of the components. The steeper the pH gradient, the less resolved two components will be that have similar isoelectric pH. Figure 9-32 shows a densitometer trace of the isoelectric focusing separation of a solution containing four different proteins: ovalbumin, horse myoglobin, ribonuclease, and cyto-chrome C. Can you figure out a way of identifying which peak corresponds to which protein? This particular run was for preparative purposes and was gel stabilized against convection. It should be noticed that the major bands in the densitometer tracing are themselves further subdivided into sharp peaks indicating that each major species has several components.

ELECTROPHORETIC LIGHT SCATTERING

This technique is yet another way in which electrophoresis can be per-formed. It is a relatively new technique and in many respects is quite sophisticated and complex; hence it will be discussed only briefly. Electro-

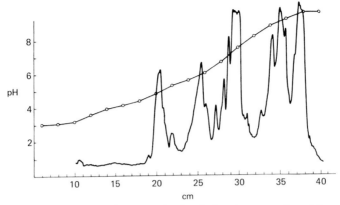

Fig. 9-32 A densitometer tracing of an isoelectric focusing separation of four proteins: ovalbumin, horse myoglobin, ribonuclease, and cytochrome C. The pH gradient of the column is superimposed. The abscissa indicates the length along the column of the various fractions. The voltage used was 400 V for 20 hr followed by 800 V for 10 hr. 500 mg of each protein were in the initial solution (Radola, 1973).

phoretic light scattering is based on the simple fact that when light is scattered off of a moving object, the frequency (or wavelength) of the scattered light is shifted a little bit with respect to the frequency of the incident light (Fig. 9-33). This is an application of the phenomena known as the Doppler effect.* The degree to which this frequency shift occurs depends

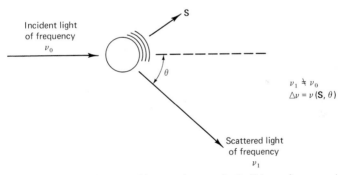

Fig. 9-33 Light scattered from an object moving at velocity **S** has a frequency (v_1) that is slightly different from that of the incident light (v_0). This frequency shift Δv is directly related to the velocity of the scattering object; i.e., the particle's speed and direction will determine the magnitude of the frequency shift. The frequency shift is also dependent on the scattering angle θ. The scattered light can be analyzed to determine the motion of the particles.

* The Doppler effect is the one responsible for the change in pitch of a train's whistle just as the train passes you. At that instant the train is changing from moving toward you to moving away from you. The apparent frequency or pitch of the whistle is dependent on the relative motion of the train and the listener. A frequency shift occurs at the moment of passing.

on the speed with which the scattering object is moving and on the angle of scatter. This frequency shift is usually very small when the light is scattered off biomolecules in a solution, and fairly elaborate electronic equipment is required to measure it; however, it can be done. In electrophoretic light scattering a solution containing one or a number of macromolecules is placed in a rectangular cell that has two electrodes implanted in it. The electrodes are used to set up an electric field to allow electrophoresis of the molecules, and a laser beam irradiates the volume between the electrodes. The molecules in the solution begin to move toward the electrode of opposite polarity and the light from the laser is scattered by these moving particles. The scattered light is then analyzed for frequency shifts from that of the original; this gives a measure of particle velocity. If the strength of the electric field is known, the electrophoretic mobilities can then be calculated.

Another nice advantage of this technique is that it can also measure the diffusion coefficients of the molecules in electrophoresis. Remember, that as the molecules move systematically toward one electrode or the other, they are also undergoing diffusion, so that any individual molecule has two types of motion associated with it. One is the preferential movement toward an electrode, and the other is the more random movement associated with diffusion. Both types of motion can be separated and analyzed in electrophoretic light scattering.

ALIGNMENT IN AN ELECTRIC FIELD

Although this section describes an electrical technique that is not as well defined as the others in this chapter, the ability to align biological components in an electrical field should at least be mentioned as a practice and as a way to achieve a spatial ordering. It was discussed previously that when placed in a uniform, homogeneous field, a dipole will align with that field. This fact can be quite valuable in a number of ways. For instance, supose it is desired to make a measurement on a solution of macromolecules in which they are all aligned in the same direction. One way of achieving this is by placing them in an electric field. This will not align them as matches in a matchbox because thermal fluctuations will disrupt complete order; but the molecules will potentially achieve much more alignment than their normal random order. Let us briefly consider just one application of this principle.

In 1965 D. Gray, R. Kilkson, and R. Deering published a paper in which they reported on their investigation to see whether or not the DNA within a bacterial cell was aligned in some orderly fashion or existed in a random array. The basis for their attack was as follows. When bacteria are irradiated with ultraviolet light, they frequently lose biological activity; they do so in part due to the interaction of the uv light with the bacterial DNA. If a single bacterium were irradiated by uv light that was polarized parallel to and perpendicular to the long axis of the bacteria, respectively, then if the DNA were aligned in some orderly fashion within the bacterium, it was argued that one type of polarization or the other should generate a larger deactivation of biological activity. This is due to the fact that ordered DNA has a preference for absorbing light that is polarized in one particular fashion. The more uv light absorbed by bacterial DNA, the higher are the chances it will have its biological activity altered. If the DNA had little or no overall order within the bacterium, then the type of polarization should not matter. But before this irradiation experiment can take place, the bacteri must first be physically lined up so that all of the bacteria themselves have essentially the same orientation. The method chosen to do this was by alignment in an electric field. What was done was to place a bacterial solution in an electric field so the bacteria would become aligned. The bacteria were then irradiated by the polarized light.

The above example is not the only case where an electric field is used to align a biological component so that some sort of physical measurement can be made. In other situations molecules have been aligned so that their differential absorption of light could be studied, or the extent to which the velocity of light differed when the light traversed the long or the short dimension of the particle. By aligning biomolecules in this fashion it is possible to examine the anisotropy of the molecules physical characteristics, i.e., to see whether the molecule has asymmetrical physical properties.

SEDIMENTATION VELOCITY

The experimental techniques discussed so far have utilized the fact that biological macromolecules are charged. However, these techniques are not the only instances where the electrical properties of biomolecules are important. There are a number of techniques not directly related to these electrical properties, but which are nevertheless influenced by them. The electrical properties of molecules must be considered by the experimenter if artifacts are to be avoided. Many times a researcher may be attempting to measure one parameter, whereas in reality a completely different effect may be the dominate influence. An example of this will now be illustrated by examining the practice of sedimentation velocity.

One of the oldest classical ways of characterizing a macromolecule is to place it in a centrigugal field and follow its motion as a function of time. The rationale for this being that the rate at which an object moves through a fluid is related to its size and shape; a centrifugal force can accelerate this motion which would normally take a very long time under normal conditions of gravity. In an experiment a solution is placed in a cell as shown in Fig. 9-34, the cell is placed in a rotor, the rotor is spun on its axis thus generating centrifugal fields in the cell up to 200,000 times that of gravity, and the motion of the molecules is observed via an optical system similar to that used in a Tiselius moving boundary electrophoresis experiment. As in electrophoresis, it is not the individual molecules that are observed but the boundary between the solvent and the solution. By following the boundary between solvent and solution as a function of time it is possible to calculate a quantity called the sedimentation coefficient S. The sedimentation value S has units of 10^{-13} sec, and the S stands for Svedberg after the developer of the analytic ultracentrifuge. The sedimentation coefficient of a molecule is defined as its rate of migration per unit of centrifugal field. (This is very analogous to the definition of electrophoretic mobility.) This sedimentation coefficient is directly related to a molecule's size and shape, and its value is commonly used in a list describing the physical characterizations of a molecule. The S value for a particular biomolecule will depend on the conditions under which the experiment is performed: solvent, viscosity, temperature, etc. A few representative values of the sedimentation coefficient for several molecules are given along with their respective molecular weights in Table 9-4. But how do the electrical properties of biomolecules affect sedimentation velocity experiments? The main way is through a phenomenon known as the primary charge effect.

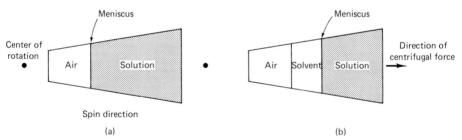

Fig. 9-34   In a sedimentation experiment a solution containing biomolecules is placed in a pie-shaped cell which is then put into a rotor. The rotor is then spun in a circle to create a centrifugal field. The field is radial and is directed toward the bottom of the cell. In (a), at the beginning of the experiment, the meniscus represents a boundary between air and solution. (b) shows the same cell a time later. Due to the centrifugal field, the solute molecules move downward to the bottom of the cell. This creates a new boundary between solution and pure solvent. It is this boundary that is actually followed optically in the sedimentation technique. The solution section shown in (b) is actually more concentrated than that initially placed in the cell.

TABLE 9-4

**Representative Values of the Sedimentation
Coefficients of Some Common Biomolecules**

Molecule	Sedimentation coefficient (S)	MW
Hemoglobin	44	68,000
Cytochrome C	1.9	15,600
Myoglobin	2	17,200
TMV	193	40,000,000

The primary charge effect occurs due to the fact that a macromolecule can sediment away from its counterions. If a protein is sedimented in a solution of low ionic strength, then because of its greater mass, the protein will experience a greater centrifugal force and will outdistance its counterions. The centrifugal forces pushing the molecules outward are directly proportional to their respective masses. The effect of this is to create an electric field that opposes the centrifugal field, and thus lowers the value of the sedimentation coefficient. Experimental work has shown that this effect can lower the sedimentation coefficient to half of its real value. As an example, consider a protein that has a net positive charge. Its counterions will be negative and will sediment relatively slowly creating a situation like that shown in Fig. 9-35. Here, the counterions will try to attract the protein, trying to draw it backward with the result that the protein will sediment more slowly than normal. This effectively reduces the value of the sedimentation coefficient. The primary charge effect can be eliminated by always performing the experiment in a solution that has a sufficiently high ionic

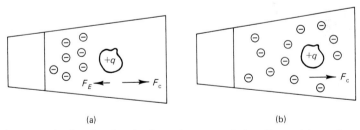

(a)   (b)

Fig. 9-35 (a) The situation that develops when a protein is sedimented in a solvent of low ionic strength. The centrifugal force F_c is opposed by an electrical force F_E between protein and counterions. The biomolecule sediments faster than the counterions because of its much greater mass. The biomolecules sediment away from the counterions. (b) The same situation, but now the solvent has an abundance of counterions, where the electrical force reducing the sedimentation velocity is no longer present. Here, the sedimentation is not obscured by the primary charge effect artifact.

strength, so that the protein is always surrounded by an abundance of counterions.

Figure 9-36 graphically illustrates the effect of ionic strength on the sedimentation coefficient. A 1% concentration (10 mg/ml) of bovine serum albumin was sedimented in solvents differing in their concentration of Br⁻ ion; the sedimentation coefficient is seen to vary sharply with salt concentration until a minimum concentration of counterions is reached, after which S is fairly constant. If the primary charge effect were not corrected for, then the sedimentation coefficient could conceivably be determined by the ionic strength of the solution and not the protein's size or shape, as it should be. This situation could easily lead to fallacious conclusions concerning the protein's characteristics.

In the above example the electrical properties of biomolecules are indirectly affecting a physical measurement that seemingly has nothing to do with electrical properties. This example is not an isolated case since there are several other experimental techniques that can be substantially influenced in a similar manner. The point to be made is that electrical properties of biomolecules can influence via indirect mechanisms.

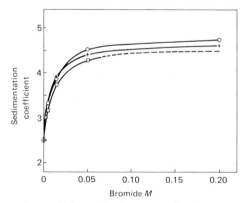

Fig. 9-36 Sedimentation coefficient of bovine serum albumin plotted vs. concentration of Br⁻. The BSA is dissolved in a solution containing 0.005 M Br⁻ + X Br⁻. The □ represent CsBr, the ○ KBr, and the + signs represent LiBr salts. (Reprinted with permission from K. O. Pedersen, *J. Phys. Chem.* **62**, 1282 (1958). Copyright by the American Chemical Society.)

SUMMARY

A number of experimental techniques in molecular biology make use of the charged nature of biological macromolecules. These techniques are used to separate, purify, and characterize biomolecules. The most noteworthy technique in this category is electrophoresis. There are several types of electrophoresis: moving boundary, disk gel, zone, electrophoretic light scattering. The electrophoretic mobility is a measure of how a biomolecule will molve in a unit electric field. Isoelectric points can also be determined via electrophoresis. Electrophoresis is not only an experimental tool but is commonly used in the hospital laboratory to aid in the detecting of disease states. SDS gel electrophoresis can be used to determine the molecular weight of proteins. Dielectrophoresis is based on the facts that biomolecules have dipole moments and that an interaction between a dipole and a large nonuniform electric field will cause a migration of the particles. Isoelectric focusing is a form of electrophoresis where the pH of the system changes as does the charge of the biomolecule. Molecules move in a pH gradient until their net charge is zero, at which point they stop. Ion exchange chromatography can separate single components from a complex mixture by relying on the differential interaction of the various components with an immobile charged phase. Isoelectric precipitation is an experimental technique used to purify components in a solution by relying on the fact that biomolecules generally have their lowest solubility at their isoelectric point.

REFERENCES

Anon. (1968). "A Guide to Ion Exchange Chromotography." Pharmacia Fine Chemicals Inc., 800 Centennial Ave., Piscataway, New Jersey.

Alberty, R. A. (1948). An introduction to electrophoresis, *J. Chem. Educ.* **25**, 426, 619.

Cann, J. R. (1969). Zone electrophoresis, *in* "Physical Principles and Techniques of Protein Chemistry, Part A (S. J. Leach, ed.). Academic Press, New York.

Cerami, A., and Peterson, C. M. (1975). *Scientific American*, April, p. 44.

Davis, B. J. (1964). Disc electrophoresis, *Ann. New York. Acad. Sci.* **121**, 404.

Dozy, A. M., and Huisman, T. H. (1969). *J. Chromatogr.* **40**, 62.

Eisenstadt, M., and Scheiberg, I. H. (1923). Dielectrophoresis of macromolecules, *Biopolymers* **12**, 249.

Gray, D., Kilkson, R., and Deering, R. A. (1965). Inactivation of oriented bacteria with polarized ultraviolet light, *Biophys. J.* **5**, 473–486.

Gray, G. W. (1951). Electrophoresis, *Scientific American*, December, p. 45.

Green, C., and Steiner, L. (1976). *J. Immunol.* **117**, 364.

Haglund, H. (1971). Isoelectric focusing in pH gradients, *in* "Methods of Biochemical Analysis," Vol. 19 (D. Glick, ed.), Wiley, New York.

Lehninger, A. L. (1970). "Biochemistry." Worth, New York.

Nakos, G., and Mortenson, L. (1971). *Biochemistry* **10**, 455.

Ornstein, L. (1964). Disc electrophoresis, *Ann. New York Acad. Sci.* **121**, 321.

Pauling, L., Itano, H., Singer, S., and Wells, I. (1949). *Science* **110**, 43.

Pedersen, K. O. (1936). *Biochem. J.* **30**, 961.

Pedersen, K. O. (1958). *J. Phys. Chem.* **62**, 1282.

Radola, B. J. (1973). *Ann. New York Acad. Sci.* **209**, 127.

Reynolds, J. A., and Tanford, C. (1970). Gross conformation of protein-sodium dodecyl sulfate complex, *J. Biol. Chem.* **245**, 5161.

Vesterberg, O. (1971). Isoelectric focusing of proteins, *in* "Methods in Microbiology," (J. R. Norris and D. W. Ribbons, eds.). Vol. 5B, Academic Press, New York.
Wachsmuth, E. D., and Pfleiderer, G. (1963). *Biochem. Z.* **336**, 545.
Ware, B. R. (1974). Electrophoretic light scattering, *Adv. Colloid Interface Sci.* **4**, 1.
Weber, K., and Osborn, M. (1969). *J. Biol. Chem.* **244**, 4406.

Appendix A

VECTOR ANALYSIS

In studying physics heavy emphasis is frequently placed on mathematics, and the usual procedure is to use as elegant and as simple mathematics as is possible. Because of this, vector analysis is routinely used in many cases. It is important that the reader be fluent in the manipulations of vectors. For those readers who are not completely at ease with vectors, this appendix is presented to give a brief survey of the information needed in order to appreciate fully the text on electrostatics. This appendix will start with a definition of a vector and a scalar; it will present a few examples of each; notation and other conventions will be discussed; and it will then proceed into the mathematics itself.

A vector is a quantity that has both a magnitude and a direction, i.e., the amount of that quantity and its direction of application are both needed to fully describe a vector quantity. Some typical examples of physical vectors are force, velocity, acceleration, electric field, and dipole moment. In all of these examples a detailed description requires one to know not only how much of the particular quantity is involved but also in what direction it is acting. A system of electrical particles has a dipole moment where the moment has a specific magnitude or strength, and it points in a specific direction. Both the strength and the direction of the dipole moment are determined by the spatial relationship of the charges. Velocity is a vector quantity because both the speed and the direction of motion must be known for a complete description. A force vector tells not only how hard the force is pushing, but also in what direction. So, a vector involves both magnitude and direction.

A scalar, on the other hand, is a quantity that has only a magnitude, and no meaningful direction associated with it, i.e., no direction is necessary to describe a scalar quantity. Examples of physical scalars are numbers, mass, time, temperature, electrical potential, and electrical potential energy. When one considers a scalar, only the magnitude or strength of the quantity is needed; direction of application makes no sense.

Symbolically, a vector is represented in this text with boldface type; often in writing one indicates a vector by placing a half arrow $\rightharpoonup$ over the letter used to represent the vector quantity. Scalars are just represented by the (italic) symbol itself with no other special markings.

Vectors are introduced into physics for two basic reasons. First, they distinguish between different types of physical quantities, and secondly because they can be manipulated mathematically in an elegant and concise fashion that has distinct advantages in many situations. Vectors and scalars are distinct not only in their definitions, but also in their algebras. The manipulations of scalars are familiar to everyone as the process of regular arithmetic. However, the mathematics of vectors is quite different; the bulk of this appendix will be used to describe vector manipulations.

COMPONENTS OF A VECTOR

Consider the situation depicted in Fig. A-1 where an arbitrary vector $\mathbf{E}$ has its tail coincident with the origin of a cartesian coordinate system. The magnitude of the vector is indicated by the length of the arrow, and the direction is as shown. It is possible to represent that part, or component, of the vector pointing exclusively parallel to the x axis by using a simple trigometric function. If $\mathbf{E}$ makes an angle θ with respect to the x axis, then the quantity $E \cos \theta$ is the component of $\mathbf{E}$ that points solely in the direction of the x axis. Similarly, $E \sin \theta$ points entirely in the direction of the y axis. E_x and E_y are referred to as the x and y components, respectively. Any

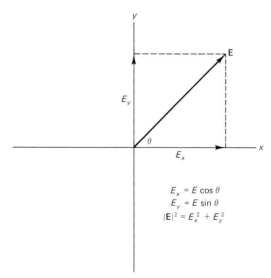

$$E_x = E \cos \theta$$
$$E_y = E \sin \theta$$
$$|E|^2 = E_x^2 + E_y^2$$

Fig. A-1   An arbitrary vector $\mathbf{E}$ is situated at an angle θ with respect to the x axis.

vector can likewise be broken down into its components along the mutually perpendicular axis of a coordinate system. Also, if a vector's components are known, it is possible to compute the magnitude of the entire vector $|\mathbf{E}| = E$. This is done for the case in Fig. A-1 by observing that E_x and E_y form a right triangle, where by Pythagoras' theorem

$$|\mathbf{E}| = E_x^2 + E_y^2 \tag{1}$$

The direction of $\mathbf{E}$ can be found by knowing the magnitude of E_x and E_y and by using a trigometric definition of $\tan \theta$. It is known that $\tan \theta = E_y/E_x$, so $\theta = \tan^{-1} E_y/E_x$.

The point to be made here is that a vector can be described in terms of its components. If a vector's components are all known, then the vector is completely described.

Unit vectors are vectors that point in the direction of each of the respective axes of a coordinate system and that have unit magnitude. For a cartesian coordinate system, the unit vectors are termed $\mathbf{i}$, $\mathbf{j}$, and $\mathbf{k}$ for the x, y, and z axes, respectively. Utilizing these unit vectors, it is possible to write the vector $\mathbf{E}$ as

$$\mathbf{E} = E_x\mathbf{i} + E_y\mathbf{j} + E_z\mathbf{k} \tag{2}$$

where E_x is the x component of $\mathbf{E}$, etc. Notice that each term on the right-hand side of Eq. (2) is composed of the product of a scalar and a vector. The direction of the component is given by the unit vector, and the magnitude is given by E_x, E_y, or E_z. It should also be remarked that vector components can be either positive or negative depending on which way the component points. As an example of this concept, consider that $\mathbf{E}$ in Fig. A-1 has a magnitude of $|\mathbf{E}| = 5$ and that $\theta = 30°$. From these data it can be calculated that

$$\mathbf{E} = 5\cos 30°\mathbf{i} + 5\sin 30°\mathbf{j} + 0\mathbf{k}$$
$$= 4.3\mathbf{i} + 2.5\mathbf{j}$$

where $4.3^2 + 2.5^2 = 5.0^2$.

If $\mathbf{A}$ is an arbitrary vector, then the unit vector $(\mathbf{a})$ that has the same direction as $\mathbf{A}$ can be computed by dividing $\mathbf{A}$ by its magnitude $|\mathbf{A}|$.

$$\text{unit vector } \mathbf{a} = \frac{\mathbf{A}}{|\mathbf{A}|} \tag{3}$$

VECTOR ALGEBRA ADDITION

The sum, or resultant, of two vectors can be obtained in several ways. Consider Fig. A-2 where we have two vectors $\mathbf{E}$ and $\mathbf{F}$, where it is desired

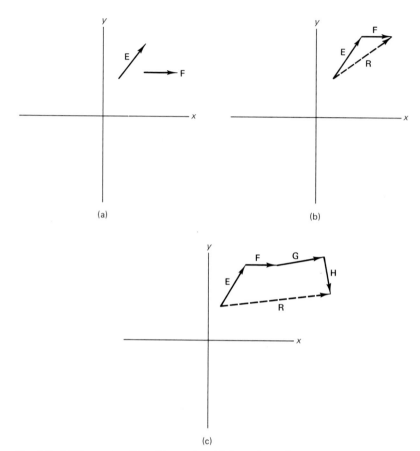

Fig. A-2 (a) Two vectors **E** and **F** are to be added together to form a resultant vector, (b) The tail of **F** is made coincident with the tip of **E** and the resultant vector **R** is formed by drawing a vector from the tail of **E** to the tip of **F**. (c) Illustration of the geometric addition method for four vectors.

to add them and come up with a single vector that will describe the combined effect of both **E** and **F**. In the geometric method of adding vectors, one vector, say **F**, is moved until its tail touches the tip of vector **E**. A vector **R** drawn from the tail of vector **E** to the tip of vector **F** now represents the sum of **E** and **F**. In this process the initial direction of vector **F** cannot be changed, otherwise the vector itself would be changed. The sum of vectors **E** and **F** is represented as

$$\mathbf{R} = \mathbf{E} + \mathbf{F} \tag{4}$$

where **R** is the sum or resultant vector. If it is desired to add three or more vectors, a similar method is used. All vectors are placed head to tail, keeping care to preserve their initial orientations and the resultant is drawn from the tail of the first to the head of the last. This method is illustrated for the addition of four vectors **E**, **F**, **G**, and **H** in Fig. A-2c.

The above method for vector addition is not very satisfactory if precise analytical results are desired. A more exact method is now described. The first step is to reduced all vectors to their x and y (z also if appropriate) components. Next, the sums of all x components and of y components are respectively calculated. These results represent the x and y components of the resultant vector **R**. The magnitude of **R** is found by remembering $|\mathbf{R}| = R_x^2 + R_y^2$ and **R**'s direction is given by $\theta = \tan R_y/R_x$. As an example, consider the vector addition of **E** and **F** where $\mathbf{E} = 2\mathbf{i} + 2\mathbf{j}$ and $F = 1.5\mathbf{i}$. We then have

$$R_x = 3.5, \qquad R_y = 2.0$$

$$|\mathbf{R}|^2 = 3.5^2 + 2.0^2 = 4.03^2$$

$$\theta = \tan^{-1}\frac{2.0}{3.5} = 29.7°$$

$$\mathbf{R} = 3.5\mathbf{i} + 2.0\mathbf{j}$$

If **E** was originally described in terms of its total magnitude and θ, then E_x and E_y would have had to be calculated. This analytical method can be used regardless of the number of vector being added.

What happens, then, when one tries to add a vector and a scalar? The answer is nothing, because it is a forbidden process. It would be like adding apples and oranges. Vectors and scalars are different and cannot be added together; however, they can be multiplied together.

VECTOR MULTIPLICATION

There are two basic ways in which vectors can be multiplied. In the dot or scalar product, a scalar results. In the cross or vector product, a vector results.

Dot Product

The scalar or dot product between two vectors and is defined as

$$\mathbf{E} \cdot \mathbf{F} = |\mathbf{E}||\mathbf{F}|\cos\theta, \qquad 0 \le \theta \le \pi \tag{5}$$

where θ is the angle separating the two vectors. This is illustrated in Fig. A-3. If the vectors are represented in terms of their components in rectangular coordinates, the dot product is given as

$$\mathbf{E} \cdot \mathbf{F} = E_x F_x + E_y F_y + E_z F_z \tag{6}$$

With these definitions, the reader should be able to see that $\mathbf{i} \cdot \mathbf{i} = \mathbf{j} \cdot \mathbf{j} = \mathbf{k} \cdot \mathbf{k} = 1.0$. As a numerical example, consider $\mathbf{E} = 5\mathbf{i} + 5\mathbf{j}$ and $\mathbf{F} = 3\mathbf{i}$. The dot product $\mathbf{E} \cdot \mathbf{F} = 15$. This can also be calculated according to Eq. (5) or $\mathbf{E} \cdot \mathbf{F} = (7.07)(3) \cos 45° = 15$. The reader should be able to understand where the 45° factor comes from after drawing the two vectors in a co-ordinate system.

The dot product is one way of ascertaining whether or not two vectors are mutually perpendicular. If they are, the dot product is zero. In this respect $\mathbf{i} \cdot \mathbf{j} = \mathbf{i} \cdot \mathbf{k} = \mathbf{j} \cdot \mathbf{k} = 0$. The dot product can also be looked upon as multiplying the magnitude of one vector times the magnitude of the other that is parallel to the first vector. In Eq. (5) the term $|\mathbf{F}| \cos \theta$ is really the component of the vector $\mathbf{F}$ that is parallel to the vector $\mathbf{E}$. Realize again that the dot product results in a scalar quantity.

Cross Product

The cross product of two vectors $\mathbf{E}$ and $\mathbf{F}$ is defined as another vector $\mathbf{H}$ that is mutually perpendicular to the original two vectors

$$\mathbf{E} \times \mathbf{F} = \mathbf{H} \qquad (7)$$

The magnitude of $\mathbf{H}$ is given by

$$|\mathbf{H}| = |\mathbf{E}| |\mathbf{F}| \sin \theta, \qquad 0 \le \theta \le \pi \qquad (8)$$

where θ is again the angle between $\mathbf{E}$ and $\mathbf{F}$. The fact that $\mathbf{H}$ is perpendicular to both $\mathbf{E}$ and $\mathbf{F}$ limits $\mathbf{H}$ to one of two possible directions; however, in the

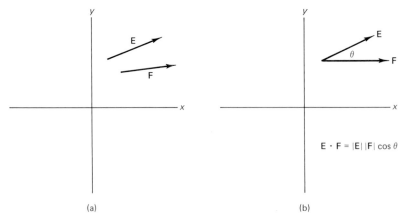

(a)   (b)

Fig. A-3 (a) Two arbitrary vectors $\mathbf{E}$ and $\mathbf{F}$ to be multiplied in a dot product. (b) The angle between $\mathbf{E}$ and $\mathbf{F}$ is θ.

definition the direction of **H** is such that **E**, **F**, and **H** all form a right-handed coordinate system. A right-handed system is defined by the fact that a right-handed screw rotated through 180° (or less) from vector **E** to **F** will advance in the direction of **H**. This spatial relationship for a vector product is shown in Fig. A-4. The following rules for a vector product hold:

$$(1) \quad \mathbf{E} \times \mathbf{F} = -\mathbf{F} \times \mathbf{E}$$

$$(2) \quad \mathbf{E} \times (\mathbf{F} + \mathbf{H}) = \mathbf{E} \times \mathbf{F} + \mathbf{E} \times \mathbf{H} \tag{9}$$

$$(3) \quad m(\mathbf{E} \times \mathbf{F}) = m\mathbf{E} \times m\mathbf{F}, \quad m \text{ is a scalar}$$

If the vectors **E** and **F** are written in terms of their components, then the cross product is given as

$$\mathbf{E} \times \mathbf{F} = \begin{vmatrix} \mathbf{i} & \mathbf{j} & \mathbf{k} \\ E_x & E_y & E_z \\ F_x & F_y & F_z \end{vmatrix} \tag{10}$$

$$= \mathbf{i}(E_y F_z - E_z F_y) - \mathbf{j}(E_x F_z - E_z F_x) + \mathbf{k}(E_x F_y - E_y F_x)$$

As a numerical example, let $\mathbf{E} = 1\mathbf{i} + 2\mathbf{j} + 3\mathbf{k}$ and $\mathbf{F} = 2\mathbf{i} + 2\mathbf{j} + 3\mathbf{k}$. We then have $\mathbf{E} \times \mathbf{F} = \mathbf{H} = 3\mathbf{j} - 2\mathbf{k}$. The reader should also be able to show that $\mathbf{H} \cdot \mathbf{E} = \mathbf{H} \cdot \mathbf{F} = 0$.

Finally, we must consider how a scalar and a vector can be multiplied. When a vector is multiplied by a scalar, it has the simple effect of increasing every component by a factor equal to the value of the scalar. The product of

$$m\mathbf{E} = mE_x\mathbf{i} + mE_y\mathbf{j} + mE_z\mathbf{k}.$$

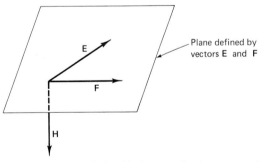

Fig. A-4 Diagram of the spatial relationship between the three vectors involved in a cross vector. Vectors **E** and **F** define a plane and **H** is perpendicular to that plane. If a right-handed screw is rotated from **E** toward **F** (through the smaller angle), it advances in the direction of **H**.

Appendix B

USEFUL CONSTANTS AND CONVERSION FACTORS

Boltzmann's constant	1.3805×10^{-16} erg
Avogadro's number	6.023×10^{23} mole^{-1}
Planck's constant	6.626×10^{27} erg sec
Electronic charge	-4.803×10^{-10} esu
	-1.602×10^{-19} C
Permittivity	8.85×10^{-12} C^2/N m^2
Coulomb constant k	8.987×10^9 N m^2/C^2

1 erg $= 2.389 \times 10^{-8}$ Cal
1 kilocalorie $= 1000$ cal
1 liter atmosphere $= 24.22$ cal
1 joule $= 2.389 \times 10^{-1}$ cal
1 Coulomb $= 3 \times 10^9$ esu
1 dyne/esu $= 300$ V/cm
1 erg/interaction $= 1.439 \times 10^{13}$ kcal/mole

INDEX